Igneous
Rocks

Igneous Rocks

Daniel S. Barker

Department of Geological Sciences
The University of Texas at Austin

Prentice-Hall, Inc., Englewood Cliffs, New Jersey 07632

Library of Congress Cataloging in Publication Data

Barker, D. S. (Daniel Stephen), (date)
  Igneous rocks.

  Bibliography: p.
  Includes index.
  1. Rocks, Igneous. I. Title.
QE461.B295   552'.1   82-7512
ISBN 0-13-450692-8   AACR2

Editorial/production supervision and interior design: Kathleen M. Lafferty
Manufacturing buyer: John B. Hall
Cover design: Diane Saxe

Cover illustration: Hydrated volcanic glass, Superstition Mountains, Arizona. Photomicrograph by Daniel S. Barker.

Printed in the United States of America

10 9 8 7 6 5 4 3 2 1

ISBN 0-13-450692-8

Prentice-Hall International, Inc., *London*
Prentice-Hall of Australia Pty. Limited, *Sydney*
Editora Prentice-Hall do Brazil, LTDA, *Rio de Janeiro*
Prentice-Hall Canada Inc., *Toronto*
Prentice-Hall of India Private Limited, *New Delhi*
Prentice-Hall of Japan, Inc., *Tokyo*
Prentice-Hall of Southeast Asia Pte. Ltd., *Singapore*
Whitehall Books Limited, *Wellington, New Zealand*

Contents

v

3

Phase Relations   *24*

4

**Estimating, Reporting, and Comparing Igneous
Rock Compositions**   *58*

5

Classification of Igneous Rocks   *86*

6

Crystallization and Textures   *99*

7 Generation and Evolution of Magma 124

8 Forms of Igneous Rock Bodies 143

9 The Effects of Volatile Components 187

10 Ultramafic Rocks   212

11 Mafic Rocks   229

12 Intermediate and Felsic Silica-Oversaturated Rocks   261

13 Silica-Undersaturated Rocks 297

14 Metasomatism 323

15 Magmatism and Tectonism 332

16 Relations of Magma to Energy and Mineral Resources   *354*

Preface

Igneous Rocks was written for undergraduate geology majors who have had a year of college-level chemistry and a course in mineralogy (not necessarily including optical crystallography) and for beginning graduate students. Geologists working in industry, government, or academia should find this text useful as a guide to the technical literature up to 1981 and as an overview of topics with which they have not worked but which may have unanticipated pertinence to their own current projects.

Rigorous development of thermodynamic principles is not attempted, but the importance of thermodynamic and kinetic models and the conclusions drawn from them are emphasized throughout to encourage the reader to dig deeper. General principles of magma genesis and modification are stressed, with few specific examples being described in any detail. Such examples are handled best (literally) in laboratory sessions concurrent with classroom work and reading. The case-history method of teaching petrology is highly effective, but as teaching materials vary widely from one school to another, teachers should draw on their own experiences with igneous rocks rather than on second-hand accounts.

Chapters 1 and 2 are reviews of introductory physical geology and mineralogy, with a perspective that should be new to most readers. These chapters provide a minimum background for nongeologist readers (such as physicists and chemists who are aware of the exciting prospects for basic and applied research in magma and its products). Chapters 3 through 9 are the traditional material of igneous petrology, each building on its predecessors. Chapters 10 through 13 apply the foregoing principles to detailed discussion of the important kinds of igneous rocks and may be most useful to advanced students and professional geologists. Chapter 14 revives a once-controversial topic to emphasize its newly recognized importance to igneous petrology. Chapter 15 exposes some of the remaining ignorance, and Chapter 16 implies some opportunities in the study of igneous rocks.

Some readers may object to excursions in Chapters 8 and 16 into structural and economic geology. However, most igneous rocks that geologists encounter are not in labeled trays, and the style, scale, and setting of their occurrence cannot be ignored. Economic aspects also deserve consideration because, as stated in Chapter 16, knowledge of igneous rocks is both a prerequisite and a byproduct of many successful mining ventures.

The list of references at the end of the book, totaling some 600 citations, is obviously incomplete. Those selected are not necessarily the definitive publications but are the most recent papers and books that can serve best as guides to older works, a few classics that have not been superseded, and, of course, the sources of all quotations, illustrations, and tabulated data. More citations are given on controversial topics, so that through independent reading the student will develop respect, but not reverence, for the literature.

Quotations from N. L. Bowen's *The Evolution of the Igneous Rocks* appear in Chapters 6 and 7 with the permission of the copyright owner, Dover Publications, Inc. Average rock compositions in Tables 10-1, 11-1, 12-1, and 13-1 from R. W. LeMaitre's 1976 paper in the *Journal of Petrology* are reproduced with permission of the Oxford University Press. Ionic radii in Tables 2-1 and 4-7 are from Brian Mason's *Principles of Geochemistry, Third Edition,* copyright 1966 by John Wiley and Sons, and from Whittaker and Muntus, *Geochimica et Cosmochimica Acta, 34,* "Ionic radii for use in geochemistry," copyright 1970, Pergamon Press, Ltd., with permission from both publishers. Tom Simkin generously provided Figure 8-20; all other photographs and photomicrographs are by the author.

Douglas Smith and J. Richard Kyle of The University of Texas at Austin and John F. G. Wilkinson of the University of New England, Armidale, N.S.W., Australia, read parts of early drafts and made valuable suggestions; they are not responsible for errors in the final version. During the years that this book was in preparation more than 400 students had to read preliminary drafts. Their criticism has been most helpful. And the comments of the following reviewers were appreciated: James Anderson of The State University of New York at Binghamton, M. E. Bickford of the University of Kansas, Russell Clemons of New Mexico State University, R. V. Fodor of North Carolina State University, and William Furbish of Duke University.

The pedagogical approach ultimately derives from examples set by many teachers, above all others A. F. Buddington, who could respond to students' questions, in a unique blend of exasperation and exultation, with "I don't know!"

Daniel S. Barker

Austin, Texas

Igneous
Rocks

1

The Role of Magma
in Geologic Processes

1.1 MAGMA

The term *magma* (Greek, "paste") was first applied in geological usage by G. P. Scrope, who introduced a pharmaceutical word for ointment to emphasize "its analogy to those compound liquids such as mud, paste, milk, blood, honey, etc., which consist of solid particles deriving a certain freedom of motion amongst one another from their intimate admixture, in greater or less proportion, with one or more perfect fluids, which act as their vehicle" (Scrope, 1825, p. 19, as quoted by Simkin, 1967, p. 66–67). Although Scrope drew the analogy to what we now call a suspension in his 1825 book, he apparently did not use the term "magma" in print until an 1872 publication.

A commonly accepted definition of magma is mobile rock material containing hot liquid, with or without suspended solids and dissolved gases, within the solid body of a planet. How hot must the hot liquid be? Hotter than the normal surface temperature of the planet, and generally between 600° and 1300°C on the earth. The "hot liquid" part of the definition is needed to exclude other mobile rock material, such as salt domes and permafrost in the earth, yet to include liquid water in other planetary bodies with much lower surface temperatures than the earth.

To answer the question of where magma comes from, we must consider the anatomy of the earth. The next two sections briefly review the distribution and movement of the earth's parts.

1.2 CORE, MANTLE, AND CRUST

Approximately 32% of the earth's mass resides in the core, slightly over 67% in the mantle, and the remaining 0.4% in the crust. This threefold division is marked by abrupt changes in seismic wave velocities at specific depths. For example, the boundary between the crust and mantle, the *Mohorovičić discontinuity* (Moho, for short), is defined by a sharp increase in seismic wave velocities in the mantle compared to the overlying crust.

Igneous petrology (the study of rocks derived from magma) is, basically, the study of the processes by which crust has segregated from the upper mantle. Volcanoes testify that the production of new crust from mantle still goes on.

Crust is divided into two types, oceanic and continental, differing in thickness, density, and composition (Figure 1-1). Oceanic crust covers approximately 61% of the earth's surface area (Figure 1-2). In spite of its greater extent and slightly higher average density, oceanic crust totals only about one third of the mass of continental crust, because the latter is so much thicker.

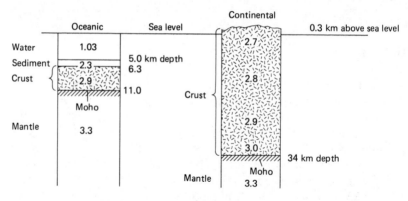

Figure 1-1. Schematic sections of average oceanic and continental crust. Numbers within columns are average densities in g/cm³. (After Hess, 1962, Fig. 2, with permission of the Geological Society of America.)

1.3 PLATE TECTONICS

The outer portion of the earth is segmented into fairly rigid plates floating on denser but weaker material. These plates move relative to each other at rates of up to 18 cm/yr (Minster and others, 1974), diverging from some plate boundaries, colliding at others, and grinding past each other along the third kind of boundary (Figures 1-2 and 1-3).

Oceanic crust is created by upwelling of magma, formed in the upper mantle, at the crests of oceanic ridges that mark *constructive plate boundaries*. As new crust is added at their edges, the plates move apart. Whether they are

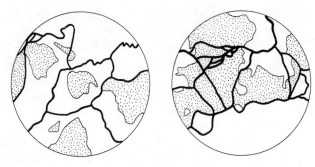

Figure 1-2. Unfamiliar views of a familiar planet. In these equal-angle polar projections, a geographic pole is at the center of each circle and the equator forms the circumference. Areas underlain by oceanic crust are white, areas underlain by continental crust are stippled, and boundaries of plates are indicated, in highly generalized fashion, by the heavy lines. The area of continental crust is considerably larger than the dry land area because of the extent of continental shelves that are below sea level but floored by continental crust. The lack of labels and the unconventional orientations of these hemisphere maps are intentional, to encourage more careful study. (Generalized from Dewey, 1975, Fig. 1, with permission of the American Journal of Science and J. F. Dewey.)

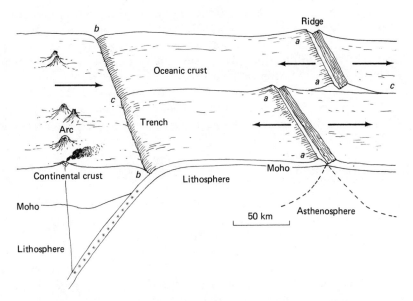

Figure 1-3. Three kinds of plate boundaries: *a-a*, constructive or spreading boundary (oceanic ridge); *b-b*, destructive or collision boundary, marked by a trench and subduction zone; *c-c*, conservative boundary (transform fault). Arrows indicate relative motions of plates. Asterisks mark seismic activity along a Benioff zone. A volcanic arc parallels the trench.

wedged apart by the magma, or are pulled apart to permit the passive rise of magma, is an unresolved question.

The total surface area of the earth remains unchanged; therefore, if crust is being continuously produced in some places, it must be destroyed in others. The destruction takes place where one plate collides with another; at such a *destructive boundary,* one of the plates curls downward into the mantle. The process of *subduction* (sinking of one edge of a plate) is marked by an oceanic trench and by an inclined zone of seismic activity called a *Benioff zone.*

Continental crust, in contrast to oceanic, is rarely subducted (apparently being more buoyant) and accumulates through geologic time. A large part of the new material added to continental crust is supplied by volcanism, commonly expressed as volcanic arcs such as the Andean chain. These arcs are parallel to the destructive plate boundaries and are located on either oceanic or continental crust, on the side under which the subducting plate sinks (Figure 1-3).

Contrast in behavior between rigid plates and nonrigid substrate is larger in scale, but more subtle, than the differences in seismic velocities that led to the distinction between crust and mantle. Consequently, another significant division of the outer portion of the earth is that between *lithosphere* and *asthenosphere.* The crust and uppermost mantle form the rigid plates of the lithosphere. The underlying, weaker, mantle is the asthenosphere (Greek, "without strength").

In Figure 1-3, the subducting plate plunges into the mantle, but the sketch ends at a depth of about 100 km. What happens at greater depth? The seismicity defining the Benioff zone, occurring within and along the top surface of the subducting plate, dies out at depths of approximately 220 km. Perhaps the lithospheric slab breaks off at this depth and sinks, perhaps it "dissolves" in the asthenosphere, or perhaps at 220 km depth the asthenosphere is too dense or too strong to allow further sinking of the slab, which levels off to rest at that depth (D. L. Anderson, 1979).

1.4 HEAT IN THE EARTH

That heat flows outward from the interior of the earth has been recognized for centuries. The sources of that heat, and the ways in which it reaches the surface, are still topics for fruitful research.

Four mechanisms are possible for transporting heat from one part of the earth to another. The first three are the familiar options of radiation, conduction, and convection. By *radiation*, heat energy is lost in the form of infrared and visible portions of the electromagnetic spectrum. The energy emitted increases (and the wavelength decreases) as the fourth power of the temperature. For example, an object heated to around 600°C gives off barely visible radiation as a dull red glow, but as the temperature is increased the luminosity rises and the color changes from red through orange to yellow. However, even deep within the earth the temperature is not sufficiently high for radiation to be an efficient means of heat transfer, and rocks are not very transparent to the radiation.

In *conduction*, thermal agitation of atoms allows heat to propagate by diffusion. Rocks are poor thermal conductors, and the remaining two heat transfer mechanisms, operating in rock material that is moving, are more efficient and rapid than radiation or conduction through a static medium.

Convection is the movement of matter, carrying heat with it, owing to density differences caused in turn by temperature differences. As the material (air, seawater, or, in the example of immediate pertinence, rock of the asthenosphere) is heated, it expands and rises, displacing cooler, more dense material, which sinks. Even at a circulation rate of only 1 cm/yr, large convection cells in the asthenosphere could easily transport a significant part of the earth's internal heat upward to the base of the lithosphere. Through the rigid outer portion of the earth (the lithospheric part of the mantle and the crust), heat transfer is accomplished by conduction, by convection of groundwater, and by the fourth mechanism, *magmatic transport*. As magma forms by melting of rock in the mantle or lower crust, heat energy is absorbed, then carried upward in the magma as it migrates toward the surface, and finally is released as the magma crystallizes or erupts as lava. Magmatic transport can be thought of as one-way, no-return convection, but is much more rapid.

On the average, the internal heat reaching the earth's surface amounts to approximately 81 ± 3 milliwatts per square meter (mW/m^2) (Sprague and Pollack, 1980). If all of this heat could be converted to electrical energy, one 100-W incandescent lamp could be powered by the average heat flowing through a square of ground 35 m on a side.

Discussion of average heat flow obscures the wide variation from one place to another, from 14 to 1600 mW/m^2 (Horai and Uyeda, 1969). Even the highest value is too low for practical extraction of energy, because conduction of heat through rock is so slow. There are places where convecting groundwater and magma convey heat toward the surface much more rapidly, and these are the sites where geothermal energy can profitably be harnessed (Chapter 16).

Heat flow is higher than average on the crests of oceanic ridges (constructive plate boundaries) and in young volcanic areas on continents, and lower than average in old ocean floor and trenches and in deep, rapidly subsiding sedimentary basins on continents. Average heat flow on continents is about 55 ± 5 mW/m^2, and that in the oceans is about 95 ± 10 mW/m^2 (Davies, 1980). Sclater and others (1980) provide a thorough inventory of heat flow variation in crustal segments of different ages. They conclude that creation of new oceanic crust through magmatism at constructive plate boundaries is responsible for at least 60% of the earth's heat loss, and that convection, not conduction, is the dominant mechanism of heat transfer elsewhere. The convecting materials that accomplish this are groundwater in the crust, seawater overlying the ridge, magma, and the asthenosphere.

The sources of the earth's internal heat are several, and each alone might account for the observed heat flow, so the problem, still unsolved in detail, is to decide the relative contribution from each source. Among the contributors is the decay of radioactive isotopes, especially of uranium, potassium, and thorium (in decreasing order of importance). One gram of basalt, if completely insulated, could be raised from 20°C to 1200°C and melted in about 300 million years if all the heat generated by decay of radioactive U, K, and Th in the basalt could be trapped within it. Although there are many other naturally occurring radioactive isotopes, ^{40}K, ^{235}U, ^{238}U, and ^{232}Th are the most abundant ones with sufficiently long half-lives, and therefore contribute most of the *radiogenic heat* now. Early in the earth's history, other isotopes must also have provided significant amounts of heat during their rapid decay.

Potassium, uranium, and thorium are strongly concentrated in continental crust; D. M. Shaw (1980) estimates that 23% of the K, and 15% each of the U and Th, in the total crust plus mantle are now in continental crust, which is only 0.5% of the total mass of crust plus mantle. Furthermore, the bulk of these three elements appears to be concentrated most strongly in the uppermost part of the continental crust (Jaupart and others, 1981). Consequently, radiogenic heat produced by these elements in the upper few kilometers accounts for about 40% of the average heat flow on continents (Pollack and Chapman, 1977). In contrast, oceanic heat flow is dominated by heat transported by magma, convecting asthenosphere, and groundwater, with little coming directly from radioactive decay in oceanic crust.

The impact of meteorites, large and small, undoubtedly provided much heat energy during the earth's early history, before this region of space was swept relatively clean of cosmogenic leftovers. The intense bombardment probably ended about 4×10^9 years ago, leaving the earth at least as heavily cratered as the moon. Gravitational compaction, and sinking of dense material toward the center to form the earth's core, were also probably important in the first stages of the earth's growth; moving 1 g from the surface all the way to the center would release 6.3×10^4 joules (J) as the heat equivalent of the decrease in potential energy.

Except for that generated by radioactive decay, heat was produced in the earth much more rapidly than it could dissipate by outward flow, and an unknown but probably large portion of the heat now reaching the surface ultimately came from early nonradiogenic sources.

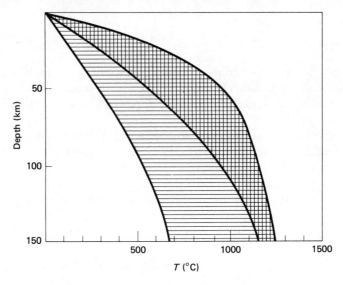

Figure 1-4. Estimated ranges of geothermal gradients under continents (horizontal ruling) and oceans (vertical ruling). (After Pollack and Chapman, 1977, Fig. 3, with permission of Elsevier Scientific Publishing Company and H. N. Pollack.)

1.5 THE GEOTHERMAL GRADIENT

If heat is flowing from the interior of the earth, the temperature must increase with depth. This rate of temperature rise, or *geothermal gradient*, has been calculated from measurements in mines and drill holes in many places. The average on the continents is about 35°C/km. However, we have penetrated only a small fraction of the total thickness of continental crust, and the observed geothermal gradient applies only to the top 10 km at the most. At greater depth, the gradient must decrease to only a few degrees per kilometer in the lower crust, and to less than 1°C/km at a depth of a few hundred kilometers into the upper mantle.

Estimates of geothermal gradients under continents and oceans (Figure 1-4) are uncertain because of our lack of precise knowledge of the distribution of K, U, and Th with depth, and our ignorance of the relative importance of the different heat transport mechanisms at different depths and times. No unique solution is possible, because the efficacy of each mechanism in turn is a function of temperature.

1.6 PRESSURE IN THE EARTH

We can be more confident about estimates of pressure as a function of depth. In the equation

$$P = dgh$$

where P is pressure, the variables are density d, gravitational acceleration g, and depth h. Density is fairly well known from seismic studies. Variation in the gravitational acceleration with depth can be calculated using classical Newtonian physics.

Within the crust, assuming an average density of 2.85 g/cm^3 and a gravitational acceleration of 9.80 m/sec^2, the pressure increase for each kilometer of depth is 279 bars or 2.79×10^7 pascals. The SI unit of pressure is the pascal (Pa), equal to 1 newton per square meter, or 1 kg m sec^{-2} m^{-2}. Previously, pressure in geological context was reported in bars, where 1 bar is 10^6 dynes per square centimeter or roughly 1 atmosphere (14.7 pounds per square inch). Geologists are adopting the pascal more reluctantly than other SI units, and the bar and kilobar (10^8 Pa) remain in wide use.

In the generation and subsequent history of magma, pressure is as important a variable as temperature, as will be evident in later chapters.

1.7 RATES OF MAGMATISM AND CRUSTAL SEGREGATION

The average annual rate at which igneous materials issue from volcanoes has been variously estimated at less than one to a few tens of cubic kilometers, worldwide. Individual production rates for persistently active volcanoes range from 0.02 (Etna and Mauna Loa) to 0.07 km^3/yr (Kilauea). During its nine years of activity, Parícutin averaged 0.16 km^3/yr. For the total volcanic output along constructive and destructive plate boundaries (each kind adding up to a length of about 54,000 km), estimated rates are 15 and 3 km^3/yr, respectively. In addition, there are volcanoes that show no proximity or clear relationship to any plate boundary; the Hawaiian volcanoes are examples. For more information, including the different ways in which the estimates are made, see Wadge (1980), Williams and McBirney (1979), and Baker and Francis (1978), and the many papers cited in those references.

Although the rate of volcanism at constructive plate boundaries exceeds that at destructive boundaries, nearly all of the oceanic crust formed at constructive boundaries eventually is returned to the mantle by subduction. New seafloor is produced at about 3 km^2/yr; the total area of oceanic crust could be generated in a little over 100 million years (Davies, 1980). In contrast, magmatism continues to augment the continents (Figure 1-5). Oceanic lithosphere is a "short-lived intermediary stage in the transport of some elements from the mantle to the [continental] crust" (DePaolo, 1980a, p. 1187).

Assuming a conservative estimate of 1 km^3/yr for the average volcanic outpouring on continents, and comparing this with the total volume of continental crust (6.5×10^9 km^3), we see that it would take considerably more time than the age of the earth (4.66×10^9 years) to derive all rocks of the continents from volcanic materials at that rate. However, an unknown but undoubtedly larger volume of magma never reaches the surface of the earth but

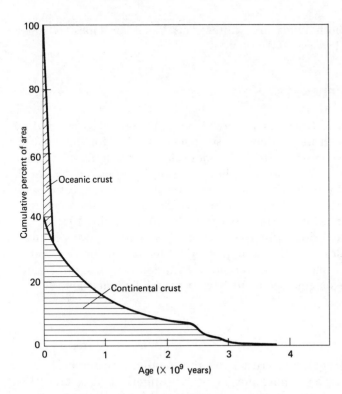

Figure 1-5. Cumulative percentage of earth's surface area occupied by *surviving* crust that formed within specific time intervals. Oceanic and continental crust are ruled. The unruled area represents crust that has been recycled by subduction and probably includes some continental, as well as oceanic, crust. (From data in Sprague and Pollack, 1980.)

solidifies at depth in the crust. Furthermore, large portions of the continents may be fragments of old oceanic crust rammed into, or shuffled under, the continents rather than subducted. Additionally, it is unlikely that the present rate of volcanism is equal to that throughout the geologic past. For one thing, generation of magma was probably more rapid in the early stages, owing to higher temperatures from gravitational compaction, meteorite impact, and rapid decay of abundant but short-lived radioactive isotopes. Indeed, there is a suggestion that the earth in its first several hundred million years was flooded with a magma ocean tens or hundreds of kilometers deep. The first crust might have formed rapidly as this ocean froze.

To what degree is the separation of crust from mantle completed? Responsible estimates differ widely. Jacobsen and Wasserburg (1979) conclude that the average growth rate of continents for the last 500 million years has been less than half the average rate for the entire history of the planet, and that some 70% of the mantle remains undepleted in constituents that could form more crust. On the other hand, D. M. Shaw (1980) estimates that only about one third of the mantle remains undepleted in crustal ingredients. In both views, there is a substantial reservoir of mantle that retains the potential for supplying additional continental crust, as well as replenishing oceanic crust. Whether the "depleted" mantle forms a layer overlying the undepleted reservoir, or whether both kinds are scattered as blocks or blobs, is still a matter for conjecture (Davies, 1981).

1.8 MAGMATISM ON OTHER PLANETS

The moon, as discussed in Chapters 11 and 12, probably once was covered by a magma ocean (for which the evidence is stronger than that for the earth's). Examination of samples returned from the moon leaves no doubt that lunar rocks are igneous, although some have been drastically modified by shock metamorphism caused by meteorite impact. Furthermore, we see features on the moon, Mars, and Mercury that must be lava flows, and other features, especially on Mars, that must be volcanoes (Greeley and Spudis, 1981).

One of the most surprising discoveries of *Voyagers 1* and *2* was the present-day and intense volcanic activity on Jupiter's satellite Io, the only other planetary body known to have active volcanoes. Eruption rates are far greater than on earth, and the magma seems to have a very different chemistry.

1.9 SUMMARY

Magmatic processes segregate the crust from the mantle, transfer heat from the earth's interior, may be prime movers of lithospheric plates, build the continents above sea level, supply the earth's atmosphere and hydrosphere, modify weather and climate, present severe localized hazards, and bring useful matter and energy from the depths up to levels accessible to mankind. Each of these effects is treated in subsequent chapters. To understand igneous rocks we must recognize and interpret their constituent minerals and be able to reconstruct the history of a magma mass from generation through completion of crystallization and cooling.

2

Igneous Minerals

2.1 COMPOSITIONS AND STRUCTURES OF MINERALS

This book assumes that the reader has had one course each in chemistry and mineralogy. Some concepts from both are reviewed in this chapter because of their importance in magmatic crystallization.

Table 2-1 lists the most abundant elements in the earth's crust. The most common valence charges and ionic radii are also given. Of all the elements in the periodic table, eight make up more than 98% of the crust. Oxygen, the most abundant element by weight, forms relatively large ions. Silicon, the second most abundant, has a much smaller ionic radius. The disparities in size and valence charge among atoms restrict the ways in which they can combine to form minerals. Rules governing the assembly of crystalline structures were deciphered in the 1930s, largely by Linus Pauling and his students. Pauling's rules follow.

(1) Anions (negatively charged ions, of which oxygen is by far the most abundant) cluster around a smaller cation (positively charged ion) to form a *coordination polyhedron*. SiO_4^{4-} and AlO_4^{5-} tetrahedra are examples. Cations show much more variety in size and charge than do anions; in Table 2-1, note that, after oxygen, only sulfur and chlorine form anions among the top 15 elements in the crust, and all three are considerably larger than most common cations. The distance from the center of the cation to the center of an anion is the sum of their ionic radii. The ratio of the ionic radius of the cation to that of the anion determines the kind of coordination polyhedron. Table 2-2

TABLE 2-1 MOST ABUNDANT ELEMENTS IN THE EARTH'S CRUST

Element	Concentration by weight (ppm)	Common valence charge	Ionic radius[a]
O	466,000	2−	1.32
Si	277,200	4+	0.48
Al	81,300	3+	0.61
Fe	50,000	3+	0.73
		2+	0.86
Ca	36,300	2+	1.08
Na	28,300	1+	1.10
K	25,900	1+	1.46
Mg	20,900	2+	0.80
Ti	4,400	4+	0.69
H	1,400	1+	—
P	1,050	5+	0.35
Mn	950	2+	0.91
S	260	2−	1.72
C	200	4+	0.16
Cl	130	1−	1.72
Total	994,290 ppm or 99.43 wt %		

Source: Data from Mason (1966, p. 45, 297–299) and Whittaker and Muntus (1970).
[a]The ionic radius in angstroms (10^{-8} cm) is for sixfold coordination.

TABLE 2-2 REGULAR COORDINATION POLYHEDRA, RADIUS RATIOS (CATION/ANION), AND COORDINATION NUMBERS

Polyhedron	Radius ratio	Coordination number
Triangle	0.15–0.22	3
Tetrahedron	0.22–0.41	4
Octahedron	0.41–0.73	6
Cube	0.73–1	8
Closest packing	1	12

lists the regular coordination polyhedra with their *coordination numbers* (the number of anions grouped around each cation) and the permitted range of values for the ratio of cation radius to anion radius. Coordination polyhedra are only rarely perfectly regular; instead, they are much more commonly distorted, with edges of slightly different lengths. Furthermore, the cation usually is not in the exact center of the polyhedron, but is pulled toward one face or corner. The distortion of polyhedra and displacement of central cations allow other coordination numbers, including 5 and 9, to be common in minerals.

(2) A cation with a charge of p^+, surrounded by m anions (where m is the coordination number), contributes p^+/m charge to each anion. Conversely, each anion with a charge of n^- receives a total contribution of n^+ from all the adjacent cations. Electrostatic attractive forces are maximized, and repulsive

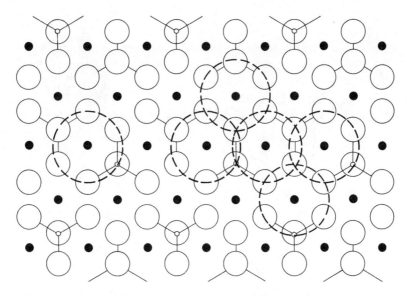

Figure 2-1. Schematic projection of the ionic structure of olivine. The smallest circles (only a few of which are visible) represent silicon cations at the centers of SiO_4^{4-} tetrahedra; the large open circles represent oxygen anions at the corners of these tetrahedra, and the medium-size dark circles represent Mg^{2+} or Fe^{2+} cations. Heavy dashed circles each connect six oxygens that are octahedrally coordinated around a divalent Mg or Fe cation, three of which, in turn, surround each oxygen anion.

forces minimized, by this arrangement. In a single SiO_4^{4-} tetrahedron, therefore, the silicon contributes a charge of 1+ to each oxygen. In the mineral olivine (Figure 2-1), the remaining charges on oxygen anions at the corners of single SiO_4^{4-} tetrahedra are balanced by Mg^{2+} or Fe^{2+} cations that are octahedrally coordinated with these oxygens.

(3) A structure is most stable if adjacent polyhedra of the same coordination number share only corners, not edges or faces. If edges or faces were shared, the distance between mutually repulsive cations at the centers of the polyhedra would be shorter. Figure 2-2, with the same scale and orientation as Figure 2-1, is a view of the olivine structure emphasizing the coordination polyhedra, rather than the individual ions. In olivine, tetrahedra do not share corners with each other, but each tetrahedron shares corners, edges, and faces with neighboring octahedra; octahedra share corners and edges, but not faces, with each other.

(4) Cations with large charges and small radii (hence, low coordination numbers) rarely will share polyhedral faces with each other. As in rule 3, the reason is that the positively charged cations at the centers of polyhedra tend to repel each other, and the higher the charge and smaller the radius, the greater the repulsion. Whenever possible, therefore, polyhedra containing small, highly charged cations are separated by polyhedra holding larger, weakly charged cations.

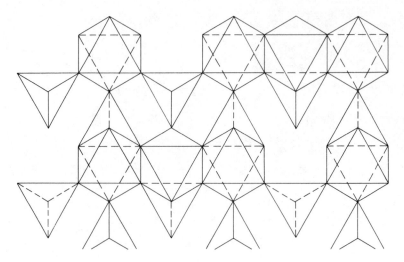

Figure 2-2. Schematic portrayal of coordination polyhedra in olivine; compare with Figure 2-1, which has the same scale and orientation. Relatively few of the tetrahedra and octahedra are shown; in reality, all the space is filled with tetrahedra and octahedra (note that a triangular face of a tetrahedron also serves as a triangular face of an octahedron).

(5) In a given crystal structure, there are only a few kinds of different cation and anion positions.

Pauling's rules describe the geometry and electrostatic charge balance of mineral structures, but do not specify the chemical elements involved. Ions can be imagined as competing for sites in crystal structures (J. A. Philpotts, 1978), and only those ions within a narrow range of charge and radius values can be successful in occupying a given niche. In addition to charge and ionic radius, two other factors must be important in determining the frequency with which one element fills a site in preference to another element. One is the abundance of the element in the available reservoir of matter from which the mineral is crystallizing, and the other is the energy required to free the element from that reservoir. The complex interrelations among the four factors make it difficult to predict with certainty which element will most frequently occupy a given site within a crystal structure. The results of experiments must take precedence over theoretical predictions.

Among the most important ionic substitutions that make *solid solutions* in many materials are Fe^{2+} for Mg^{2+}, K^+ for Na^+, and Fe^{3+} for Al^{3+}. These are predictable from the similarities of charge and ionic radius in each pair. In addition, Si^{4+} in fourfold coordination can be replaced by Al^{3+}, Fe^{3+}, P^{5+}, and Ti^{4+}, and all but the last require additional ionic substitution to balance the charges. The most frequent of these tandem substitutions are $Ca^{2+}Al^{3+}$ for Na^+Si^{4+}, $2Al^{3+}$ for $Mg^{2+}Si^{4+}$, and Na^+Al^{3+} for $Ca^{2+}Mg^{2+}$.

This discussion has assumed that ions behave as rigid spheres of constant radius, but these assumptions are not strictly correct. The radius of a cation is influenced by coordination number (the higher the coordination number, the larger the ionic radius), and the electron clouds around some kinds of ions can be distorted from spherical symmetry. Burns (1970) has taken these deviations from ideal behavior into account.

As indicated in Table 2-1, only a dozen or so elements make up the greatest bulk of the earth's crust. Pauling's rules imply that there is a restricted number of ways in which these few elements are allowed to combine to build the crystal structures of minerals. Chemical studies further indicate that different kinds of ions compete for the same positions within the same structure. The result of these combined tendencies is that, although more than 2500 mineral species are known, only about 20 are at all abundant in the crust and mantle. Of the vast majority of minerals that are rare, some, such as cassiterite, SnO_2, are chemically stable but an essential ingredient is not abundant. Others, such as oldhamite, CaS, contain fairly abundant elements but are not stable relative to other minerals under most conditions in the earth's crust.

The most abundant minerals in the crust and uppermost mantle are the silicates, in which the crystal structures are dominated by SiO_4^{4-} tetrahedra that are arranged in patterns according to the degree to which adjacent tetrahedra share corner oxygen anions. Liebau (1980) gives a clear and brief classification of silicate structures.

Partly because oxygen is the most abundant anion, but mostly as the result of tradition, the analytical chemist usually reports the composition of a rock or mineral in terms of weight percentages of oxides. The oxides SiO_2, TiO_2, Al_2O_3, Fe_2O_3, FeO, MnO, MgO, CaO, Na_2O, K_2O, P_2O_5, CO_2, and H_2O account for at least 95% of the chemical composition of nearly every rock, be it igneous, metamorphic, or sedimentary. However, it is very rare that all these oxides occur in significant amounts within a single mineral.

The *rock-forming minerals* (so called because they are abundant and widespread, in contrast to minerals that are found in relatively few places and in small amounts) have compositions that can be expressed in terms of these oxides. Silicates are the most abundant rock-forming minerals, but the 13 oxides listed above also serve to describe the compositions of oxides, carbonates, phosphates, and hydroxides. Indeed, the only rock-forming minerals requiring chemical constituents not included in the 13 oxides are sulfates and halides (most common in sedimentary rocks) and sulfides, zircon, monazite, chromite, and tourmaline (all usually making up less than 1% of an igneous rock).

2.2 COMPONENTS, PHASES, AND SYSTEMS

A *system* is any portion of the universe that is considered for study. It may be any size (as long as it is macroscopic; that is, it contains a large number of atoms), may be any shape, and may be homogeneous or may differ markedly from one point to another. A *phase* is any part of a system that is mechanically separable from the other parts. In rocks, each mineral constitutes a phase, which may be separated from the others by its density, magnetic and electrostatic properties, or by its size and shape. In addition to the solid phases, rocks may contain one or more liquid phases (for example, water and petroleum). Gases mix completely to form one homogeneous phase. To sum up, all minerals are phases, but not all phases are minerals. Furthermore, minerals are poorly behaved phases for the most part, because a mineral may have different compositions within different parts of the same crystal, owing to solid solution, or the composition may vary from one crystal to another within the rock sample that you have selected as the system.

Components are chemical formula units necessary to describe the compositions of all possible phases in a system. For simplicity, we choose components so that the minimum number of components characterizes all the phases. For example, consider a rock made up of the following three phases:

Quartz, SiO_2
Plagioclase, a solid solution of $NaAlSi_3O_8$ and $CaAl_2Si_2O_8$
Alkali feldspar, a solid solution of $NaAlSi_3O_8$ and $KAlSi_3O_8$

These three phases could be described by different sets of components. We could use the six elements O, Si, Al, Ca, Na, and K as components. Or, noticing that the silicates can be expressed as simple combinations of oxides,

$$NaAlSi_3O_8 = 0.5(Na_2O \cdot Al_2O_3 \cdot 6SiO_2)$$

$$KAlSi_3O_8 = 0.5(K_2O \cdot Al_2O_3 \cdot 6SiO_2)$$

$$CaAl_2Si_2O_8 = CaO \cdot Al_2O_3 \cdot 2SiO_2$$

we could choose the five oxides as components. Alternatively, we could select as components SiO_2 and the three feldspar endmembers, so that all phases can be described in terms of the four components SiO_2, $NaAlSi_3O_8$, $KAlSi_3O_8$, and $CaAl_2Si_2O_8$. The last choice fully describes the compositions of all the phases, and uses the smallest number of components.

One reason for selecting the smallest number of components necessary and sufficient to describe the compositions of all phases is so that we can portray mineral compositions graphically. If we can characterize all phases in a system with only two or three components, we can show the compositional relations by one- or two-dimensional diagrams (that is, lines or planes). For a

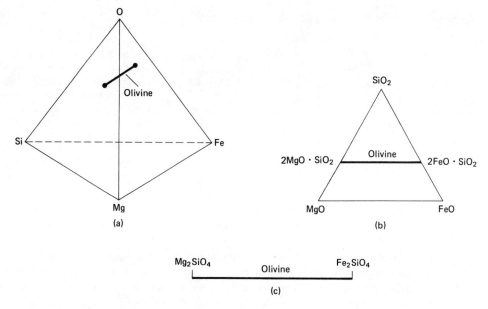

Figure 2-3. Three ways of expressing olivine compositions: (a) in terms of four elements, with a heavy line representing olivine running through the tetrahedron from the back face to the left front face; (b) in terms of three oxides; (c) in terms of two endmembers.

system containing four components, we need a three-dimensional figure. Examples follow.

Consider the mineral olivine, a solid solution involving the endmembers Mg_2SiO_4 and Fe_2SiO_4. We could represent the composition of olivine by the elemental components O, Si, Mg, and Fe, plotted in a three-dimensional tetrahedron (Figure 2-3a), or by the three oxide components SiO_2, MgO, and FeO in an equilateral triangle (Figure 2-3b), or in terms of the two endmembers on a line (Figure 2-3c). If we are dealing with a system containing olivine as the only phase, the third is the logical choice. Because Mg^{2+} and Fe^{2+} are completely interchangeable in the olivine solid solution, olivine compositions fall on an unbroken line between the compositions of the two endmembers.

If a phase A is made up of two components B and C, then A, B, and C lie on a straight line; B and C are endpoints of the line, and A lies somewhere between them, its exact position depending on the proportions of components B and C in the formula of A (Figure 2-4). The relative amounts of B and C

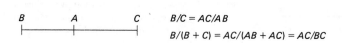

$B/C = AC/AB$

$B/(B + C) = AC/(AB + AC) = AC/BC$

Figure 2-4. Graphical relationship of a reaction $B + C = A$. The equations show two ways of expressing the proportions of B and C at the point A.

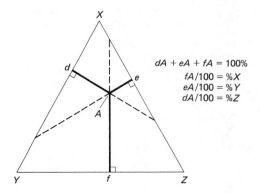

$$dA + eA + fA = 100\%$$
$$fA/100 = \%X$$
$$eA/100 = \%Y$$
$$dA/100 = \%Z$$

Figure 2-5. Representing the composition of a phase A in terms of three components X, Y, and Z. The lines dA, eA, and fA are drawn through point A and perpendicular to the sides of the triangle. According to plane geometry, regardless of the position of point A, $dA + eA + fA =$ the height of the triangle, which is defined as 100%.

are inversely proportional to their distances from A (this relationship is called the *lever rule*). The amount of B times the length AB is equal to the amount of C times the length AC.

If the composition of a phase must be described by three components, we can represent the components as the apices of a triangle (an equilateral triangle, purely for convenience of measurement, but any triangle will work) and still apply the lever rule. If the height of the triangle is taken as 100%, the composition of a phase can be plotted or read in the triangle in terms of the three components (Figure 2-5).

Of course, it makes a difference whether we plot compositions in weight percent or mole (atomic) percent. For example, we can express $2MgO + SiO_2 = Mg_2SiO_4$ in weight percent (Figure 2-6a), so that Mg_2SiO_4 is made up of 42.8% SiO_2 and 57.2% MgO, or in mole percent (Figure 2-6b), so that Mg_2SiO_4 contains 33.3% SiO_2 and 66.7% MgO.

Table 2-3 summarizes the compositions of important minerals in igneous rocks. Only the more important endmembers of some solid solutions are listed; manganese olivine and barium feldspar are examples of those deleted. To emphasize that many endmembers, especially of silicates, are combinations of oxide components in simple rational proportions, compositions are expressed both as a conventional formula and as oxides. Figure 2-7 shows many of the simpler endmembers plotted in terms of three components, to accustom the reader to the graphical representation of mineral compositions.

These plots also provide insights concerning compatible and incompatible phases that will be useful in the next chapter. For example, in Figure

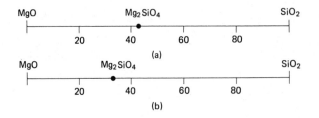

Figure 2-6. Mg_2SiO_4 plotted in terms of the components MgO and SiO_2: (a) in weight percent; (b) in mole (atomic) percent.

18

TABLE 2-3 GENERALIZED COMPOSITIONS AND SELECTED ENDMEMBERS
OF SIGNIFICANT MINERALS IN IGNEOUS ROCKS

A. SILICATES
1. Single tetrahedra and rings
  Olivine, X_2SiO_4, where $X = Mg$, Fe^{2+}, Ca, Mn, Ni
   Forsterite, Mg_2SiO_4 or $2MgO \cdot SiO_2$
   Fayalite, Fe_2SiO_4 or $2FeO \cdot SiO_2$
  Monticellite, $CaMgSiO_4$ or $CaO \cdot MgO \cdot SiO_2$
  Zircon, $ZrSiO_4$ or $ZrO_2 \cdot SiO_2$
  Sphene (titanite), $CaTiSiO_5$ or $CaO \cdot TiO_2 \cdot SiO_2$
  Garnet, $X_3Y_2Si_3O_{12}$, where $X = Mg$, Fe^{2+}, Ca, Mn; $Y = Al$, Fe^{3+}, Cr
   Almandine, $Fe_3Al_2Si_3O_{12}$ or $3FeO \cdot Al_2O_3 \cdot 3SiO_2$
   Pyrope, $Mg_3Al_2Si_3O_{12}$ or $3MgO \cdot Al_2O_3 \cdot 3SiO_2$
   Grossularite, $Ca_3Al_2Si_3O_{12}$ or $3CaO \cdot Al_2O_3 \cdot 3SiO_2$
  Zoisite, $Ca_2Al_3Si_3O_{12}(OH)$ or $4CaO \cdot 3Al_2O_3 \cdot 6SiO_2 \cdot H_2O$
  Epidote, $Ca_2Fe^{3+}Al_2Si_3O_{12}(OH)$ or $4CaO \cdot Fe_2O_3 \cdot 2Al_2O_3 \cdot 6SiO_2 \cdot H_2O$
  Melilite, $X_2YZ_2O_7$, where $X = Ca$, Na; $Y = Mg$, Al; $Z = Si$, Al
   Akermanite, $Ca_2MgSi_2O_7$ or $2CaO \cdot MgO \cdot 2SiO_2$
   Gehlenite, $Ca_2Al_2SiO_7$ or $2CaO \cdot Al_2O_3 \cdot SiO_2$
   Sodium melilite, $NaCaAlSi_2O_7$ or $Na_2O \cdot 2CaO \cdot Al_2O_3 \cdot 4SiO_2$
2. Single chains
  Orthopyroxene, $XSiO_3$, where $X = Mg$, Fe^{2+}
   Enstatite, $MgSiO_3$ or $MgO \cdot SiO_2$
   Ferrosilite, $FeSiO_3$ or $FeO \cdot SiO_2$
  Clinopyroxene, XYZ_2O_6, where $X = Ca$, Na; $Y = Mg$, Fe^{2+}, Fe^{3+}, Al, Cr; $Z = Si$, Al
   Diopside, $CaMgSi_2O_6$ or $CaO \cdot MgO \cdot 2SiO_2$
   Hedenbergite, $CaFeSi_2O_6$ or $CaO \cdot FeO \cdot 2SiO_2$
   Jadeite, $NaAlSi_2O_6$ or $Na_2O \cdot Al_2O_3 \cdot 4SiO_2$
   Acmite, $NaFeSi_2O_6$ or $Na_2O \cdot Fe_2O_3 \cdot 4SiO_2$
   Calcium Tschermak's "molecule," $CaAl_2SiO_6$ or $CaO \cdot Al_2O_3 \cdot SiO_2$
  Wollastonite, $CaSiO_3$ or $CaO \cdot SiO_2$
3. Double chains
  Amphiboles, $W_{0-1}X_2Y_5Z_8O_{22}(OH)_2$, where $W = Na$
   $X = Ca$, Na, Mg, Fe^{2+}
   $Y = Mg$, Fe^{2+}, Fe^{3+}, Al, Ti
   $Z = Si$, Al
4. Layers
  Micas, $X_2Y_{4-6}Z_8O_{20}(OH)_4$, where $X = K$, Na
   $Y = Al$, Mg, Fe^{2+}, Fe^{3+}, Ti
   $Z = Si$, Al
  Talc, $Mg_3Si_4O_{10}(OH)_2$ or $3MgO \cdot 4SiO_2 \cdot H_2O$
  Serpentine, $Mg_3Si_2O_5(OH)_4$ or $3MgO \cdot 2SiO_2 \cdot 2H_2O$
  Chlorite, $X_{12}Y_8O_{20}(OH)_{16}$, where $X = Mg$, Fe^{2+}, Fe^{3+}, Al; $Y = Si$, Al
5. Frameworks
  Feldspars, $XYSi_3O_8$ and $WY_2Si_2O_8$, where $X = Na$, K, Rb
   $W = Ca$, Ba, Sr
   $Y = Al$, Fe^{3+}
   K-feldspar, $KAlSi_3O_8$ or $K_2O \cdot Al_2O_3 \cdot 6SiO_2$
   Albite, $NaAlSi_3O_8$ or $Na_2O \cdot Al_2O_3 \cdot 6SiO_2$
   Anorthite, $CaAl_2Si_2O_8$ or $CaO \cdot Al_2O_3 \cdot 2SiO_2$
  Quartz, tridymite, cristobalite, SiO_2

TABLE 2-3 *(Continued)*

      Feldspathoids
        Nepheline, $NaAlSiO_4$ or $Na_2O \cdot Al_2O_3 \cdot 2SiO_2$
        Kalsilite, $KAlSiO_4$ or $K_2O \cdot Al_2O_3 \cdot 2SiO_2$
        Leucite, $KAlSi_2O_6$ or $K_2O \cdot Al_2O_3 \cdot 4SiO_2$
        Analcime, $NaAlSi_2O_6 \cdot H_2O$ or $Na_2O \cdot Al_2O_3 \cdot 4SiO_2 \cdot 2H_2O$

B. **NONSILICATES**
1. Oxides
      Corundum, Al_2O_3
      Rutile, TiO_2
      Ilmenite, $FeTiO_3$ or $FeO \cdot TiO_2$
      Hematite, Fe_2O_3
      Perovskite, $CaTiO_3$ or $CaO \cdot TiO_2$
      Spinels, YZ_2O_4, where Y = Mg, Fe^{2+}, Zn, Mn, Ni; Z = Al, Fe^{3+}, Cr
        Magnetite, Fe_3O_4 or $FeO \cdot Fe_2O_3$
        Ulvöspinel, Fe_2TiO_4 or $2FeO \cdot TiO_2$
        Hercynite, $FeAl_2O_4$ or $FeO \cdot Al_2O_3$
        Spinel, $MgAl_2O_4$ or $MgO \cdot Al_2O_3$
        Chromite, $FeCr_2O_4$ or $FeO \cdot Cr_2O_3$
2. Carbonates
      Calcite, $CaCO_3$ or $CaO \cdot CO_2$
      Dolomite, $CaMg(CO_3)_2$ or $CaO \cdot MgO \cdot 2CO_2$
      Siderite, $FeCO_3$ or $FeO \cdot CO_2$
3. Halides
      Fluorite, CaF_2
4. Phosphates
      Apatite, $Ca_5(PO_4)_3(OH, F, Cl)$ or $10CaO \cdot 3P_2O_5 \cdot (F_2, (OH)_2, Cl_2)$
      Monazite, $(Ce, La, Th)PO_4$
5. Sulfides
      Pyrite, FeS_2
      Pyrrhotite, $Fe_{1-x}S$
      Sphalerite, ZnS
      Chalcopyrite, $CuFeS_2$

Figure 2-7. *(Opposite page)* Mole percent plots of endmember compositions from Table 2-3 that can be represented by three oxide components. Lines indicate solid solutions in orthopyroxene, olivine, magnetite–ulvöspinel, and ilmenite– hematite. Other solid solutions require more than three oxide components. (a) $MgO-FeO-SiO_2$. *q,* quartz; *en,* enstatite; *fo,* forsterite; *fa,* fayalite. (b) $MgO-CaO-SiO_2$. *q,* quartz; *en,* enstatite; *di,* diopside; *wo,* wollastonite; *ak,* akermanite; *fo,* forsterite. (c) $FeO-CaO-SiO_2$. *q,* quartz; *fs,* ferrosilite; *hd,* hedenbergite; *wo,* wollastonite; *fa,* fayalite. (d) $FeO-Al_2O_3-SiO_2$. *q,* quartz; *fs,* ferrosilite; *alm,* almandine; *fa,* fayalite; *herc,* hercynite; *cor,* corundum. (e) $MgO-Al_2O_3-SiO_2$. *q,* quartz; *en,* enstatite; *py,* pyrope; *fo,* forsterite; *spin,* spinel; *cor,* corundum. (f) $CaO-Al_2O_3-SiO_2$. *q,* quartz; *wo,* wollastonite; *an,* anorthite; *gross,* grossularite; *ca tsch,* calcium Tschermak's "molecule"; *gehl,* gehlenite; *cor,* corundum. (g) $Na_2O-Al_2O_3-SiO_2$. *q,* quartz; *ab,* albite; *jd,* jadeite; *ne,* nepheline; *cor,* corundum. (h) $K_2O-Al_2O_3-SiO_2$. *q,* quartz; *kf,* K-feldspar; *lc,* leucite; *ks,* kalsilite; *cor,* corundum. (i) $MgO-H_2O-SiO_2$. *q,* quartz; *en,* enstatite; *talc,* talc; *fo,* forsterite; *serp,* serpentine; *per,* periclase (MgO); *bruc,* brucite $(Mg(OH)_2)$. (j) $CaO-TiO_2-SiO_2$. *q,* quartz; *wo,* wollastonite; *sphene,* sphene; *pv,* perovskite; *ru,* rutile. (k) $FeO-Fe_2O_3-TiO_2$. *ru,* rutile; *il,* ilmenite; *usp,* ulvöspinel; *mt,* magnetite; *hm,* hematite.

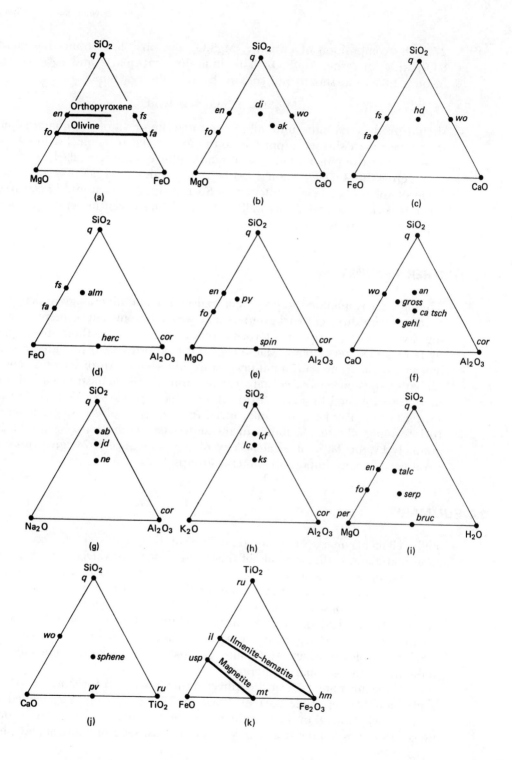

2-7a, the composition of enstatite, $MgSiO_3$, lies on a line connecting quartz, SiO_2, with forsterite, Mg_2SiO_4. This indicates that quartz and forsterite are incompatible, and should not coexist, because the reaction

$$Mg_2SiO_4 + SiO_2 = 2MgSiO_3$$

should produce enstatite until all of the forsterite or all of the quartz (whichever phase is in shorter supply) is consumed by the reaction, leaving a mixture of forsterite plus enstatite or of quartz plus enstatite. Similarly, in Figure 2-7j, perovskite, $CaTiO_3$, should not be found with quartz because the composition of sphene, $CaTiSiO_5$, intervenes. Such diagrams are useful in classifying igneous rocks (Chapter 5), as well as in predicting which minerals we should expect to find with others.

2.3 FURTHER READING

This chapter is intended to provide a transition from mineralogy to petrology. Students with different backgrounds may welcome some supplementary reading. Systematic description of rock-forming minerals is authoritatively given by Deer and others (1966). MacKenzie and Guilford (1980) provide an excellent visual aid to mineral identification in thin section. In order of ascending difficulty, the concepts of crystal structure and crystal chemistry touched upon here are explained by Frye (1974), Mason (1966, p. 75–95), Fyfe (1964), and Burns (1970). Papike and Cameron (1976) lucidly apply these principles to rock-forming silicates. Volumes in the series *Reviews in Mineralogy*, published annually by the Mineralogical Society of America since 1974, give thorough coverage of many important mineral groups.

2.4 SUMMARY

Eight elements make up more than 98% of the earth's crust. Oxygen, the most abundant, is the only one of these with a negative valence charge, and is also one of the largest of the eight. The other seven balance the negative charge of oxygen by their placement in the crystal structure. Pauling's rules summarize the constraints on combinations of ionic charges and sizes, but ions are not strictly rigid spheres of constant volume. One element may proxy for another of similar charge and size occupying a given site in a crystal structure, producing solid solution. Most minerals that are abundant in igneous rocks show this compositional variation.

A system is any part of the universe that is selected for study. A phase is a mechanically separable part of a system. If a rock sample is defined as a system, each mineral in the sample is a phase, whether it forms a large continuous mass or is scattered as many small crystals. One or more liquid phases and a gas phase are commonly present in rock systems. A component is a chemical formula unit (element, oxide, or more complex combination) needed to express the chemical composition of a phase. Many rock-forming minerals can be expressed as simple combinations of oxide components, and may be shown graphically to advantage, allowing prediction of phase incompatibility.

3

Phase Relations

In Chapter 2, the concepts of components, phases, and systems were introduced after a discussion of the constraints that limit compositions and crystalline structures of minerals. Two additional factors strongly influence the mineralogy of igneous rocks. These are temperature and pressure, discussed in Chapter 1. This chapter, in turn, describes the ways in which components combine to form phases under varying conditions of temperature and pressure.

3.1 ONE-COMPONENT SYSTEMS

Any system in which the compositions of all phases can be expressed by only one component is obviously the simplest kind of system, so simple that it has little geological significance. Very few igneous rocks even remotely resemble one-component systems. Nevertheless, systems with a single component are the best starting place for considering more complex systems, and they do illustrate polymorphism, a phenomenon they share with multicomponent systems. Figure 3-1 is a generalized diagram showing the relative stabilities of phases in a one-component system. By definition, the composition of each phase is fixed (if it could vary, more than one component would be needed to describe the variation). We therefore need no dimensions for representing compositions, and one dimension or axis can be assigned to variation in temperature and the other to variation in pressure.

At low pressures and high temperatures, vapor is stable relative to liquid

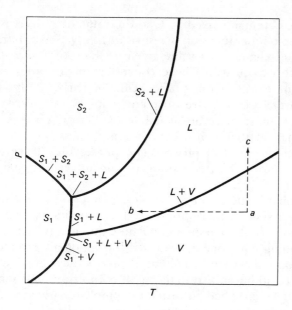

Figure 3-1. Phase relations in a generalized one-component system. P, pressure; T, temperature; L, liquid; V, vapor; S_1 and S_2, solid polymorphs.

or solid. If the pressure is increased at constant temperature (path *a-c*), or if the temperature drops at constant pressure (path *a-b*), the system moves out of the field in which only vapor is stable and crosses the curve labeled $L + V$. Along that curve, liquid coexists with vapor. At b and c in the field above the curve, only liquid is stable. At the point labeled $S_1 + L + V$, three phases coexist. The pressure–temperature diagram of a one-component system is thus made up of fields, curves, and points. In a field, only one phase is stable; along a curve (the boundary between two fields), two phases coexist, and at a point (the intersection of three curves or the place where three fields touch), three phases make up the system at equilibrium. The phrase "at equilibrium" requires a digression concerning the restriction of an assemblage of three phases to a point and of two phases to a curve.

These restrictions are summarized by the *phase rule*,

$$F + P = C + 2$$

where P is the number of phases, C the number of components, and F the number of degrees of freedom. *Degrees of freedom* are the attributes of a system (temperature, pressure, and composition, for example) that can be changed independently without destroying a phase or creating a new phase. Consider the situation in the field labeled V in Figure 3-1. The composition is fixed because this is a one-component system, but both temperature and pressure can vary independently without destroying the vapor and forming a liquid phase. Applying the phase rule, P equals 1 and C equals 1, so F equals 2; there are two degrees of freedom corresponding to the variables temperature and pressure. Exactly on the curve labeled $L + V$, on the other hand, the number of phases is two, C remains fixed at 1, so the value of F decreases to 1. This means that either temperature or pressure can be varied arbitrarily, but the other variable must change to a value given by the coordinates of the curve, or else one of the two phases will be converted to the other as the system "falls off" the curve into a one-phase field. Similarly, at the point labeled $S_1 + L + V$, three phases are present but C equals 1, so $F + P = C + 2 = 3$, and F must therefore equal zero; there is no degree of freedom. This means that, at that point, neither temperature nor pressure may vary without loss of one or two phases. The phase rule, developed a century ago by Josiah Willard Gibbs, is a powerful aid in reading and constructing phase diagrams.

Derivation of the phase rule requires simple algebra and a few definitions. For each phase in a system, there are $(C - 1)$ compositional variables; specifying the proportions of all components but one automatically fixes the proportion of the remaining component, because all component proportions must add up to unity. So, in a system containing P phases there are $P(C - 1)$ compositional variables. In addition, two state variables (pressure and temperature, for example) must be specified in order to completely describe a system. Therefore, the total number of variables required to describe a system containing C components and P phases is $P(C - 1) + 2$. Now F, the number of independent variables (degrees of freedom) is equal to the total number of variables, $P(C - 1) + 2$, minus the number of restrictions. Next we have to calculate the number of restrictions.

At equilibrium, the chemical potential of a component is the same in all phases in a system. For our purposes, we can symbolize chemical potential, without defining it, as u_c^p, the chemical potential of component c in phase p. We can build an array expressing the equality of the chemical potential of each component in all phases, with a vertical column for each phase and a horizontal row for each component.

$$u_1^1 = u_1^2 = u_1^3 = \cdots = u_1^P$$

$$u_2^1 = u_2^2 = u_2^3 = \cdots = u_2^P$$

$$\cdot \qquad \cdot \qquad \cdot \qquad \cdot \qquad \cdot$$

$$u_C^1 = u_C^2 = u_C^3 = \cdots = u_C^P$$

There are $(P - 1)$ restrictions in each row (count the equal signs), and there are C rows, so the total number of restrictions is $C(P - 1)$. Therefore F, the total number of variables minus the total number of restrictions, is

$$F = P(C - 1) + 2 - C(P - 1)$$

$$F = PC - P + 2 - PC + C$$

$$F = C + 2 - P \quad \text{or} \quad F + P = C + 2 \quad \text{the phase rule.}$$

A little reflection will show that a diagram such as Figure 3-1, with the phase rule as a guide, is a remarkably concise way of expressing information. The diagram is an idealization, showing phase relations at equilibrium. *Equilibrium* is the condition under which no spontaneous change can occur; the rate of change in the system is zero. Figure 3-1 indicates that we can never, at equilibrium, find four phases coexisting in this system, and that S_2 is never in equilibrium with V. It might be possible, starting with the system in the field of S_2, to decrease the pressure very rapidly at constant temperature, so that the system moves into the field of V before all of S_2 is converted to L. But this process of rapid change contradicts our requirement that the system remain in equilibrium.

Phase diagrams are constructed with the assumption of equilibrium, and there is indirect evidence in many igneous rocks, particularly in those that cooled slowly, that the phases were in equilibrium with each other, at least over part of the crystallization and cooling history. If, however, the phases of igneous rocks continuously reequilibrated during crystallization and cooling to surface temperature, many of the mineralogical and textural features that record a high-temperature origin would be erased.

As an example of high-temperature history, consider *polymorphism*. In Figure 3-1, two solids, S_1 and S_2, are shown, each with its own field of stability. Because these are two distinct phases, they must be mechanically separable, and therefore differ in physical properties such as density, although they have the same composition. Polymorph S_2 is stable at higher temperatures and pressures than S_1, and the two can coexist in equilibrium only at temperature–pressure combinations lying on the line labeled $S_1 + S_2$. Assume that the hypothetical system corresponds to a real rock, and that the temperature and pressure at the surface of the earth are near the lower left corner of Figure 3-1, in the field in which S_1 is stable. If we find S_2 in the rock at the surface, it is obviously not in equilibrium with S_1, but persists metastably.

Metastable polymorphs, far outside their equilibrium fields in terms of pressure and temperature, are commonly found. Some polymorphic transitions are extremely sluggish, requiring millions of years to be completed under surface conditions. Alluvial diamonds have survived in Precambrian sedimentary rocks for over 2×10^9 years without changing to graphite, the more stable polymorph at the surface. Other polymorphic changes are extremely rapid and cannot be "quenched" to preserve a metastable phase.

When we recognize that a phase is metastable, we are faced with two alternative interpretations; either the phase formed stably but has not "kept up" with changes in temperature and pressure since its formation, or the phase formed metastably and never has been in equilibrium. Examples of both are shown by the one-component system SiO_2. Figure 3-2 is a pressure–temperature diagram showing the stability fields of five solid polymorphs and SiO_2 liquid. There are other polymorphs, at very high and low pressures, but these are omitted for clarity. Note that the tridymite–cristobalite transition

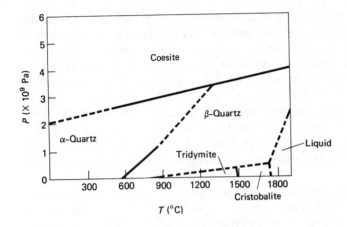

Figure 3-2. The system SiO_2. Dashed lines indicate greater uncertainty in location of boundaries between stability fields. (After Fig. 4-23A of W. G. Ernst, *Earth Materials*, © 1969, p. 88. Adapted by permission of Prentice-Hall, Inc., Englewood Cliffs, N. J.)

is much more strongly dependent on temperature than on pressure (being nearly parallel to the pressure axis), in contrast to the cristobalite–beta quartz transition, which is more nearly parallel to the temperature axis.

Alpha quartz is the most common form of silica on the earth's surface. Coesite, a very high-pressure form, has been produced by meteorite impacts in quartz-bearing rocks, and then persists metastably. Tridymite and cristobalite, high-temperature, low-pressure forms, are found in volcanic rocks (as expected, if they crystallized in their stability fields and survived metastably during rapid cooling). In contrast, beta quartz, easily synthesized by heating alpha quartz above 573°C at atmospheric pressure, inverts instantaneously to alpha quartz upon cooling. No one has yet quenched the beta quartz structure at temperatures outside its stability field.

The great differences in ease of polymorphic transition illustrated by the forms of SiO_2 reflect their differences in crystal structure. For example, the change from beta quartz to alpha quartz, or the reverse, involves only a slight rotation of SiO_4^{4-} tetrahedra and consequently only a small energy change in this so-called *displacive transformation*. On the other hand, tridymite and cristobalite differ more markedly from alpha quartz, necessitating the breaking and reestablishment of oxygen-sharing bonds between tetrahedra during this *reconstructive transformation*, and consequently greater energy changes. As a result, transformation to alpha quartz may not occur in quickly cooled rocks, so tridymite and cristobalite survive.

To show the stability field for coesite, a compressed scale was necessary on the pressure axis. Consequently, the stability field for SiO_2 vapor cannot be shown, because it lies at very low pressure and high temperature, in the lower right corner of the diagram. As a matter of fact, the "triple point" where liquid, vapor, and solid coexist in equilibrium has yet to be experimentally located for any silicate (Yoder, 1980, p. 23).

Finally, note that SiO_2 melts at a very high temperature (over 1700°C at atmospheric pressure) unattainable in the earth's crust except momentarily by meteorite impact or nuclear explosion. When we find a mass composed entirely of quartz, 10 m long, we can be certain that it did not crystallize in the one-component system SiO_2, but that other components were present but escaped. The influence of additional components is the next topic.

3.2 TWO-COMPONENT SYSTEMS

Systems characterized by two or more components have one attribute that is lacking in one-component systems, namely compositional variation. For a system with two components, the two must total 100% of the system. Therefore, specifying the amount or percentage of one component and the total mass of the system automatically fixes the mass of the other component. We can represent variations in the proportions of two components by points on a line, as in Figures 2-4 and 2-6. The line consumes one dimension of the phase diagram. Simultaneously to portray variations in temperature, pressure, and composition in two-component systems, we need three-dimensional phase diagrams (Figure 3-3). Such a temperature–pressure–composition diagram can be visualized as an infinite number of temperature–pressure diagrams, each representing a fixed ratio of component 1 to component 2, stacked side by side along the composition axis. These diagrams are, to say the least, cumbersome, and we generally sacrifice some information by taking temperature–composition slices through the three-dimensional diagrams at a constant pressure (horizontal planes, in the orientation of Figure 3-3), and customarily neglect the vapor phase.

Remember that Gibbs's phase rule for the general case is $F + P = C + 2$, involving two state variables in the last term, but by fixing the pressure we lose one of the state variables; for a two-component system at constant pressure, $F + P = C + 1 = 3$.

The simplest example of a two-component system is shown in Figure 3-4. All liquids, and all solids, are completely immiscible; L_1 is pure liquid A, L_2 is pure liquid B, and the crystalline phases M and N show no solid solution

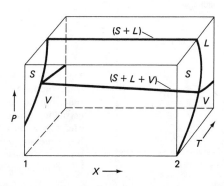

Figure 3-3. Generalized phase diagram for a two-component system, expressed in terms of temperature, pressure, and the compositional variable X, the ratio of component 1 to component 2. S, L, and V are solid, liquid, and vapor.

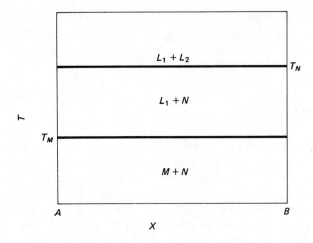

Figure 3-4. Two-component system with complete immiscibility.

but consist entirely of pure components A or B, respectively. T_M and T_N are the melting temperatures of phases M and N. Unfortunately, such simple phase relations, involving complete immiscibility between liquids, are rare in systems involving the rock-forming minerals. Progressing to greater complexity, and therefore more closely approaching reality, the next simplest system is one in which all liquid compositions are completely miscible, but the two solid phases show no solid solution (Figure 3-5).

Such a phase diagram shows three kinds of stability fields, one in which only solid phases are stable, one in which a liquid is the only phase, and one in which a solid coexists with a liquid. It is the solid plus liquid field that is most important in igneous petrology. The boundaries of these kinds of fields require a specific vocabulary.

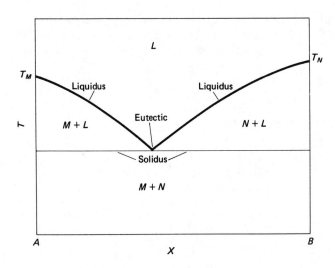

Figure 3-5. Two-component system with complete miscibility in the liquid state, but no solid solution in crystalline phases. As in Figure 3-4, phase M is pure component A and has a melting temperature T_M, and phase N is pure component B and has a melting temperature T_N.

The *liquidus* is the curve representing the highest temperature at which crystals are in equilibrium with liquid. Above the liquidus temperature for a given composition, the system at equilibrium contains no solid phase.

The *solidus* is the curve representing the lowest temperatures at which crystals are in equilibrium with liquid. Below the solidus temperature for a given composition, the system at equilibrium contains no liquid.

A *eutectic* is an invariant point (that is, a site of zero degrees of freedom) representing the temperature and composition at which two solid phases and a liquid coexist at equilibrium. The eutectic point is one kind of intersection between liquidus and solidus.

Before proceeding to other types of phase diagrams for two-component systems, we should examine the rules for interpreting such diagrams. These rules are not based as much on chemistry as on common sense, plane geometry, and algebra.

(1) If two phases are in equilibrium, they must be at the same temperature. On a temperature–composition diagram of any two-component system, a line perpendicular to the temperature axis (and therefore parallel to the composition axis) is a line of constant temperature, and all phases that are in equilibrium must lie on that line.

(2) As long as the system is *closed* (meaning that matter cannot pass into or out of the system) its bulk composition remains constant and must lie on a line perpendicular to the composition axis (parallel to the temperature axis). The number of phases, and their proportions, may vary, but the compositions of all phases, taken in their proportions, must add up to the bulk composition. If only one phase is present, its composition must coincide with the bulk composition of the system.

(3) The lever rule of Chapter 2 applies in determining the proportions of phases in a given bulk composition, just as it does in determining the proportions of components in a phase.

(4) When a phase crystallizes from a liquid, and has a composition different from that of the liquid, the liquid becomes depleted in the component that is preferentially incorporated in the crystals. As crystallization proceeds, the composition of the remaining liquid moves away from the composition of the solid.

(5) Combining rules 2, 3, and 4, if the bulk composition of the system is fixed, but the liquid composition changes, the ratio of liquid to solid must change. Conversely, if the proportion of liquid to solid changes, the liquid must change composition.

These rules should be clear after examining Figure 3-6 and the following discussion. Start with bulk composition Q (25% B) at 1100°C (point a). The entire system consists of liquid with the composition 75% A, 25% B. This

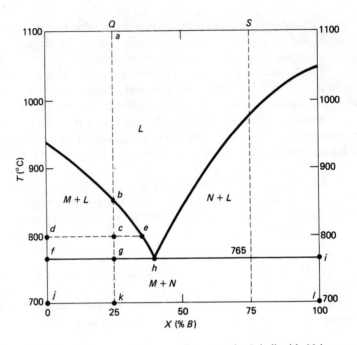

Figure 3-6. Two-component system with a eutectic. *L* is liquid, *M* is pure *A*, and *N* is pure *B*.

liquid persists as the only phase in the system during cooling, until a temperature of 850°C is reached at point *b*. At this temperature, on the liquidus for composition *Q*, phase *M* begins to crystallize. *M* has the composition 100% *A*. As the temperature drops, more of *M* crystallizes, depleting the remaining liquid in component *A* and driving its composition toward *B* along the liquidus curve *b-e*.

At 800°C, the bulk composition (point *c*) is made up of phase *M* (100% *A*, at point *d*) and liquid *L* (65% *A*, 35% *B*, at point *e*). The proportion of *M* to *L*, according to the lever rule, is given by the ratio of line *c-e* to line *d-c*, or roughly 70% *L*, 30% *M*. A check on this calculation is simple, and should always be made after using the lever rule; the percentage of *B* in *M* times the proportion of *M*, plus the percentage of *B* in *L* times the proportion of *L*, equals the percentage of *B* in the bulk composition of the entire system. In this example, [0% *B* in *M* times 30*M*] + [35% *B* in *L* times 70*L*] = 24.5% *B*, close enough to the bulk composition *Q* (25% *B*).

The system continues to cool, with the proportion of *M* to *L* increasing, until the eutectic temperature, 765°C, is reached. The liquid has been driven so far toward *B* that it has reached the eutectic composition (point *h*), and has become saturated with respect to phase *N*. Phases *M* and *N* now crystallize simultaneously until all the liquid is used up. Until the last drop of liquid disappears, the temperature remains at 765°C; according to Gibbs's phase rule when the pressure is held constant, $F + P = C + 1$, and $C = 2$ and $P = 3$ (the three phases are L, M, and N), so there are no degrees of freedom ($F = 0$) and the temperature cannot change until one of the three phases disappears, in this case L. The crystallization of M and N releases heat to maintain this temperature, until all the liquid has crystallized.

At the eutectic temperature, the proportions of L, M, and N are changing, and the lever rule cannot be applied when three phases lie on the same equal-temperature line (also called an *isotherm*). During this crystallization in a closed system at the eutectic temperature, the bulk composition of the system remains at point g, and phases M, L, and N are, respectively, at points f, h, and i.

After the last drop of liquid of composition 60% A, 40% B (point h) freezes to $M + N$, the number of phases is reduced to 2 and the system now has one degree of freedom, and can cool. The 765°C isotherm is the solidus, and below this temperature the system consists of phases $M + N$ in the proportions 75% M, 25% N, as dictated by the bulk composition. For example, at 700°C, the bulk composition is at point k and phases M and N are at points j and l, respectively.

To summarize the behavior of bulk composition Q upon cooling, liquid alone is present from 1100 to 850°C, crystals of M start to form at 850°C and increase in their ratio to L from 850 to 765°C, where the liquid disappears and N crystallizes. Below 765°C, the system consists of 75% M and 25% N.

If the system in Figure 3-6 happened to have the bulk composition coinciding with that of the eutectic h (60% A, 40% B), upon cooling it would remain entirely liquid until the eutectic temperature was reached at 765°C, at which M and N would crystallize simultaneously.

The relations for bulk compositions Q, h, and S are summarized in a different way in Figure 3-7, where the temperatures and the proportions of the phases are shown, but not their compositions. If we make the very approximate assumption that elapsed time is proportional to decreasing temperature in a cooling system, Figure 3-7 conveys the important information that in a two-component system one phase may crystallize before another, after it, or entirely simultaneously with it, depending on the bulk composition of the system.

One other point demands consideration here. In a simple two-component system such as that in Figure 3-6, the consequences of increasing the temperature will be the reverse of cooling, if (and only if) equilibrium is maintained among all phases. That is, regardless of whether we start heating a bulk composition Q, h, or S, the phase assemblage will consist of crystalline M and N until the eutectic temperature of 765°C is reached, at which the first droplet of liquid will form. This first liquid will have the eutectic composition h, even if the bulk composition of the system is 99% A, 1% B. Figure 3-7, showing the proportions of phases as a function of temperature, applies whether the temperature is increasing or decreasing, if the phases are all in

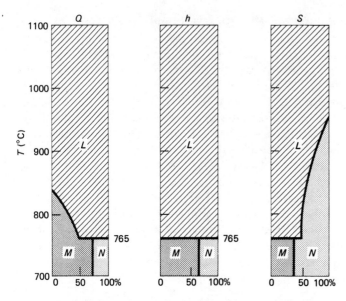

Figure 3-7. Proportions of phases as a function of temperature for bulk compositions Q, h, and S of Figure 3-6.

equilibrium. The compositions of individual phases, shown in Figure 3-6, are independent of the direction of temperature change. The effects of disequilibrium will be taken up later.

Having considered the rules for reading phase diagrams of two-component systems, we can now continue toward more complicated examples, comforted with the knowledge that the rules will remain the same.

The next example (Figure 3-8) involves no solid solution and limited immiscibility between liquid compositions over the range from X' to X''. The compositions of the two liquids approach each other as the temperature rises, and above a critical temperature T_c only one liquid is present over the entire

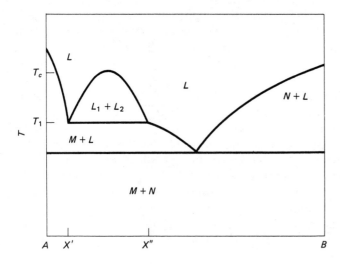

Figure 3-8. Two-component system with no solid solution, a eutectic, and limited liquid immiscibility.

range of compositions from A to B. If the system is poorer in component B than X', or richer in B than X'', no liquid immiscibility will occur at any temperature. Upon slow cooling of a system with a bulk composition lying between X' and X'', the two immiscible liquids diverge in composition until, at temperature T_1, the liquid with composition X' has become so rich in component A that phase M (pure A) begins to precipitate, while the proportion of liquid with composition X'' increases (apply the lever rule at two temperatures, one slightly above T_1 and the other slightly below, to demonstrate this).

A further complication, but a minor one, involves polymorphism (Figure 3-9). Inversion defines a boundary between stability fields for the two polymorphs (M and M' in this example). Note in Figure 3-9b that the liquidus curve must show an inflection at the inversion temperature.

Another variation on the theme, and an extremely common one, results from limited solid solution (Figure 3-10). Note that M and N can tolerate large amounts of components B and A, respectively, in their crystalline struc-

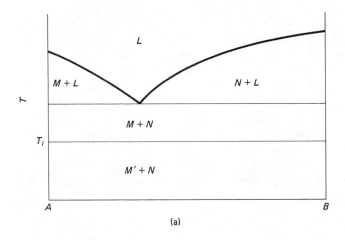

(a)

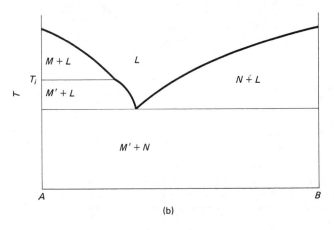

(b)

Figure 3-9. Two-component systems with a eutectic, polymorphism, and no solid solution. M' is the low-temperature polymorph of M. In (a), inversion takes place at temperature T_i below the solidus. In (b), inversion takes place above the solidus.

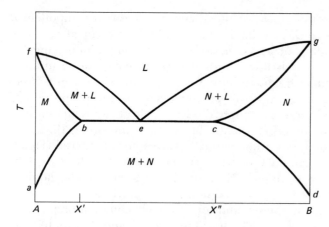

Figure 3-10. Two-component system with a eutectic and with limited solid solution. The liquidus is *f-e-g*, the solidus is *f-b-e-c-g*, and the solvus is *a-b-e-c-d*.

tures, and that at the eutectic temperature a bulk composition between X' and X'' crystallizes to two solid phases b and c with the compositions X' and X'', the limits of solid solution at this temperature. A *solvus* is defined as the curve showing the limit of solid solution. On one side of the solvus, a solid solution is stable as a single phase, but on the other side it unmixes into two phases. In Figure 3-10, the solvus is the curve a-b-e-c-d, and is truncated by the solidus from b to c. As the extent of solid solution increases, the solvus contracts (or drops to lower temperatures), and points b and c in Figure 3-10 pull closer to each other and to point e, the eutectic, as shown in Figure 3-11a. If the solvus detaches from the solidus (Figure 3-11b), point e is no longer a eutectic but is

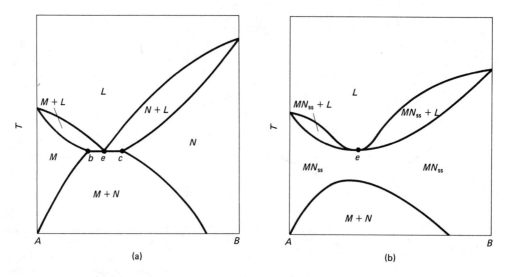

Figure 3-11. (a) Nearly complete solid solution at near-solidus temperatures in a two-component system; compare with Figure 3-10. (b) Complete solid solution (MN_{ss}) just below the solidus in a two-component system. Point e is a minimum.

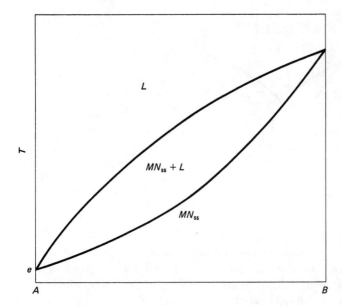

Figure 3-12. Complete solid solution with a minimum (e) at one extreme composition, in a two-component system.

called a *minimum*, and is still a point of intersection of liquidus and solidus. A minimum, like a eutectic, is invariant (has no degrees of freedom), but for a different reason. At a eutectic in a two-component system, three phases coexist at constant pressure, so $F = C - P + 1 = 2 - 3 + 1 = 0$. At a minimum, only two phases are present (solid and liquid), but both must have the same composition. The system is said to be "compositionally degenerate," being part of a two-component system but a part that can be completely described in terms of only one component, because both phases have fixed and identical compositions. Therefore, $F = C - P + 1 = 1 - 2 + 1 = 0$.

Between the solidus and the solvus in Figure 3-11b, there is complete solid solution; components A and B can enter the crystal structure of the solid phase in any proportions. In many systems exhibiting complete solid solution, the minimum is at one end of the compositional range (Figure 3-12), as in the olivine and plagioclase solid solutions.

We have now considered all kinds of two-component systems except two, both of which depend on the existence of an intermediate compound, not a solid solution but a stoichiometric combination of the two components. In the first possibility, the intermediate compound melts *congruently,* that is, to a liquid with the same composition as the compound. In effect, a congruently melting compound divides the two-component system into two back-to-back two-component systems, sharing the composition of the intermediate compound as a common component of both (Figure 3-13). The intermediate compound acts as a fence preventing a liquid in the system A-O from changing composition into the system O-B. The vertical line representing the composition of phase O (with the formula AB) in Figure 3-13 is really a one-component system; upon heating, a system with its bulk composition exactly on that

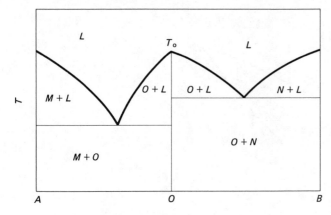

Figure 3-13. Two-component system with a congruently melting intermediate compound O (formula AB), melting at T_0. There is no solid solution in this example.

line will remain pure crystalline O until the melting temperature T_O is reached, although any bulk composition falling a little to one side of that line will begin to melt at one of the two eutectics, at much lower temperatures. Note that in this system M and N are incompatible phases; at equilibrium, one or the other coexists with phase O, but the assemblage $M + N$ cannot exist.

In the one remaining possibility (Figure 3-14), the intermediate compound melts *incongruently*, breaking down to another solid phase with a different composition, plus a liquid which must also have a different composition from that of the intermediate compound. In Figure 3-14, the point p is called a *peritectic*; at that point, three phases (O, N, and a liquid of specific composition p) coexist, so $F = 0$. Peritectics, eutectics, and minima are the kinds of invariant points shown in temperature–composition phase diagrams of two-component systems.

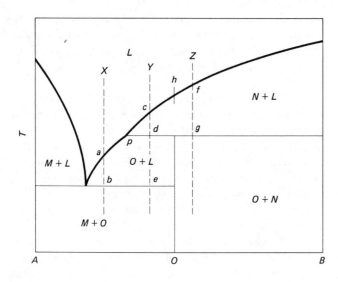

Figure 3-14. Two-component system with an incongruently melting intermediate compound O melting to produce solid phase N plus liquid p. There is no solid solution.

Incongruent melting is shown by some rock-forming minerals (potassium feldspar and magnesian orthopyroxene among them), so we must examine this phenomenon in detail. In Figure 3-14, bulk compositions X, Y, and Z are indicated. Starting at high temperature with a system of bulk composition X, only liquid is present until the liquidus is reached at point a, where crystals of O begin to precipitate. At point b, the solidus is reached, and the last bit of liquid (with the composition of the eutectic between M and O) crystallizes to phase M plus additional O. Below the solidus, the system now contains roughly equal proportions of crystalline phases M and O.

Composition Y remains entirely liquid upon cooling until the liquidus is reached at point c. Then crystals of N begin to precipitate. At point d, at the peritectic temperature, all the crystals of N that have formed previously now react with the liquid, which has in the meantime become so depleted in component B that it has attained the peritectic composition p. The temperature now remains constant, just as it does at a eutectic, until all of phase N is consumed by reaction with liquid p, leaving some of the liquid, while a new phase O begins to crystallize. After all the N is used up at point d, O continues to crystallize from the liquid until the solidus is encountered at point e. But point e also lies on the temperature of the eutectic between M and O, so the system again halts at this second temperature until the last drop of liquid (of eutectic composition) crystallizes to yield M plus additional phase O. Below the solidus, the system consists of approximately 85% O plus 15% M.

Bulk composition Z cools until the liquidus is reached at point f, where N begins to crystallize. At point g, the system reaches the peritectic temperature, and liquid of composition p reacts with the crystals of N. However, unlike the behavior with bulk composition Y, there is not enough liquid to react with all the crystals of N and destroy them, so the liquid itself is consumed in the reaction at the peritectic, and below the solidus the system is made up of approximately 90% O and 10% N.

Finally, let us heat a system with bulk composition identical to that of the intermediate compound O. The system will contain only one phase, crystalline O, until the incongruent melting temperature is reached. As more heat is added to the system, the temperature does not rise until all crystals of O have decomposed into approximately 70% liquid p and 30% crystals of N. After all the O has disappeared, the temperature can rise, and the liquid now moves from composition p toward the vertical line through the composition O, while crystals of N dissolve in the liquid and finally disappear as the liquid reaches the bulk composition of the system at point h. Above the temperature of h, there is only one phase, liquid with the composition of O. Below the solidus and above the liquidus, a system with bulk composition O could be treated as a one-component system, but in the region where both crystals and liquid are present, two components are needed. This contrasts with the one-component behavior of the congruently melting intermediate compound in Figure 3-13 over the entire temperature range.

Earlier it was noted that a congruently melting intermediate compound acts as a "fence," preventing the migration of a liquid composition from one side to the other. In contrast, an intermediate compound that melts incongruently acts as a one-way turnstile, allowing the compositional change of liquid over the dividing line in one direction, but not in the other, during cooling.

In Figure 3-14, bulk composition *Y* is particularly significant because it shows that a crystallizing phase (in this example, *N*) may be the first to crystallize from a liquid upon cooling, but is absent from the final subsolidus assemblage of phases. Such intermittent or aborted crystallization of a phase has been found in many igneous rocks. It is helpful to look at the relations in Figure 3-14 with a temperature–phase proportion diagram, just as we compared Figures 3-6 and 3-7. Figure 3-15 does this for compositions *X, Y, O,* and *Z* of Figure 3-14. In bulk compositions *Y* and *O*, phase *N* was the first to crystallize during cooling of the liquid, but *N* was not present at equilibrium below the peritectic temperature. Inspection of Figure 3-14 shows that phase *N* was "doomed" from the beginning, because bulk composition *Y* lies between the compositions of phases *M* and *O*, and bulk composition *O* corresponds to the single subsolidus phase *O*. Component *B*, which makes up 100% of phase *N*, is simply not sufficiently abundant to stabilize this phase below the peritectic for these bulk compositions.

All two-component systems have phase diagrams like those shown in the preceding figures, or combinations of them. So far, no "real" systems with geologically significant components and phases have been introduced as examples. The main reason is that most "real" systems involve combinations of the idealized relations that have been illustrated, such as solid solution, incongruent melting, and polymorphism, and quite commonly they cannot adequately be described in terms of only two components. Figures 3-16 to 3-18 are examples of real, significant, and complicated systems.

Figure 3-16, the simplest, shows relations in the system $NaAlSi_3O_8$–$CaAl_2Si_2O_8$, the plagioclase solid solution series. The solidus and liquidus form

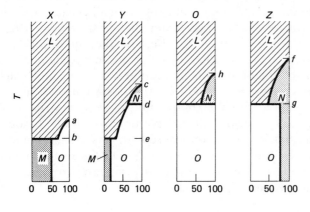

Figure 3-15. Proportions of phases as a function of temperature for bulk compositions *X, Y, O,* and *Z* in Figure 3-14. Lowercase letters correspond to those in Figure 3-14.

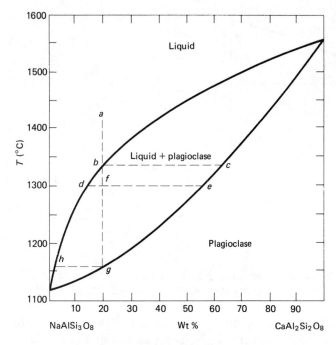

Figure 3-16. The system $NaAlSi_3O_8$–$CaAl_2Si_2O_8$ at 1 atm pressure. (After Bowen, 1913, Fig. 1, with permission of The American Journal of Science.)

a simple loop like that of Figure 3-12. Anorthite, $CaAl_2Si_2O_8$, melts at about 1550°C but albite, $NaAlSi_3O_8$, melts at a little over 1100°C. Compositions between these extremes show a melting interval (the temperature span between liquidus and solidus). The diagram offers a useful review of the rules for reading compositions and proportions of phases, and carries the additional incentive of dealing with the single most abundant rock-forming mineral in the earth's crust.

Start with a bulk composition a (20% anorthite) at 1400°C, well above the liquidus. Cool it in a closed system. When the system reaches 1340°C (point b), it intersects the liquidus; the first crystals of plagioclase (c) form with the composition 62% anorthite. If cooling is slow enough to maintain equilibrium, the crystals continuously react with the liquid, so that the compositions of coexisting crystals and liquid lie at the two points of intersection of the solidus and liquidus with an isotherm. For example, when the system has cooled to 1300°C (f), the coexisting liquid and solid are respectively at d and e, and therefore contain 14 and 55% anorthite. The proportions of liquid and solid are given by the relative lengths of the line segments ef and df, or roughly 90% liquid and 10% solid. When the system has cooled to intersect the solidus at g, all the plagioclase has reequilibrated with the remaining liquid until it has achieved the composition 20% anorthite. When this composition is reached, the composition of the crystalline phase equals the composition of the entire system and there can be no more liquid, so the last trace of liquid (h) crystallizes as the plagioclase arrives at 20% anorthite. There are sub-

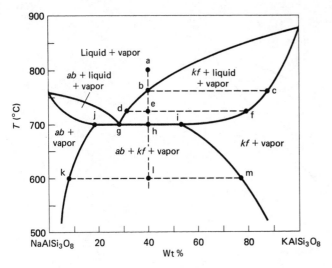

Figure 3-17. The system $NaAlSi_3O_8$–$KAlSi_3O_8$ with H_2O at 5×10^8Pa. *ab*, albite-rich solid solution; *kf*, K-feldspar-rich solid solution. (After Morse, 1970, Fig. 2, with permission of Oxford University Press and S. A. Morse.)

solidus complexities in this system, involving at least three regions of unmixing under solvus curves, but they are still poorly understood and need not concern us here.

Figure 3-17, showing relations in the system $NaAlSi_3O_8$–$KAlSi_3O_8$ in the presence of water at a pressure of 5×10^8 Pa, appears at first glance to be like Figure 3-10. However, each field in the diagram contains one more phase than Gibbs's phase rule should allow if the system contains only two components. The explanation is that the liquid and vapor phases require an additional component, H_2O. The diagram can only show the ratio of the two components $NaAlSi_3O_8$ and $KAlSi_3O_8$ in the solid and liquid phases at a given bulk composition and temperature, but cannot indicate how much water is dissolved in the liquid (Chapter 9 will discuss this), and tells us nothing about the composition of the vapor, which we can correctly surmise is mostly H_2O but also contains Na, K, Al, and Si in unknown proportions, which will vary with temperature.

A liquid in this system containing $NaAlSi_3O_8$ and $KAlSi_3O_8$ in the ratio 60:40 at 800°C (point *a*) will cool to intersect the liquidus at *b* with precipitation of potassium-rich alkali feldspar at *c*. Cooling of the system to *e* results in abundant liquid *d* and subordinate feldspar *f*. What are the compositions and proportions of *d* and *f*? Check your results for agreement with the bulk composition.

Further cooling eventually brings the system to the solidus (the horizontal line *jghi*) at approximately 700°C. Just above the solidus, the system contains liquid *g* and feldspar *i* in roughly equal amounts. At the solidus, the liquid reacts with feldspar *i* to form a new phase, the sodium-rich feldspar *j*. Just below the solidus, the system contains roughly twice as much of feldspar *i* as feldspar *j* (plus the omnipresent vapor). With slow cooling below 700°C, both feldspar phases change composition along the limbs of the solvus. The potassium-rich phase can tolerate more sodium in solid solution than the sodium-rich phase can accommodate potassium, at a given temperature. By the time the temperature has slowly dropped to 600°C, the system contains nearly equal proportions of feldspars k and m, both of distinctly different compositions from the feldspars that were present at higher temperatures.

As the third example, Figure 3-18 shows phase relations in the system $KAlSi_2O_6$–SiO_2 at atmospheric pressure. The outstanding features of this diagram are the intermediate compound $KAlSi_3O_8$ (K-feldspar) that melts incongruently to $KAlSi_2O_6$ (leucite) plus liquid, the eutectic between K-feldspar and SiO_2, and the polymorphic inversion of cristobalite to tridymite upon cooling through 1470°C. Because of the incongruent melting, K-feldspar appears on the liquidus only over a short span from about 40 to 54% SiO_2. Less-silica-rich compositions will first crystallize leucite, but whether this phase will persist in the subsolidus assemblage depends on whether the bulk composition lies between $KAlSi_2O_6$ and $KAlSi_3O_8$ or between $KAlSi_3O_8$ and SiO_2. Note that leucite and a silica polymorph (tridymite, or quartz at a lower temperature than shown) are incompatible.

The bulk composition a, containing 30% silica, cools as liquid from 1600°C to the liquidus at b (about 1420°C). As leucite crystallizes, the remaining liquid becomes more rich in silica. At about 1150°C, the liquid reaches point c and reacts with the leucite to form K-feldspar. In passing through the peritectic at c, the system changes from one containing approximately 75%

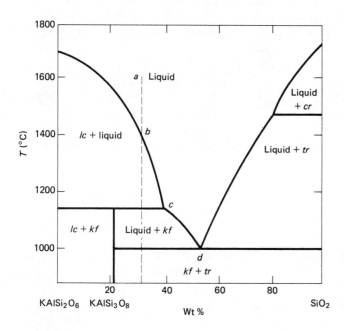

Figure 3-18. The system $KAlSi_2O_6$–SiO_2 at 1 atm pressure. *cr*, cristobalite; *tr*, tridymite; *lc*, leucite; *kf*, K-feldspar. (After Schairer and Bowen, 1947, Fig. 3, with permission of the Geological Society of Finland.)

liquid c plus 25% leucite to one containing about 50% liquid c plus 50% K-feldspar. In other words, roughly equal amounts of leucite and liquid are consumed in making K-feldspar at the peritectic. From 1150 to about 990°C, more K-feldspar forms at the expense of liquid, which ultimately reaches the eutectic at point d. Here the assemblage of 30% liquid and 70% K-feldspar is converted to the subsolidus assemblage 12% tridymite plus 88% K-feldspar.

3.3 THREE-COMPONENT SYSTEMS

Although it is possible to visualize four-component systems by plotting compositions within tetrahedra and holding both temperature and pressure constant, no quantitative treatment is attempted here for systems containing more than three components. Most rocks are assemblages of phases that require at least four components to describe completely the compositional variations in all phases. Three-component systems are obviously a closer approximation to geological reality than are two-component systems.

According to Gibbs's phase rule, $F + P = C + 2$, when $C = 3$, $F = 5 - P$. In a one-phase system of three components, $F = 4$. The four degrees of freedom correspond to temperature, pressure, and two compositional variables (the third compositional variable being constrained by the other two because all three components sum to unity). To show phase relations in a three-component system in only three dimensions, we must surrender one degree of freedom, for example by fixing pressure as constant, and drawing a different diagram for each value of pressure that we wish to consider. Remember, from Figure 2-5, that we can represent three compositional variables (two of which are independent) by a triangular coordinate system, confining the compositional variation to two dimensions. Then we can add the third dimension, and third degree of freedom, as the vertical axis of a triangular prism with a height proportional to temperature (pressure being constant) or, more rarely, to pressure at constant temperature. Figure 3-19 gives an example. More commonly than attempting a three-dimensional diagram, we represent the third dimension as contour lines, as on a topographic map. In the example of Figure 3-20, the contours are isotherms representing the shape of the liquidus surface.

Every ternary (three-component) system is bounded by three binary (two-component) systems. Each binary temperature–composition diagram is a cross section along one edge of the ternary temperature–composition prism. The "valleys" (extensions of binary eutectics into the ternary system) are "bottomed" by *univariant lines* along which three phases are in equilibrium. These meet at an *invariant point* (such as the ternary eutectic d in Figure 3-20, where $M + N + P + L$ coexist and $F = 0$).

Most ternary phase diagrams show the shape of the liquidus surface, not of the solidus which underlies the liquidus and intersects it at certain invariant

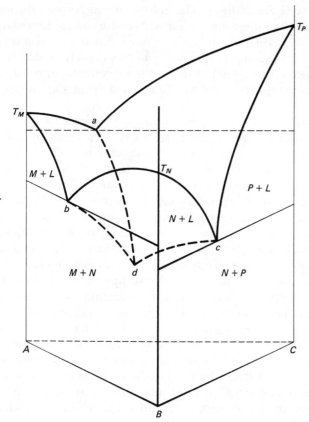

Figure 3-19. Three-component system at constant pressure, represented as a triangular prism with temperature as the vertical dimension. T_M, T_N, and T_P are the melting temperatures of the phases M, N, and P (pure components A, B, and C, respectively). The liquidus surface is outlined by the heavy curves. Points a, b, and c are eutectics in the two-component systems A-C, A-B, and B-C, respectively. Point d is a eutectic in the three-component system A-B-C and is an invariant point where M, N, P, and liquid of composition d coexist at equilibrium.

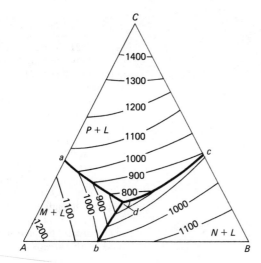

Figure 3-20. The same three-component system as in Figure 3-19, represented as an equilateral triangle with isotherms (in degrees Celsius) on the liquidus surface. Points a, b, c, and d are the same eutectics as in Figure 3-19.

45

points and along certain univariant lines. The solidus is much more difficult to locate precisely than the liquidus by experiment because, using the polarizing microscope and the X-ray diffractometer, it is much easier to detect a few crystals in abundant glass (indicating that the system was quickly cooled from a temperature slightly below the liquidus) than it is to detect a trace of isotropic, amorphous glass among many crystals (quenched from slightly above the solidus).

Ternary systems may include intermediate compounds which may themselves be binary (lying on one edge of the triangle) or ternary (lying within the triangle). Before we can go further, more vocabulary is needed. A *conjugation line* is a straight line connecting the compositions of two phases. A *liquidus field* is an area on the liquidus surface representing all compositions from which a specific phase is the first to crystallize. Here $F = 4 - P$, and we have one solid phase and a liquid in equilibrium, so liquidus fields are divariant ($F = 2$). Boundary curves or *cotectics* are junctions of the liquidus fields. On them, two solid phases and a liquid are in equilibrium, so cotectics are univariant. Whenever a cotectic separating the liquidus fields of two phases crosses the conjugation line that joins the compositions of these two phases, the point of intersection is a temperature maximum on the cotectic. Figure 3-21 illustrates this principle, which is important because it limits the compositional change allowed a cooling liquid. A liquid falling within the triangle *A-B-BC* in the figure can never, while cooling, migrate into the triangle *A-BC-C*. The last drop of liquid in a crystallizing system with a bulk composition in the triangle *A-B-BC* will have the composition of the eutectic *y*, and cannot climb over the temperature maximum at *x* into the triangle *A-BC-C*. This restriction is important in prohibiting certain compositional gradations among igneous rocks.

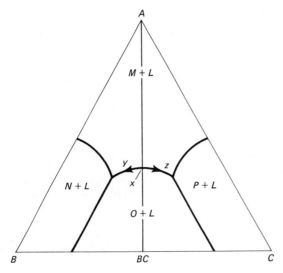

Figure 3-21. Ternary system *A-B-C* with an intermediate binary compound *BC*. Phases *M, N, O,* and *P* are pure *A, B, BC,* and *C,* respectively. The curve *y-x-z* is the cotectic representing *M + O + L*; points *y* and *z* are eutectics where respectively *N + O + M + L* and *M + O + P + L* coexist. The line *A-BC* is the conjugation line joining the compositions of phases *M* and *O*; it intersects the cotectic *M + O + L* at point *x*. Point *x* is consequently a temperature maximum on the cotectic, along which temperatures decrease in the directions of the arrows.

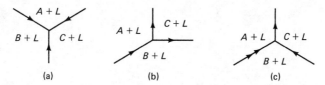

Figure 3-22. Three kinds of invariant points in ternary systems. (a) Ternary eutectic; three cotectics intersect at a temperature minimum. (b) Solid decomposition; three cotectics intersect, and the temperatures decrease along two away from the invariant point. Either phase A or B dissolves as C crystallizes at the invariant point. (c) Peritectic; three cotectics intersect, and the temperature along one decreases away from the invariant point, after phase B is resorbed into the liquid. Resorption of B takes place along the cotectic $A + B + L$, and is completed at the peritectic.

Because the migration of liquid compositions is so important, some labeling conventions have been established for cotectics, and these are shown in Figure 3-22. An arrow points in the direction in which the liquid composition moves with falling temperature. A double-headed arrow indicates that a solid phase is being resorbed into the liquid. There are three kinds of invariant points in ternary systems, at which three solid phases and a liquid coexist. These are also summarized in Figure 3-22.

As with binary systems, there are rules for estimating the proportions and compositions of phases in equilibrium in ternary systems.

1. The composition of the aggregate already crystallized is given by the intersection of a binary edge with a line drawn through the bulk composition of the system and through the composition of the liquid (Figure 3-23).

2. A tangent to the cotectic, extended to a binary edge, shows the relative amounts of solid phases crystallizing at the temperature of the point of tangency.

These rules should be clarified by the figure; start with bulk composition x. Above 1400°C for this composition, only liquid is stable. At 1400°C, a trace of phase P (100% C) begins to crystallize. As the temperature decreases, more P crystallizes, extracting component C, and only that component, from the liquid. The composition of the liquid therefore moves in a straight line directly away from C. At 1300°C, the liquid has the composition y; by the lever rule, the proportion of P already crystallized is given by xy/Cy and the proportion of liquid remaining is given by Cx/Cy.

Cooling proceeds with crystallization of more P and movement of the liquid composition toward point z. At z, a trace of phase N joins $P + L$ in equilibrium. Phase N is 100% component B. Now, with both P and N crystallizing along the cotectic $P + N + L$, the liquid must move down the

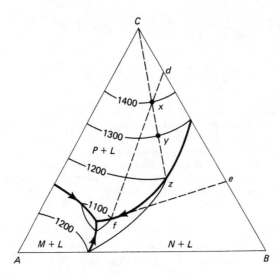

Figure 3-23. Ternary system with a ternary eutectic. Phases *M, N,* and *P* are pure components *A, B,* and *C,* respectively. Line *f-e* is tangent to the cotectic *P + N + L* at point *f* at 1100°C.

cotectic in the direction of the arrow, away from both *C* and *B,* because both components are now being extracted from the liquid.

At point *f* (1100°C), the relative amounts of *P* and *N* already crystallized are given by the lengths of the line segments *Bd* and *Cd,* respectively (rule 1), and the relative amounts of *P* and *N* crystallizing at the moment that the liquid is at 1100°C (point *f*) are given by the lengths of the line segments *Be* and *Ce,* respectively, where *fe* is the tangent to the cotectic at *f* (rule 2).

The liquid then proceeds down the cotectic to the ternary eutectic, where phase *M* begins to crystallize along with *P* and *N.* The temperature remains constant at this invariant point until all the liquid has crystallized. Then the temperature drops, and the final assemblage consists of the phases *M + N + P* in the proportions given by the initial bulk composition *x,* if the system has remained closed.

So far, we have considered only ternary systems with a eutectic and with no solid solution. To appreciate the more complex and realistic situations, we start with a ternary system containing a binary compound that melts congruently (Figure 3-24). In the bounding binary system *A-C,* there is a compound *AC* that melts congruently. In effect, the conjugation line *AC-B* divides the ternary system into two simple eutectic systems *A-AC-B* and *B-AC-C.* Point *d,* the intersection of the conjugation line with the cotectic $E_1 - E_2$ (representing the equilibrium assemblage *AC + N + L*) is a temperature maximum or "thermal divide." For the final liquid to reach $E_2,$ the bulk composition must lie in the triangle *A-AC-B.*

As examples, trace the crystallization histories of bulk compositions *x, y,* and *z* in Figure 3-24. All three lie in the liquidus field of phase *N,* so crystallization of *N* drives the liquids away from *x, y,* and *z,* respectively, to points *a, b,* and *c* on cotectics. At point *a, P* begins to crystallize with *N,* and the liquid

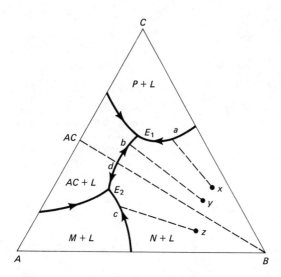

Figure 3-24. Ternary system A-B-C with a binary compound AC. Phases M, N, and P are pure components A, B, and C, respectively; AC is an intermediate compound that melts congruently. AC-B is a conjugation line.

moves to E_1, ultimately crystallizing to the final solid assemblage $P + N + AC$. However, for composition y, at point b phase AC begins to crystallize with N, and phase P does not form until the liquid reaches E_1; the final assemblage is again $P + N + AC$, but the order of crystallization was different, and the final assemblage is slightly richer in AC and N than was composition x. Composition z crystallizes N first, and the liquid moves to c, where M joins $N + L$ at equilibrium. M and N crystallize, driving the liquid to E_2, and the final assemblage is $M + AC + N$.

Often we need to know the proportions of phases in a subsolidus assemblage, but the presence of intermediate compounds has divided the equilateral triangle of the phase diagram into two or more scalene triangles. Even in the absence of an equilateral triangular coordinate system, we can still read the composition of any point in terms of phases or components at the apices of a scalene triangle. Figure 3-25 shows how. From each apex of the triangle, draw a line (dashed in the figure) to the point representing the composition sought, and extend the line (heavy and solid in the figure) to the opposite side of the triangle. The ratio of the length of a solid line segment to the total length of all three solid line segments is the percentage of the component or phase toward which the line segment points. This method will also work, of course, for equilateral triangles, but is more general than that shown in Figure 2-5.

If an intermediate binary compound melts incongruently, its participation in a ternary system will be that shown in Figure 3-26a. Note that as a consequence of incongruent melting, the liquidus field of phase AB does not overlap the composition AB. This lack of congruence is the easy way to recognize the existence of an incongruently melting compound on such a diagram.

The behavior of a ternary intermediate compound introduces more geo-

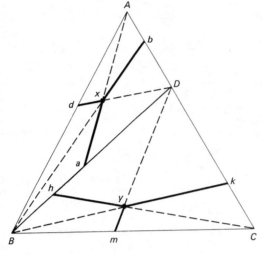

Figure 3-25. Bulk compositions of points x and y can be expressed as percentages of A, B, and D or of B, C, and D, respectively, as well as in percentages of A, B, and C. For point x, $\% A = ax/(ax + bx + dx)$, $\% B = bx/(ax + bx + dx)$, and $\% D = dx/(ax + bx + dx)$. For point y, $\% B = ky/(my + hy + ky)$, $\% C = hy/(my + hy + ky)$, and $\% D = my/(my + hy + ky)$.

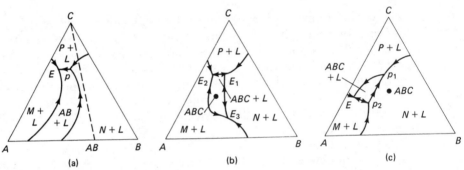

Figure 3-26. Ternary systems with intermediate compounds. (a) Binary compound AB melts incongruently. (b) Ternary compound ABC melts congruently. (c) Ternary compound ABC melts incongruently.

metrical complexity but no new principles. Figure 3-26b shows a congruently melting ternary compound ABC which effectively divides the ternary system into three ternary eutectic systems A-ABC-B, C-ABC-B, and A-ABC-C. This situation is comparable to that in the corresponding binary system (Figure 3-13). In Figure 3-26b the conjugation lines A-ABC, B-ABC, and C-ABC have been omitted for clarity. You can place a straightedge on the figure to trace the position of each conjugation line and confirm that the rule concerning temperature maxima on cotectics has been obeyed.

In Figure 3-26c, the ternary compound ABC melts incongruently; note that the liquidus field of ABC does not overlap the composition ABC. The ABC liquidus field is bounded by three cotectics which terminate in a eutectic (point E) and two peritectics (p_1 and p_2).

Solid solution has the same effects on ternary as on binary systems; in

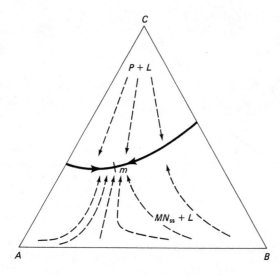

Figure 3-27. Ternary system A-B-C with complete solid solution between components A and B in the phase MN_{ss}. Phase P is pure component C. Point m is a minimum on the cotectic $P + MN_{ss} + L$.

general, the number of liquidus fields is reduced. Because the composition of a solid solution phase may change continuously during crystallization, the composition of the remaining liquid does not migrate in a straight line across the liquidus field of the solid solution, but follows a curve controlled by the ratio of components in the crystals forming at a given temperature. Figure 3-27 shows a simple example of solid solution between two of the three components, to be compared with the ternary eutectic of Figure 3-20. The invariant point has changed from a eutectic to a minimum, point m. Phase P is pure component C, so liquid compositions in the liquidus field of P will migrate toward the cotectic following straight lines (dashed in the figure) radiating from C. In contrast, liquid compositions in the liquidus field of the solid solution MN_{ss} will follow the dashed curves, because the ratio A/B in the crystals changes. For each bulk composition in the liquidus field of MN_{ss}, there is a unique path which the liquid follows during equilibrium crystallization. These paths cannot be predicted, but can only be reconstructed using experimental data. If the bulk composition and the composition of the solid solution that is crystallizing at a specific temperature are known, the liquid composition at that temperature can be calculated.

As in the analogous binary system of Figure 3-12, in a ternary system the invariant minimum may lie at one extreme on the cotectic, as shown in Figure 3-28. In that system, the phase compositions that coexist at various temperatures are shown in Figure 3-29. Take bulk composition b, starting above the liquidus temperature. Upon cooling, phase P begins to crystallize, and the liquid composition moves in a straight line away from C until it reaches the $P + MN_{ss} + L$ cotectic. At a temperature slightly below that at which the cotectic was reached, the coexisting phases L_1, x_1, and P are indicated by the dashed three-phase triangle connecting their compositions. Bulk composition

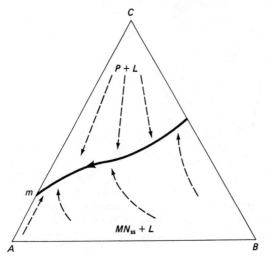

Figure 3-28. Ternary system A-B-C with complete solid solution between components A and B in the phase MN_{ss}. Phase P is pure component C. Point m is the minimum on the cotectic $P + MN_{ss} + L$, and actually lies in the binary system A-C. Dashed curves show composition trends of liquids during crystallization.

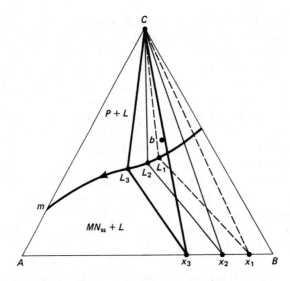

Figure 3-29. Phases coexisting at three temperatures in the ternary system of Figure 3-28. Compositions x_1, x_2, and x_3 represent the MN_{ss} phase coexisting with liquids L_1, L_2, and L_3, respectively. Phase P is pure component C.

b is close to the P-L_1 side of that triangle, indicating that L_1 and P are more abundant than MN_{ss} with composition x_1 at that temperature. As the system continues to cool and crystallize, the three-phase triangle migrates to the left, as shown by the light and heavy triangles drawn with unbroken lines. When the liquid composition has reached L_3 and the solid solution composition has reached x_3, the bulk composition b is almost on the P-x_3 side of the three-phase triangle. With a further slight drop in temperature, the triangle moves a little more to the left, point b will fall on the side of the triangle connecting the two solid phases, and the last of the liquid will disappear. The final subsolidus phase assemblage is indicated by a straight line from C through the bulk composition b and extended to the base of the triangle; the lever rule can be used to find the proportions of the two phases P and MN_{ss}.

3.4 FRACTIONAL FUSION AND CRYSTALLIZATION CONTRASTED WITH EQUILIBRIUM FUSION AND CRYSTALLIZATION

So far in this chapter, we have assumed that each system remained closed (the bulk composition was constant during cooling or heating) and that equilibrium was maintained. Now we examine the consequences of open systems and of departures from equilibrium, conditions that are geologically probable.

Presnall (1969) has summarized the contrasts and similarities between equilibrium and nonequilibrium processes in systems containing liquid and crystals. The following is a modification and amplification of his discussion.

In treating closed systems at equilibrium, we assumed that crystals and liquid remain in complete communication with each other at all times. There are, however, four extreme possibilities. In equilibrium crystallization, removal of heat causes crystals to form; these continue to react with the remaining liquid, if the phase relations so dictate (as in systems involving incongruently melting compounds). In equilibrium fusion, addition of heat produces a liquid that continually reacts with the remaining crystals so that its composition immediately reflects changes in the proportions and compositions of surviving crystalline phases. In *fractional crystallization*, the crystals are isolated from the liquid as soon as they form. In *fractional fusion*, the liquid is immediately isolated from the remaining solid phases and does not react further with them. The mechanisms by which crystals and liquid become isolated need not concern us here, but will be described in Chapter 7.

It is extremely unlikely that removal or isolation of any phases in a system could be instantaneous and totally efficient, so fractional crystallization and fusion are unattainable ideals, just as are equilibrium crystallization and fusion. In natural processes, crystallization and fusion must occur in ways intermediate between equilibrium and perfect fractionation. Nevertheless, it is much easier to deal with the extreme models, so they will be treated here, but we must remember that these are the limiting possibilities, not the strict analogues of magmatic processes.

Figure 3-30 shows a ternary system with a ternary eutectic and no solid solution. We will compare the four extreme processes for one bulk composition, point x in the figure (40% A, 40% B, 20% C).

Take the example of equilibrium crystallization first. Bulk composition x cools to its liquidus temperature of 1000°C, where a trace of phase N (100% B) begins to form. With further cooling, more N separates, and the composition of the remaining liquid migrates away from B toward point s. As s is reached at 900°C, the system contains 77% liquid (of composition s) and 23% crystals of N. Because s lies on the $N + M + L$ cotectic, phase M (100% A) now

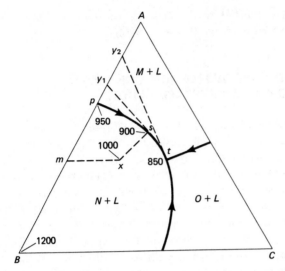

Figure 3-30. Ternary system A-B-C with a ternary eutectic at point t and no solid solution. Crystalline phases M, N, and O are pure components A, B, and C, respectively. Numbers are liquidus temperatures for the indicated points. (After Presnall, 1969, Fig. 1, with permission of the American Journal of Science and D. C. Presnall.)

begins to crystallize, and the liquid composition moves down the cotectic, eventually reaching the ternary eutectic point t at 850°C. Just as the eutectic is reached, the system contains 50% liquid (now of composition t), plus 20% M and 30% N. At this eutectic temperature the remaining liquid crystallizes, and the final subsolidus assemblage, as required by bulk composition x, is 40% M, 40% N, and 20% O. Changes in the proportions of phases, as a function of temperature, are shown in Figure 3-31a.

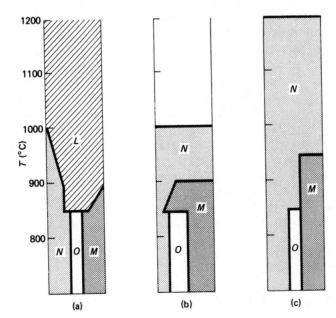

Figure 3-31. Phase proportions as functions of temperature for the system in Figure 3-30. (a) Equilibrium crystallization and fusion. (b) Fractional crystallization; proportions among accumulated crystals. (c) Fractional fusion; proportions among solid residue. Initial bulk composition x is assumed.

Equilibrium fusion is exactly the reverse of equilibrium crystallization. Upon heating bulk composition x, the first liquid, with composition t, forms at 850°C. After all of phase O has dissolved in the liquid, the temperature increases and the liquid composition moves up the $M + N + L$ cotectic until the point s is reached at 900°C. From 850 to 900°C, phase M has been dissolving in the liquid, and the last of M disappears at point s. With further heating from 900 to 1000°C, crystals of N dissolve in the liquid, which reaches the composition x as the liquidus temperature of 1000°C is attained and the last crystals of N disappear. Figure 3-31a applies to equilibrium processes, whether the temperature rises or falls.

Now consider perfect fractional crystallization of the same bulk composition x. As the system cools from the liquidus temperature of 1000°C, N crystallizes and the liquid changes composition from x to s. At 900°C on the $N + M + L$ cotectic at s, phase M begins to separate along with N. From 900 to 850°C, the proportions of solid phases change from 72% M, 28% N (given by point y_1 and the lever rule) to 85% M, 15% N (point y_2). At the eutectic t at 850°C, the remaining liquid crystallizes to yield a subsolidus assemblage of 40% O, 40% M, 20% N, as required by the composition of the eutectic liquid. In this model the crystals are separated from the liquid as rapidly as they form; Figure 3-31b shows the proportions of phases among the accumulated crystals (not the entire system including liquid) at any temperature. On Figure 3-30, although the composition of the liquid changed continuously from x to s to t, the composition of the separated crystals followed a discontinuous path, remaining first at B, jumping from B to y_1, gradually moving to y_2, and finally jumping from y_2 to t.

Perfect fractional fusion is not the reverse of perfect fractional crystallization. Start again with bulk composition x. The assemblage remains unchanged as the temperature rises, until the eutectic temperature (850°C) is reached. At this temperature, liquid of composition t (40% A, 20% B, 40% C) forms, and is removed immediately, leaving a solid residue. The melting and instantaneous removal are repeated until all of phase O (pure C) has been consumed in liquid t. Because C is an essential component of t, no more of that liquid can form after all the C has been removed. At this stage about half of the mass of the original system has been removed as liquid, and the remaining solid assemblage has a bulk composition that has migrated from x away from t to point m (40% M, 60% N). Further heating will not generate a liquid until the temperature of p, the eutectic in the binary system A-B, is reached at 950°C. As liquid of composition p (65% A, 35% B) is removed as fast as it forms, the bulk composition of the solid residue changes from m to pure B. When about 63% of the system has been removed as liquid p, the remaining solid is pure B, which will not melt until 1200°C. Therefore, in contrast to fractional crystallization, fractional fusion produces liquids in a discontinuous path (in this example, liquid compositions are t, then p, and finally B, with gaps during which heating produces no liquid), but the solid residue changes composition along the continuous path from x to m to B. The proportions of phases (in the solid residue only) at a given temperature are summarized in Figure 3-31c. Fractional fusion produces liquids of more restricted compositional range than does fractional crystallization, equilibrium crystallization, or equilibrium fusion. This is an important conclusion, because fractional fusion probably is the closest approximation to the mechanism by which magmas are generated.

3.5 FURTHER READING

More complete and detailed instructions for reading phase diagrams are provided by Ehlers (1972), Cox and others (1979), and especially by Morse (1980), in a book that is recommended reading as a sequel to this one. N. L. Bowen's 1928 classic, still in print but out of date, is unsurpassed for clarity and insight. Yoder (1979) has edited a symposium volume that looks at Bowen's book with a half century of hindsight.

Papers by Presnall (1969), Roeder (1974), and Morse (1976) offer high-level presentation of specific topics. Levin and others (1964, with a 1969 supplement) have compiled thousands of phase diagrams for systems involving inorganic materials of mutual interest to geologists and to the ceramics industry. Their book is prefaced by a clear and practical treatment of rules for reading phase diagrams.

3.6 SUMMARY

This chapter has dealt largely with idealized examples of phase relations in one-, two-, and three-component systems, with few examples involving actual mineral phases. Other chapters provide additional examples. The major points made in this introduction to phase relations are:

1. Compositions and proportions of phases coexisting at equilibrium can be predicted using graphical construction.
2. The composition of a liquid is controlled by the compositions and amounts of phases that have crystallized from it, or from which it is forming by fusion, and by the initial bulk composition of the system.
3. During cooling and crystallization, liquid compositions tend to migrate toward invariant points (eutectics, peritectics, and minima).
4. "Thermal barriers" imposed by the presence of intermediate compounds restrict the compositional variation allowed liquids in a system.
5. There is a general tendency for liquidus temperatures to fall as more components are added to a system.

6. Equilibrium fusion is the reverse of equilibrium crystallization.
7. Perfect fractional crystallization produces continuous changes in liquid composition, but discrete jumps in crystal compositions.
8. Perfect fractional fusion produces discrete jumps in liquid compositions, but continuous changes in the composition of the remaining solids.

4

Estimating, Reporting, and Comparing Igneous Rock Compositions

4.1 MODAL ANALYSIS

In the preceding chapters it was shown that the bulk composition of any system (including an igneous rock) can be expressed in two equally valid ways: the proportions of phases or of components. Consider first the phases.

The composition of a rock expressed as percentages of minerals is the *mode*. Modes can be estimated in several ways with different degrees of confidence. A glance at a sample of coarse-grained rock may indicate that pink K-feldspar and glassy quartz are present in roughly equal amounts, that white plagioclase is in smaller amount than either of these, and that dark biotite makes up an even smaller portion of the rock. Such an estimate, qualitative though it may be, is the mode of the rock.

A better estimate of the mode is obtained by counting points or measuring lengths of lines on the surface of an outcrop, a detached sample, or a thin section. The choice of a particular scale depends on the grain size of the rock and the degree of precision required.

The standard petrographic thin section (25 × 44 mm) contains a slice of rock 30 μm (0.03 mm) thick. For a rock with a density of 2.7 g/cm^3, the thin section contains only 0.1 g of rock, an extremely small fraction of the rock sample. The petrologist must decide whether one thin section per sample will suffice for estimation of the mode. If the rock is heterogeneous or coarse-grained, obviously more than one thin section will be needed.

There are three methods of estimating the relative proportions of phases (minerals) in a rock:

1. By recalculation from a chemical analysis of a sample of the rock, if the composition of each phase is known.
2. By crushing the rock fine enough that each fragment is a particle of only one phase, and counting the grains or separating and weighing them.
3. By measuring the relative areas occupied by the phases on flat surfaces (usually thin sections).

The first two methods are tedious, expensive, and prone to large errors. The third may be tedious, but requires less time and equipment. By way of an example of relative area measurement, consider the problem of determining the percentage of holes (porosity) in Swiss cheese. One way to obtain the percentage is to cut many thin slices and measure the area occupied by holes in each slice. The ratio of total hole area to cheese area in all the slices would be an excellent estimate of the volume ratio of holes to cheese. One randomly chosen slice should provide a good estimate of that ratio if the slice is large compared to the maximum diameter of holes.

We can estimate areas by counting points on a symmetrical grid or by measuring lengths of closely spaced parallel lines (Figure 4-1). There are many pieces of apparatus, of varying degrees of sophistication, for counting points or measuring lines on rock surfaces.

(a)   (b)

Figure 4-1. Three grains of mineral A are exposed in the total area T. (a) Using a symmetrical grid of points (shown as larger dots where they fall within A) to estimate the areal proportions of A and T; % A = number of points in A divided by total number of points = 4/19 = 21%. (b) Using closely spaced parallel lines (shown as heavier lines within A) to estimate the areal proportions of A and T; % A = length of lines in A divided by total length of lines = 55 mm/275 mm = 20%.

It is worthwhile to practice estimation of modes in thin section without resort to any device other than the crosshairs in the microscope ocular. Such modes are not precise, but with care and practice one can determine whether a given phase makes up 5, 10, 25, or 50% of a rock. There are two common methods of modal estimation using the ocular crosshairs (Figure 4-2). One is the quadrant method, in which the observer mentally "crowds" all areas of a particular phase into one quadrant of the microscope field. If that phase could fill the quadrant exactly, its modal abundance would be 25% in this field of view. The second technique uses the linear intercepts of the crosshairs over a given phase. For example, if the east–west crosshair has half of its total

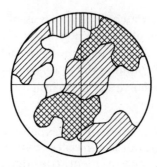

Figure 4-2. Microscopic field of view, with ocular crosshairs superimposed, of a rock containing three phases (one crosshatched, one with parallel ruling, and the third unornamented).

length within quartz grains, and the north–south crosshair has one fourth of its total length within quartz, the modal percentage of quartz in this field is between 25 and 50. Modal estimates based on the ocular crosshair methods must be repeated for many separate fields on a thin section and averaged to arrive at a reasonably accurate mode.

In addition to errors in sampling and estimation, there are several other causes of inaccurate modes. For example, we must only consider the "outcrop areas" of minerals on the upper surface of a thin section. Sections are not infinitesimally thin, and we do see into them. When grain diameters are small relative to section thickness, or when boundaries between grains are inclined from our line of sight, in transmitted light we mistakenly measure the maximum cross-sectional area of the grain, rather than the area on the top surface of the section. This source of error is particularly troublesome with opaque grains and with phases having strong color or high relative relief.

Heterogeneity and preferred orientation present problems. For greatest efficiency, the section should be cut in a plane perpendicular to any linear or planar features. Chayes (1956, p. 16–30) should be consulted before attempting to measure modes of rocks with strongly developed directional features.

Another obvious source of error is misidentification of phases. It is essential to make a complete petrographic study of a rock before estimating the mode. Selective staining of certain minerals, and combinations of reflected with plane- and cross-polarized transmitted light, will assure more accurate identification of phases during modal analysis.

The problem of tabulation remains. How should phases be grouped in recording the mode? Most equipment for modal analysis allows only a limited number of phases to be counted separately. All constituents in excess of 10% (by visual estimate) should be distinguished. Others may be significant enough, in some studies, to require separate listing, or they may simply be lumped as "others" or as "opaque accessories" and "nonopaque accessories." The smaller the percentage of a phase, the less the precision with which this percentage can be estimated in thin section. For phases making up less than 5% of a rock, modal variation of ±3% may not be true variation but statistical scatter.

With practice, and with familiarity with the rock, a precise mode can be measured in about 30 minutes. Modes are usually presented in volume percent but, with knowledge of the approximate specific gravities of all phases, the numbers can be recalculated into weight percent. The use of either volume or weight percent should be specified. For more detail on techniques of modal analysis, refer to Hutchison (1974, p. 47–65). Table 4-1 lists two examples of modes, and Figure 4-3 illustrates some fields of view with varying percentages of phases for reference and practice.

Minerals of igneous rocks are broadly divisible into two groups, *felsic* and *mafic*. "Felsic" (from *fel*dspar and *si*lica) minerals are quartz (and its polymorphs), feldspars, feldspathoids, muscovite, and corundum. Felsic minerals

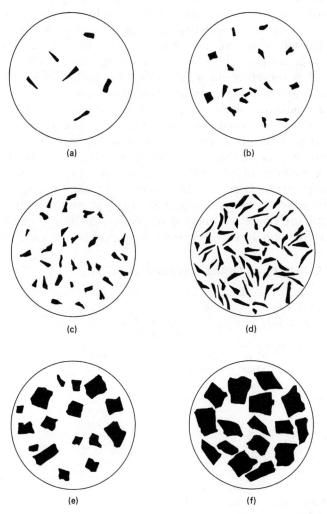

(a)

(b)

(c)

(d)

(e)

(f)

Figure 4-3. Circular fields with the dark constituent covering a measured percentage of the total area: (a) 2%; (b) 5%; (c) 10%; (d) and (e) 25%; (f) 50%.

are sometimes called the "light" or "leucocratic" constituents. "Mafic" (from *ma*gnesium and *f*errous iron) minerals, also called "dark" or "melanocratic" constituents, are all those not included in the definition of felsic. The terms "light" and "dark" are misleading, because plagioclase can be darker than orthopyroxene in the same hand sample of rock, and many mafic minerals are colorless in thin section, as are the felsic constituents. The term "mafic" is not ideal, either, because many minerals included in this group by default (not being felsic) contain neither magnesium nor ferrous iron as essential elements. Nevertheless, the terms "mafic" and "felsic" are widely used and will be retained in this book.

The term *color index*, used in Table 4-1, refers to the total of the modal percentages of mafic minerals in a rock. Glass, not being a mineral, is considered neither felsic nor mafic, and does not enter the color index. As discussed in Chapter 5, color index is important for classifying igneous rocks. For now it is sufficient to know that *ultramafic rocks* are those in which the color index is at least 90, and *mafic rocks* are those in which this total exceeds 50. One of the examples in Table 4-1 is therefore mafic, the other extremely felsic.

The maximum amount of compositional data concerning any rock would be conveyed by a tabulation of the proportions of all phases and the composition of each phase, in other words, by the mode and a chemical analysis of each mineral and, if present, glass. The compositions of phases in rocks can be estimated, with widely varying degrees of confidence, by several methods. These include measurement of physical properties that correlate with chemical composition (refractive indices, specific gravity, optic axial angle, extinction angles, X-ray diffraction lattice spacings, and so on), chemical analysis of separated fragments of the phase, or analysis of portions of thin sections using the electron probe microanalyzer. This instrument, informally called a

TABLE 4-1 EXAMPLES OF MODES FOR IGNEOUS ROCKS (VOL %)

Alkali basalt, Pilot Knob, Austin, Texas		Porphyritic rhyolite, Llano County, Texas	
Olivine	12.0	Quartz	42.0
Clinopyroxene	48.4	Alkali feldspar	54.8
Plagioclase	17.6	Biotite	2.0
Opaque oxides	7.2	Others[a]	1.2
Glass	13.6	Total	100.0
Nepheline + analcime + zeolites	1.2	Color index	3.2
Total	100.0		
Color index	67.6		

[a]Zircon, sphene, opaque oxides, apatite, fluorite, calcite, muscovite, epidote.

"probe," uses a high-energy electron beam to excite characteristic X-rays from spots as small as a few micrometers in diameter in polished thin sections. The intensities and wavelengths or energies of the X-rays provide quantitative estimates of the kinds and abundances of elements in the analyzed spots. Probe analyses are advantageous because they can be made on very small portions of a thin section without destroying the textural relations among the phases, which can be very informative (Chapter 6). Widespread use of the probe and of modal analysis (which can be accomplished using the probe to identify each grain encountered on a grid of points) now permit more complete and exact information on the compositions and proportions of phases. The bulk composition of the system (rock sample) can then be calculated from the mode and the compositions of all phases.

However, for many tens of thousands of rock samples, these data are lacking and only bulk compositions are known, having been determined by chemical analysis of a powdered (presumably representative) portion of a crushed (presumably representative) sample of rock. We now turn to these so-called "whole-rock" analyses.

4.2 CHEMICAL ANALYSIS

The compositions of rocks can be expressed in terms of components without reference to the kinds or amounts of phases. Customarily, chemical analyses of rocks are reported in weight percentages of simple oxide components. In part, this format reflects the abundance of oxygen as the principal anion, but it also reflects historical precedent, dating from the time when many chemical constituents during analysis were isolated and weighed as oxides.

Because one of the purposes of making a chemical analysis is to allow comparison with analyses of other rocks, the data should always be presented in a standard order to facilitate such comparison. A widely accepted standard format is shown in Table 4-2, for the same rocks as the modes in Table 4-1. Other ways of expressing chemical data include recalculating analyses into atomic proportions of cations and anions, but these have not been widely adopted.

In Table 4-2, the alkali basalt shows the oxide components that are commonly reported for igneous rocks. Note that the analysis discriminates ferrous and ferric iron. If only "total iron" is reported (as is common in many recently published data), the information is less useful because it disregards the oxidation state of iron in mafic phases and glass.

H_2O^+ is water expelled when the rock powder is heated above 110°C, and is often termed "structural" or "bound" water, on the assumption that it all represents water or hydroxyl entering the crystal structure of hydrous minerals (such as amphiboles and micas) in the rock. In contrast, H_2O^-, water driven off upon heating a sample from room temperature to 110°C, is called "absorbed" or "hygroscopic" water, assumed to be loosely bound to surfaces

TABLE 4-2 EXAMPLES OF CHEMICAL ANALYSES OF IGNEOUS ROCKS (WT %)

Alkali basalt, Pilot Knob, Austin, Texas[a]		Porphyritic rhyolite, Llano County, Texas[b]	
SiO_2	42.80	SiO_2	75.20
TiO_2	3.57	TiO_2	0.35
Al_2O_3	10.15	Al_2O_3	12.27
Fe_2O_3	2.71	Fe_2O_3	0.63
FeO	9.90	FeO	1.89
MnO	0.17	MnO	0.06
MgO	13.10	MgO	0.27
CaO	12.27	CaO	0.81
Na_2O	2.16	Na_2O	3.05
K_2O	0.36	K_2O	5.00
H_2O^+	2.26	H_2O^+	0.32
H_2O^-	0.58	H_2O^-	0.02
CO_2	0.11	CO_2	0.10
P_2O_5	0.62	P_2O_5	0.06
Total	100.76	BaO	0.04
		F	0.25
		S	0.01
		Less O = (F + S)	0.11
		Total	100.22

Note: The corresponding modes are in Table 4-1.

[a]G. K. Hoops, analyst; from Barker and Young, 1979, with permission of the Texas Academy of Science.

[b]S. S. Goldich, analyst; from Goldich, 1941, Table 2, no. 5, *Journal of Geology,* with permission of The University of Chicago Press.

and filling cracks and fluid inclusions, but not playing a role in the crystal structures of specific phases. H_2O^- and, to some degree, CO_2 contents reflect the degree of postmagmatic alteration or weathering of the sample (but some CO_2 is a primary magmatic constituent). Occasionally, one finds igneous rock analyses which report "loss on ignition" (abbreviated L.O.I.), the weight loss upon heating a sample to very high temperatures. "Loss on ignition" therefore includes H_2O^+, H_2O^-, CO_2, and some other easily volatilized constituents, and may also be affected by oxidation of iron from ferrous to ferric during heating. L.O.I. is a warning that an analysis may be imprecise with respect to other components.

Depending on economic possibilities or mineralogical peculiarities of the rock, other components may be reported in weight percent, as for the rhyolite in Table 4-2. Sulfur is given as S if the rock contains a sulfide mineral, or as SO_2 or SO_3 if a sulfate mineral is present. The geologist must examine the rock in thin section before submitting a portion for analysis, and must tell the analyst if a sulfide or sulfate is present. It also helps the analyst to know in advance if large amounts of hydrous minerals, carbonates, and particularly insoluble or unusual minerals are present. Other constituents reported in weight percent, if warranted by the mineralogy, include ZrO_2, SrO, BaO, Cr_2O_3, NiO, Cl, and F.

Because the assumption is made that all cations are bound only to oxygen, in reporting an analysis as weight percentages of oxide components, some correction must be made if Cl, F, or S is significant. Before summation of an analysis, as in the rhyolite in Table 4-2, the equivalent amount of oxygen for the other anions is calculated and subtracted. After this correction, the sum of all constituents in an acceptable rock analysis should fall between 99.5 and 101.0 wt %. For further discussion of rock analysis, consult Washington (1917, p. 9–26), still one of the clearest accounts, and Fairbairn and others (1951).

4.3 THE CIPW NORM

Rock analyses, expressed as weight percentages of oxides, may be difficult to compare with each other because of the large number of components involved (at least 12), and the wide variation in abundance of each component. To make comparison easier, several schemes have been proposed for combining components into a smaller number of variables analogous to the compositions of mineral phases. The most generally used of these calculations is the *CIPW norm* developed in the late nineteenth century by the American petrologists Cross, Iddings, Pirsson, and Washington.

In the calculation of the CIPW norm, oxide components are combined into arbitrary compounds according to a rigidly prescribed order, originally thought to be the order of crystallization of the analogous minerals in most magmas. These arbitrary compounds or *normative constituents* have the chemical formulas of simple endmembers in the major rock-forming mineral groups (Table 2-3), with the important exception that no normative constituent is hydrous. The CIPW norm, expressed in weight percent, thus resembles the mode of an igneous rock composed entirely of anhydrous crystalline phases in which there were no solid solutions. Table 4-3 lists the normative constituents, their abbreviations, formulas, and the molecular weights of these and of their simple components (called "oxide" components for brevity). The procedure for calculating the CIPW norm follows, based on Washington (1917, p. 1162–1165), with clarifications suggested by Kelsey (1965). These 28 steps are the complete set of instructions, covering all possibilities, but all will be applicable only in extremely rare instances. It will soon be obvious that norm calculation is closer to accounting than to science; one must carefully monitor each expenditure of each component to avoid overdrafts and misallocations.

TABLE 4-3 CIPW NORMATIVE CONSTITUENTS AND THEIR OXIDE COMPONENTS

Oxide components		Normative constituents			
Formula	Mol. wt.[a]	Name	Abbrev.	Mol. wt.[a]	Formula
SiO_2	60.09	Quartz	q	60.09	SiO_2
TiO_2	79.90	Corundum	c	101.96	Al_2O_3
Al_2O_3	101.96	Orthoclase	or	556.70	$K_2O \cdot Al_2O_3 \cdot 6SiO_2$
Fe_2O_3	159.69	Albite	ab	524.48	$Na_2O \cdot Al_2O_3 \cdot 6SiO_2$
FeO	71.85	Anorthite	an	278.22	$CaO \cdot Al_2O_3 \cdot 2SiO_2$
MnO	70.94	Leucite	lc	436.52	$K_2O \cdot Al_2O_3 \cdot 4SiO_2$
MgO	40.30	Nepheline	ne	284.12	$Na_2O \cdot Al_2O_3 \cdot 2SiO_2$
CaO	56.08	Kaliophilite	kp	316.34	$K_2O \cdot Al_2O_3 \cdot 2SiO_2$
Na_2O	61.98	Acmite	ac	462.03	$Na_2O \cdot Fe_2O_3 \cdot 4SiO_2$
K_2O	94.20	Sodium metasilicate	ns	122.07	$Na_2O \cdot SiO_2$
P_2O_5	141.95				
		Potassium metasilicate	ks	154.29	$K_2O \cdot SiO_2$
CO_2	44.01				
Cr_2O_3	151.99	Wollastonite	wo	116.17	$CaO \cdot SiO_2$
NiO	74.70	Diopside	di		
BaO	153.34	Diopside	$di\text{-}di$	216.56	$CaO \cdot MgO \cdot 2SiO_2$
SrO	103.62	Hedenbergite	$di\text{-}hd$	248.11	$CaO \cdot FeO \cdot 2SiO_2$
Cl	35.45				
		Hypersthene	hy		
F	19.00				
		Enstatite	$hy\text{-}en$	100.39	$MgO \cdot SiO_2$
S	32.06				
		Ferrosilite	$hy\text{-}fs$	131.94	$FeO \cdot SiO_2$
SO_3	80.06				
ZrO_2	123.22				
		Olivine	ol		
		Forsterite	$ol\text{-}fo$	140.69	$2MgO \cdot SiO_2$
		Fayalite	$ol\text{-}fa$	203.79	$2FeO \cdot SiO_2$
		Calcium orthosilicate	cs	172.25	$2CaO \cdot SiO_2$
		Magnetite	mt	231.54	$FeO \cdot Fe_2O_3$
		Ilmenite	il	151.75	$FeO \cdot TiO_2$
		Hematite	hm	159.69	Fe_2O_3
		Sodium carbonate	nc	105.99	$Na_2O \cdot CO_2$
		Titanite	tn	196.07	$CaO \cdot TiO_2 \cdot SiO_2$
		Perovskite	pf	135.98	$CaO \cdot TiO_2$
		Rutile	ru	79.90	TiO_2
		Apatite	ap	336.22	$3CaO \cdot P_2O_5 \cdot \frac{1}{3}CaF_2$
		Calcite	cc	100.09	$CaO \cdot CO_2$
		Pyrite	pr	119.97	FeS_2
		Thenardite	th	142.04	$Na_2O \cdot SO_3$
		Fluorite	fr	78.08	CaF_2
		Zircon	zr	183.31	$ZrO_2 \cdot SiO_2$
		Halite	hl	58.44	$NaCl$
		Chromite	cm	223.84	$FeO \cdot Cr_2O_3$

[a]Molecular weights are based on the 1975 scale of atomic weights, in which $^{12}C = 12.000$, as tabulated in Robie and others (1978, P. 4).

(1) Convert the weight percentage of each oxide component listed in the analysis to a "molecular proportion" (m.p.) by dividing the weight percent by the molecular weight of that component (listed in Table 4-3). Using SiO_2 as an example, dividing 75.20 wt % by 60.09, the m.p. of SiO_2 is 1.2515. Do this for each component, rounding each result to four decimal places.

(2) Add the m.p.'s of MnO and, if reported, NiO, to that of FeO.

(3) Add the m.p.'s of BaO and SrO to that of CaO.

(4) If ZrO_2 is reported in the analysis, assign it all to the normative constituent zircon (zr) by writing down the m.p. of $ZrO_2 = zr$. Because the formula for zircon can be written $ZrO_2 \cdot SiO_2$, a m.p. of SiO_2 equal to that of ZrO_2 must also be assigned to zircon. Keep track of the amount of SiO_2 allotted to all normative constituents by adding the m.p. of SiO_2 assigned in each step to a running total Y. So, for this step, $Y = zr$.

(5) Assign the m.p. of all P_2O_5 to apatite (ap). Subtract 3.33 times the m.p. of P_2O_5 from the m.p. of CaO in the analysis. If fluorine is reported, and if the m.p. of F is equal to or greater than two thirds the m.p. of P_2O_5, subtract two thirds of the m.p. of P_2O_5 from the total m.p. of F. If the m.p. of F is less than two thirds of that of P_2O_5, use all the F for ap. These manipulations conform to the formula for normative apatite, $3CaO \cdot P_2O_5 \cdot \frac{1}{3}CaF_2$. Record the m.p. of ap = the m.p. of P_2O_5. A common mistake is to add the m.p.'s of all components to arrive at that of the normative constituent; do not do it. The m.p. of any normative constituent is equal to that of any oxide component of which one formula unit is used, in this case P_2O_5.

(6) Any F remaining after step 5 is assigned to fluorite (fr), CaF_2. The m.p. of fr equals 0.5 times that of F remaining after step 5. The m.p. of CaO available for subsequent steps is reduced by an amount equal to fr.

(7) Assign all Cl to halite (hl) by recording hl = the m.p. of Cl. Reduce the m.p. of Na_2O available for subsequent steps by an amount equal to one half of hl.

(8) Depending on the identity of the modal sulfur-bearing phases, SO_3 reported in the analysis can be retained or converted to S. If it is kept as SO_3 (because a sulfate mineral has been identified in the sample), assign it all to thenardite (th). From the m.p. of Na_2O remaining after step 7, subtract an amount equal to th.

(9) If sulfur is reported as S, or if SO_3 is converted to S because modal sulfide is present, assign all S to normative pyrite (pr). The m.p. of pr equals one half that of S. From the m.p. of FeO subtract an amount equal to pr.

(10) If the rock contains the modal feldspathoid cancrinite, CO_2 is all assigned to sodium carbonate (nc). Set nc equal to the m.p. of CO_2, and reduce the m.p. of Na_2O remaining from step 8 by an amount equal to nc.

If the rock does not contain modal cancrinite but does contain modal calcite (by far the more common situation), all CO_2 is assigned to calcite (*cc*). The m.p. of *cc* is equal to that of CO_2. The amount of CaO remaining after step 6 is reduced by an amount equal to *cc*.

A problem of interpretation arises here; perhaps the calcite is a primary magmatic mineral, but in most igneous rocks it is almost certainly a secondary alteration product. Was the required calcium derived from primary magmatic phases in the rock, or was it introduced from outside, along with the CO_2? There is no easy solution to this dilemma, which is a significant problem because, in rocks containing more than a trace of CO_2, a significant amount of CaO must be assigned to calcite, with consequent reduction in the amounts of normative anorthite, diopside, and wollastonite. Customarily, if the modal calcite is thought to be secondary, CO_2 is ignored, and *cc* is not listed in the norm. This logic assigns all the CaO to other normative constituents on the assumption that CO_2 entered the rock to combine with CaO already residing there. This is probably the better alternative for rocks in which calcite occurs in patches replacing plagioclase or other magmatic minerals, but is less appropriate for rocks in which calcite lines cavities and fractures, having precipitated from fluids that migrated into the rock, probably bringing calcium as well as CO_2.

(11) Assign the m.p. of all Cr_2O_3, if reported, to chromite (*cm*), so that *cm* equals Cr_2O_3. From the amount of FeO remaining after step 9, subtract an amount equal to *cm*.

At this stage all of the less abundant and infrequently reported components have been combined with others (MnO, NiO, SrO, BaO) or have been assigned to normative constituents (ZrO_2, P_2O_5, F, Cl, S, SO_3, CO_2, and Cr_2O_3).

(12) If the m.p. of FeO remaining after step 11 is greater than that of TiO_2, set the m.p. of ilmenite (*il*) equal to that of TiO_2 and subtract the same amount from FeO. There is no TiO_2 remaining for other normative constituents.

If the m.p. of FeO remaining after step 11 is equal to or less than that of TiO_2, set the m.p. of *il* equal to that of FeO, and subtract the same amount from TiO_2. There is no FeO remaining for other normative constituents.

(13) If the m.p. of Al_2O_3 is greater than that of K_2O, the m.p. of provisional orthoclase (*or'*) is set equal to that of K_2O, and an equivalent amount is subtracted from Al_2O_3. Because the formula for *or* is $K_2O \cdot Al_2O_3 \cdot 6SiO_2$, add $6or'$ of silica to Y from step 4. Now $Y = zr + 6or'$.

In the far less common situation that the m.p. of Al_2O_3 is equal to or less than that of K_2O, set *or'* equal to the m.p. of Al_2O_3 and subtract the same amount from K_2O. There is no more Al_2O_3 available to form other normative constituents, but the excess K_2O must be assigned, by making potassium metasilicate (*ks*). To keep track of the assigned silica, add $6or' + ks$ to Y.

(14) If the m.p. of Al_2O_3 remaining after step 13 exceeds that of Na_2O, make provisional albite (*ab'*) in an amount equal to that of Na_2O, and subtract the same amount from Al_2O_3. There is no more Na_2O available for other normative constituents. Add $6ab'$ to Y.

If the m.p. of Al_2O_3 after step 13 is equal to or less than that of Na_2O, set *ab'* equal to the amount of Al_2O_3, and subtract the same amount from the Na_2O remaining after steps 7 and 10. Al_2O_3 is now exhausted. Add $6ab'$ to Y.

(15) If the m.p. of Na_2O remaining from step 14 exceeds that of Fe_2O_3, set acmite (*ac*) equal to the m.p. of Fe_2O_3 and subtract the same amount from Na_2O. This uses up all Fe_2O_3. The Na_2O still remaining is assigned to sodium metasilicate (*ns*). Add $4ac + ns$ to Y.

If the m.p. of Na_2O remaining from step 14 is equal to or less than that of Fe_2O_3, the amount of *ac* is equal to that of Na_2O, which becomes exhausted at this point. Deduct the same m.p. from Fe_2O_3, and add $4ac$ to Y.

(16) If Al_2O_3 exceeds CaO, let anorthite (*an*) equal the amount of CaO, and deduct the same amount from Al_2O_3. All CaO has now been assigned. The Al_2O_3 remaining after formation of *an* is all assigned to corundum (*c*). Add $2an$ to Y.

If Al_2O_3 is less than or equal to CaO, *an* equals the amount of Al_2O_3; deduct the same amount from CaO. All Al_2O_3 has been assigned, and no other aluminous constituents can be calculated. Add $2an$ to Y.

(17) If CaO exceeds TiO_2 remaining from step 12, make provisional titanite (*tn'*), $CaO \cdot TiO_2 \cdot SiO_2$ (titanite is an older name for sphene). The m.p. of *tn'* equals that of TiO_2. Deduct the same amount from CaO, and add *tn'* to Y.

If CaO is less than or equal to TiO_2, *tn'* equals the m.p. of CaO; deduct the same amount from TiO_2, and add *tn'* to Y. CaO is now used up. The remaining TiO_2 is allotted to rutile (*ru*).

(18) If Fe_2O_3 exceeds FeO, the amount of magnetite (*mt*) equals that of FeO, and the same amount is deducted from Fe_2O_3. All FeO is now assigned. The remaining Fe_2O_3 is hematite (*hm*).

If FeO exceeds Fe_2O_3 or equals it, *mt* equals the m.p. of Fe_2O_3, and an amount equal to *mt* is subtracted from FeO. All Fe_2O_3 has been assigned.

(19) Using the m.p.'s of MgO and FeO remaining after step 18, calculate the ratio MgO/(MgO + FeO). For convenience call this ratio M.

(20) If the m.p. of CaO remaining after step 17 is larger than the sum of MgO + FeO remaining after step 18, calculate provisional diopside (*di'*) as two endmembers, *di'-di* and *di'-hd*, respectively $CaO \cdot MgO \cdot 2SiO_2$ and $CaO \cdot FeO \cdot 2SiO_2$, in proportions dictated by the ratio M of step 19. In

other words, the m.p. of *di'-di* equals M times (MgO + FeO); *di'-hd* equals $(1 - M)$ times (MgO + FeO). The m.p.'s of *di'-di* and *di'-hd* thus computed are subtracted from CaO. The remaining CaO is allotted to provisional wollastonite (*wo'*). All MgO and FeO are now assigned. Add $2(di'\text{-}di + di'\text{-}hd) + wo'$ to Y.

If the m.p. of CaO remaining after step 17 is less than or equal to the sum of MgO and FeO remaining after step 18, set the sum of *di'-di* + *di'-hd* equal to the m.p. of CaO. The m.p.'s of the two end-members are then M times CaO and $(1 - M)$ times CaO. All CaO has now been assigned. The remaining MgO and FeO are assigned to provisional hypersthene (*hy'*); the m.p.'s of the endmembers are *hy'-en* equals remaining MgO and *hy'-fs* equals remaining FeO. Add $2(di'\text{-}di + di'\text{-}hd) + (hy'\text{-}en + hy'\text{-}fs)$ to Y.

(21) The "running total" of assigned silica Y, with contributions added in steps 4, 13, 14, 15, 16, 17, and 20, must now be compared with the m.p. of SiO_2 calculated in step 1. This is the moment of truth. If the amount of SiO_2 is greater than Y, the excess silica is assigned to normative quartz (*q*). The norm calculation is now complete, and the provisional constituents calculated in steps 13, 14, 17, and 20 are no longer provisional; their m.p.'s need not be changed, and the primes are removed from their abbreviations. Convert the m.p.'s of all constituents to weight percentages by multiplying each by the appropriate molecular weight (given in Table 4-3).

If, as less commonly happens, the m.p. of silica available in step 1 is less than the total encumbered as Y, the provisional constituents must be reallocated to eliminate the silica deficit. Define this deficit D equal to Y minus the m.p. of silica in step 1. The steps for reducing D to zero must be taken in the following order.

(22) If D is less than half of (*hy'-en* + *hy'-fs*), calculate normative olivine with the endmembers *ol-fo* ($2MgO \cdot SiO_2$) and *ol-fa* ($2FeO \cdot SiO_2$). The sum of (*ol-fo* + *ol-fa*) equals D, and the endmembers are in the proportions dictated by the ratio M in step 19. *Ol-fo* equals M times D, and *ol-fa* equals $(1 - M)$ times D. Then calculate *hy-en* and *hy-fs* in that same ratio M, with their sum equal to (*hy'-en* + *hy'-fs*) − $2D$. The silica deficit has now been reduced to zero by conversion of some of the provisional hypersthene into olivine. All the normative constituents can now be converted to weight percent as in step 21.

If, on the other hand, D in step 21 is greater than half of (*hy'-en* + *hy'-fs*), set (*ol-fo* + *ol-fa*) equal to one half of (*hy'-en* + *hy'-fs*), and calculate the olivine endmembers using the ratio M. The remaining silica deficit D_1 is equal to D minus $0.5(hy'\text{-}en + hy'\text{-}fs)$, and at least one more step is needed.

(23) If D_1 is less than the m.p. of *tn'* (step 17), set *tn* equal to *tn'* minus D_1 and set perovskite (*pf*) equal to D_1. The silica deficit is now zero.

If D_1 is larger than *tn'*, set *pf* equal to *tn'*, and *tn* becomes zero; in other words, convert all titanite to perovskite. The remaining silica deficit D_2 equals D_1 minus tn'.

If, through shortage of TiO_2 or CaO, it was impossible to make tn' in step 17, step 23 must be skipped.

(24) If D_2 (or D_1, if step 23 must be omitted) is less than four times the m.p. of ab', make normative nepheline (ne) such that ne equals $D_2/4$ and ab equals ab' minus $D_2/4$. By this conversion of some provisional albite to nepheline, the silica deficit has been reduced to zero.

If D_2 is at least four times ab', convert all provisional albite to nepheline by setting ne equal to ab' and ab equal to zero. The remaining silica deficit D_3 equals D_2 minus $4ab'$.

(25) If D_3 is less than $2or'$, set leucite (lc) equal to $D_3/2$ and or equal to or' minus $D_3/2$. The silica deficit is zero.

If D_3 is equal to or greater than $2or'$, set lc' equal to or' and or equal to zero. All orthoclase has been converted to leucite, and the remaining silica deficit D_4 equals D_3 minus $2or'$.

(26) If D_4 is less than one half wo', set cs equal to D_4 and wo equal to wo' minus $2D_4$. The silica deficit is now zero.

If D_4 is equal to or greater than one half wo', set cs equal to one half wo' and wo equal to zero. D_5 equals D_4 minus $0.5wo'$.

(27) If D_5 is less than (di'-di + di'-hd), add an amount equal to $D_5/2$ to the amounts of cs and (ol-fo + ol-fa) already calculated; set (di-di + di-hd) equal to (di'-di + di'-hd) minus D_5. The silica deficit is now zero.

If D_5 is equal to or greater than (di'-di + di'-hd), add an amount equal to one half (di'-di + di'-hd) to the amounts of cs and (ol-fo + ol-fa) already calculated. Then di-di and di-hd equal zero. The remaining silica deficit D_6 equals D_5 minus (di'-di + di'-hd).

(28) Set kp equal to $D_6/2$ and lc equal to lc' minus $D_6/2$.

This is the end of the CIPW calculation. Norm calculation is a tedious routine, ideally suited for computer processing. Before one can write a program for norm computation, or communicate with a programmer who will write such a program, considerable practice should be acquired. Two worked examples appear in Table 4-4. In later chapters appear many analyses and norms which may be checked for practice. Those norms were calculated by computer, and the numbers will differ slightly from results of hand calculations due to differences in rounding.

The significance of many of the normative constituents is obscured by purely algebraic manipulation in the steps just enumerated, so additional comments may be useful.

The presence or absence of certain constituents in the norm is highly significant. When quartz appears, the SiO_2 content of the analyzed sample must be high enough that some silica remains in excess of Y at the end of step

TABLE 4-4 CALCULATION OF CIPW NORMS FROM ROCK ANALYSES IN TABLE 4-2

Alkali Basalt

Oxide	Wt %	Mol. wt.	Mol. prop.	
SiO_2	42.80	60.09	0.7123	Y:[13]0.0228[14]0.2316[16]0.3534 [20]0.8977[21]D=0.1854
TiO_2	3.57	79.90	0.0447	[12]il0.0447
Al_2O_3	10.15	101.96	0.0995	[13]0.0957[14]0.0609[16]an0.0609
Fe_2O_3	2.71	159.69	0.0170	[18]mt0.0170
FeO	9.90	71.85	0.1378 ⎫	[12]0.0955[18]0.0785
			⎬ 0.1402	[19]di'-hd'0.0274
MnO	0.17	70.94	0.0024 ⎭	[22]fa0.0256
				[19]M=0.8055
MgO	13.10	40.30	0.3251	[19]di'0.1133[22]fo0.1059
CaO	12.27	56.08	0.2188	[5]0.2041[10]0.2016[16]0.1407
Na_2O	2.16	61.98	0.0348	[14]ab'0.0348[24]ne0.0135 ab0.0213
K_2O	0.36	94.20	0.0038	[13]or'0.0038
CO_2	0.11	44.01	0.0025	[10]cc0.0025
P_2O_5	0.62	141.95	0.0044	[5]ap0.0044

	Mol. prop.	Mol. wt.	CIPW norm (wt %)
or	0.0038	556.70	2.12
ab	0.0213	524.48	11.17
an	0.0609	278.22	16.94
ne	0.0135	284.12	3.84
di-di	0.1133	216.56	24.54
di-hd	0.0274	248.11	6.80
ol-fo	0.1059	140.69	14.90
ol-fa	0.0256	203.79	5.22
mt	0.0170	231.54	3.94
il	0.0447	151.75	6.78
ap	0.0044	336.22	1.48
cc	0.0025	100.09	0.25
H_2O			2.84
			Total 100.82

20. Such a rock with normative quartz is defined as *silica-oversaturated*. In contrast, if olivine, perovskite, nepheline, leucite, calcium orthosilicate, or kaliophilite is present (in other words, if the calculation must be carried beyond step 21 because Y exceeds the available silica), the rock is *silica-undersaturated*.

Corundum in the norm indicates that Al_2O_3 exceeds the molecular proportions required to combine with other components to form normative analogues of feldspar and feldspathoids, and suggests that either modal mica should be present in the sample or the rock contains glass from which alkalis have been leached.

Normative orthoclase, albite, and anorthite require little comment; they are the chief repositories of Al, as well as of K, Na, and Ca, in most igneous rocks. An excess of Na_2O over that required to combine with Al_2O_3 and SiO_2

TABLE 4-4 *(Continued)*

Porphyritic Rhyolite				
Oxide	Wt %	Mol. wt.	Mol. prop.	
SiO_2	75.20	60.09	1.2515	Y:[13]0.3186[14]0.6138 [16]0.6230[20]0.6482[21]q0.6033
TiO_2	0.35	79.90	0.0044	[12]il0.0044
Al_2O_3	12.27	101.96	0.1203	[13]0.0672[14]0.0180[16]0.0134 c0.0134
Fe_2O_3	0.63	159.69	0.0039	[18]mt0.0039
FeO	1.89	71.85	0.0263 ⎫ 0.0271	[9]0.0268[12]0.0224
MnO	0.06	70.94	0.0008 ⎭	[18]0.0185[20]fs0.0185
MgO	0.27	40.30	0.0067	[19]M=0.2659 en0.0067
CaO	0.81	56.08	0.0144 ⎫ 0.0147	[5]0.0134[6]0.0069
BaO	0.04	153.34	0.0003 ⎭	[10]0.0046[16]an0.0046
Na_2O	3.05	61.98	0.0492	[14]ab0.0492
K_2O	5.00	94.20	0.0531	[13]or0.0531
CO_2	0.10	44.01	0.0023	[10]cc0.0023
P_2O_5	0.06	141.95	0.0004	[5]ap0.0004
F	0.25	19.00	0.0132	[5]0.0129[6]fr0.0065
S	0.01	32.06	0.0003	[9]pr0.0003

	Mol. prop.	Mol. wt.	CIPW norm (wt %)
q	0.6033	60.09	36.25
c	0.0134	101.96	1.37
or	0.0531	556.70	29.56
ab	0.0492	524.48	25.80
an	0.0046	278.22	1.28
hy-en	0.0067	100.39	0.67
hy-fs	0.0185	131.94	2.44
mt	0.0039	231.54	0.90
il	0.0044	151.75	0.67
ap	0.0004	336.22	0.13
cc	0.0023	100.09	0.23
fr	0.0065	78.08	0.51
pr	0.0003	119.97	0.03
H_2O			0.34
		Total	100.28

Note: Numbers in brackets indicate steps in the calculation.

in forming normative albite and nepheline is indicated by acmite. If there is still an excess of sodium after combining Na_2O with Fe_2O_3 and SiO_2 to form acmite, sodium metasilicate is calculated. These two indicators of high sodium can appear in norms of both silica-oversaturated and silica-undersaturated rocks.

Normative diopside is analogous to Ca-rich clinopyroxene; its presence indicates that CaO is in excess of that required to combine with Al_2O_3 to form anorthite, and therefore that corundum and diopside cannot appear in the same norm. The presence of wollastonite indicates that CaO is even too high to combine with sufficient MgO and FeO in diopside. Excess MgO and FeO are allotted to hypersthene, the normative analogue of orthopyroxene, if the rock is silica-oversaturated, to hypersthene plus olivine if the rock is only slightly silica-undersaturated, or to olivine without hypersthene if it is more strongly silica-undersaturated.

Silica undersaturation is also indicated by normative perovskite, and to an extreme degree by calcium orthosilicate. Normative potassium metasilicate is similar to sodium metasilicate, and does not necessarily indicate silica-undersaturation, only abnormally high potassium. Normative rutile can appear only if TiO_2 is in excess of the FeO and CaO required to form ilmenite and titanite or perovskite. Sodium carbonate is calculated only if the feldspathoid cancrinite is in the mode. The other constituents, pyrite, thenardite, fluorite, zircon, halite, and chromite, usually are not present in large amounts and indicate only that their essential ingredients were reported in the analysis.

The CIPW norm was initially developed as a basis for igneous rock classification (Chapter 5), but the classification was never widely adopted. Nevertheless, the norm itself has remained a useful means of comparing rocks with each other and with phase diagrams. Because the norm calculation takes alternate branches, depending on the ratios of some components, some normative constituents are incompatible. "Forbidden" combinations are summarized in Table 4-5. This list also serves as a guide to incompatibilities among modal phases in igneous rocks, although there are exceptions to be mentioned later in this chapter.

Just as modal constituents are classified as felsic and mafic, normative constituents are divided into two groups, salic and femic. *Salic* (from *silicon*

TABLE 4-5 INCOMPATIBLE CONSTITUENTS IN THE CIPW NORM OF ANY ROCK

q	never with *ol, pf, ne, kp, lc, cs*
c	never with *ac, wo, di, ne, kp*
or	never with *kp*
ab	never with *lc, cs, kp*
an	never with *ac, ns, kp*
lc	never with *q, ab, hy*
ne	never with *q, hy*
kp	never with *q, c, ab, an, or, hy*
ac	never with *c, an*
ns	never with *c, an*
wo	never with *c, hy, ol*
di	never with *c*
hy	never with *wo, ne, lc, cs, pf, kp*
ol	never with *q, wo*
cs	never with *q, ab, hy*
pf	never with *q, hy*

and *al*uminum) constituents include, as one might expect, quartz, corundum, orthoclase, albite, anorthite, nepheline, leucite, and kaliophilite, analogous to the modal felsic phases. However, the inventors of the CIPW norm also included zircon, halite, thenardite, and sodium carbonate in the salic constituents. The *femic* group (from *fe*rrous iron and *m*agnesium, analogous to mafic) includes all constituents except those listed above. Many petrologists have been dissatisfied by the lack of closer correspondence between the felsic and salic and the mafic and femic groups of modal and normative constituents. Partly to overcome this disadvantage, Thornton and Tuttle (1960) defined the *differentiation index* (D.I.) as the sum of *q* + *ab* + *or* + *ne* + *kp* + *lc*, normative analogues of the modal felsic group excluding *an* and *c*. Because of the incompatibilities arising from the way the norm is calculated (Table 4-5), no more than three of these constituents can appear in any norm (Figure 4-4). The major difference between D.I. and the sum of modal felsic phases is the omission of anorthite; one of several justifications for deleting the anorthite component from the D.I. is that this permits all contributors to the D.I. to be plotted on a two-dimensional figure of the system $NaAlSiO_4$–$KAlSiO_4$–SiO_2 (Figure 4-4), a system of great importance discussed later.

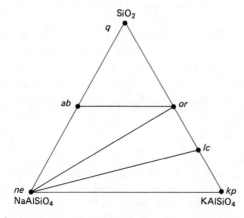

Figure 4-4. Normative constituents of the Thornton–Tuttle differentiation index plotted in mole percent in the system $NaAlSiO_4$–$KAlSiO_4$–SiO_2. The combinations *q* + *ab* + *or*, *ne* + *ab* + *or*, *ne* + *or* + *lc*, and *ne* + *lc* + *kp* are the only ones permitted that involve three normative constituents.

It must be emphasized that, no matter how elegant the calculation, a norm is not primary information, but a chemical analysis is, and must always be presented. Given an analysis, one can compute the norm. However, it is not possible to recalculate the rock composition in oxide weight percentages if only the norm is available.

How close a correspondence should we expect between mode and norm? Obviously, the mode can contain complex hydrous phases such as amphiboles and micas that have no simple normative analogues. Potassium in micas shows up in the norm only in *or*, *lc*, or *kp*, and other components are distributed among many normative constituents. For amphiboles, the situation is even more complex. Table 4-6 shows an analysis of a mineral treated as if it were a rock; contrast the norm with the mode, which is 100% amphibole. Some am-

TABLE 4-6 CHEMICAL ANALYSIS AND CIPW NORM (WT %) FOR HORNBLENDE FROM A GABBRO, ARGYLLSHIRE, SCOTLAND

SiO_2	48.40	*or*	1.00
TiO_2	1.08	*ab*	16.10
Al_2O_3	7.27	*an*	10.79
Fe_2O_3	2.44	*di*	36.11
FeO	9.53	*hy*	8.75
MnO	0.10	*ol*	19.65
MgO	15.86	*mt*	3.54
CaO	11.25	*il*	2.05
Na_2O	1.90	H_2O	2.08
K_2O	0.17	Total	100.07
H_2O^+	1.71		
H_2O^-	0.37		
Total	100.08		

Source: Nockolds, 1941, p. 490, with permission of the Geological Society of London.

phiboles yield quartz in the norm, others nepheline. This one is only slightly silica-undersaturated, so that olivine appears in the norm but hypersthene survives.

Another reason for differences between mode and norm lies in the assumption in the CIPW norm calculation that neither endmember of olivine can coexist with quartz; hypersthene forms instead until either the excess quartz or olivine is consumed. In real igneous rocks, however, iron-rich orthopyroxene is unstable relative to fayalite plus quartz. Figure 4-5 contrasts

Figure 4-5. Diagrams, in mole percent, contrasting normative and modal assemblages of quartz, olivines, and orthopyroxenes. (a) Subsolidus phase relations in the system SiO_2–Mg_2SiO_4–Fe_2SiO_4. The iron endmember of orthopyroxene, $FeSiO_3$ (ferrosilite), is not stable relative to quartz plus fayalite. (After Smith, 1971, Fig. 1, with permission of the American Journal of Science and D. Smith.) (b) The analogous assemblages in the CIPW norm. Labeled fields are: 1, quartz + orthopyroxene in (a), *q* + *hy* in (b); 2, orthopyroxene + olivine in (a), *hy* + *ol* in (b); 3, quartz + iron-rich orthopyroxene + iron-rich olivine; 4, quartz + iron-rich olivine.

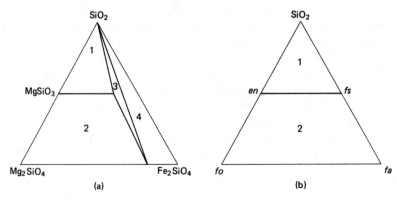

the possible modal and normative assemblages involving quartz, orthopyroxene, and olivine.

Solid solutions of Ti- and Al-bearing components in clinopyroxenes, the
wide range of solid solution between Na and K in high-temperature alkali
feldspars, solid solution in the magnetite–ulvöspinel and ilmenite–hematite
series, and the presence of glass, also cause discrepancies between norm and
mode. With all of these causes in mind, examine the differences between
modes and norms in the two analyzed rocks of Tables 4-1 and 4-4.

4.4 VARIATION DIAGRAMS

It is usually very difficult to detect similarities and differences by scanning
rock analyses or norms arranged as columns and rows of numbers. Graphical
methods aid the eye and imagination during comparison. In many areas, igneous rocks of approximately the same age have different compositions,
largely attributable to varying histories of fusion or crystallization. Properly
shown on graphs, the compositional variation should trace out curves analogous to the paths on phase diagrams. The ideal diagram for showing trends
in rock compositions will show all the significant variations, and only the significant ones, while arranging the rocks in a genetic sequence, say from parent liquid through successive steps of crystallization to the final residual
liquid. The "ideal" diagram has not yet been devised.

Because the oxide component commonly exhibiting the greatest numerical variation is SiO_2, an early type of variation diagram (Harker, 1909) used a
plot of weight percentages of other oxide components versus SiO_2 (Figure
4-6). In this example from Parícutin volcano, Mexico, there is strong evidence
that all the analyzed lava samples are genetically related. All issued from the
same small volcano and, during its nine years of nearly continuous activity,
the SiO_2 content increased from 55 wt % in 1943 to over 60% at the close of
the eruptive activity in 1952 (Wilcox, 1954). The general decrease in FeO,
MgO, and CaO with increasing SiO_2, and the increasing K_2O, are features
common to most such plots. These trends can reflect changes in the proportions and compositions of phases, but do not necessarily give unambiguous
information on how one rock sample is related to another. In the first place,
remember that we are dealing with percentages that total a fixed sum, 100.
Silica is the most abundant oxide component and as it increases, the sum of
the others must decrease. Furthermore, with increasing silica content, the
amounts of mafic phases decrease, leading to general declines in FeO, MgO,
and CaO. Solid solutions in mafic minerals show a strong tendency for increasing substitution of Fe^{2+} for Mg^{2+} as the color index of the rocks decreases. Problems of interpreting variation diagrams based on percentage data
adding to a constant sum have been treated mathematically by Chayes (1971).

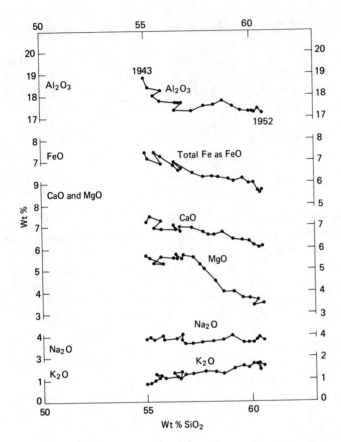

Figure 4-6. Silica variation diagram for lavas of Parícutin volcano, Mexico. Data points are connected in order of eruption of the analyzed lavas; activity began in 1943 and ended in 1952. (Simplified from Wilcox, 1954, Fig. 101.)

For specific suites of igneous rocks, other types of variation diagrams may be useful. For example, Wright (1971, 1974) used MgO as a parameter against which to plot other oxides in Hawaiian lavas. Figure 4-7 is one such diagram, in which CaO is compared with MgO for lavas of the 1959 eruption of Kilauea and for important minerals (olivine, clinopyroxene, orthopyroxene, and plagioclase) in the lavas. The rock compositions fall on or close to a line

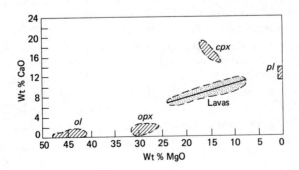

Figure 4-7. CaO versus MgO for lavas and minerals of the 1959 eruption of Kilauea, Hawaii. "Lavas" are whole-rock compositions; *ol, opx, cpx,* and *pl* are the ranges of CaO and MgO contents in olivine, orthopyroxene, clinopyroxene, and plagioclase, respectively. (After Wright, 1971, Fig. 2.)

extending between olivine and plagioclase. This suggests, by analogy with the crystal–liquid composition relations in phase diagrams discussed in Chapter 3, that the chemical variation in these Kilauea lavas can be explained by precipitation of varying amounts of olivine (driving the residual liquids along a line away from olivine) or of plagioclase (driving them away from plagioclase), or of both. Extraction of either clinopyroxene or orthopyroxene cannot account for the observed CaO–MgO relations. This hypothesis should be checked using other oxide pairs (FeO versus Al_2O_3, for instance), because if all the chemical variation in the lavas is caused by fractional crystallization of olivine or plagioclase, the same relations between rock and fractionated mineral compositions should show on all the diagrams. On these graphs of one oxide versus another, the lever rule can be applied to estimate the proportions of liquid and solid phases required for a specific bulk composition.

Another widely adopted variation diagram employs three components, recalculated to sum 100% in a triangular diagram (Figure 4-8). In one version MgO, total iron as FeO, and total alkalis ($Na_2O + K_2O$) are plotted on an AFM diagram. AFM is an acronym for alkalis–ferrous iron–magnesium. Another shows the relative abundances of Ca, Na, and K. Because this plot discriminates the Na and K that are lumped at one corner of an AFM diagram, and includes the Ca that is ignored on AFM, the Ca–Na–K diagram should be used in conjunction with the AFM diagram. For extended discussion of the applications and limitations of AFM diagrams, see Barker (1978).

Implicit in the compositional trends shown on both triangular diagrams in Figure 4-8 are variations in the ratios Mg/(Mg + Fe), K/(K + Na), and Na/(Na + Ca). These variations reflect variable compositions of the major rock-forming minerals. During progressive crystallization of magma, there is a strong tendency for Mg to be incorporated in ferromagnesian phases relative to ferrous iron, and the remaining liquid thus becomes enriched in iron relative to magnesium. Similarly, crystallization of plagioclase preferentially depletes the liquid in Ca relative to Na and K.

If a large number of rock analyses, representing a broad variety of rocks in a restricted geographic region, is plotted on an AFM diagram, one or the other of two distinct trends is likely to emerge (Figure 4-9), one leading to stronger iron enrichment than the other. The difference between these two trends can be better appreciated if we examine the relations between bulk compositions and phase compositions on such a diagram. All the felsic minerals contain negligible iron and magnesium; silica polymorphs and the anorthite component of plagioclase also contain negligible alkalis, so these are ignored in an AFM plot. Alkali feldspars and the feldspathoids have compositions that fall at the A corner of the diagram. The remaining, mafic, phases will plot along the F-M edge if they contain negligible alkalis (olivines, orthopyroxenes, many clinopyroxenes, and opaque oxides), or within the triangle (alkali-bearing amphiboles, clinopyroxenes, and micas). Now the bulk composition of a rock, represented by a point in the AFM triangle, is a sum-

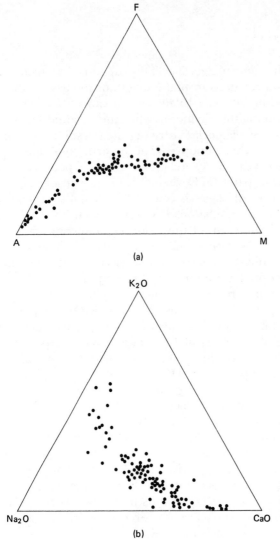

Figure 4-8. (a) AFM and (b) CaO–Na₂O–K₂O diagrams for intrusive rocks of the Mount Stuart Batholith, Washington, in weight percent. Each dot represents one rock analysis. A, Na₂O + K₂O; F, total iron calculated as FeO; M, MgO. (From Erikson, 1977, Fig. 6, with permission of Springer-Verlag and E. H. Erikson, Jr.)

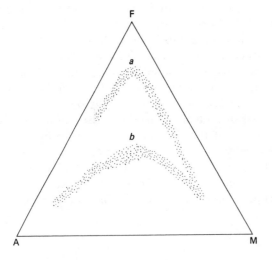

Figure 4-9. Bulk compositions of rocks in two suites (*a* and *b*) plotted on an AFM diagram. Suite *a* shows much more iron enrichment, relative to alkalis and magnesium.

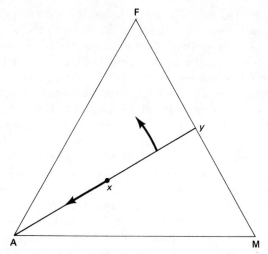

Figure 4-10. AFM diagram showing the factors that control bulk composition points on such a diagram. Point *x* represents the bulk composition, and point *y* represents the average Fe/Mg ratio of all mafic phases in the rock. Arrows indicate the directions in which line A*xy* and point *x* move with progressive crystallization of magma.

mation of the proportions and compositions of all phases. The bulk composition therefore lies on a line radiating from the A apex toward the average Fe/Mg ratio of all the mafic phases, and the position of the bulk composition point on this line reflects the proportion of mafic to felsic phases (that is, the color index). With progressive crystallization of magma, the Fe/Mg ratio in residual liquids (and in the mafic minerals that crystallize from them) tends to increase, and the color index tends to decrease. The line A*xy* in Figure 4-10 therefore sweeps toward the F corner, pivoting around A, at the same time that the bulk composition *x* slides closer to A along the line A*xy*. The two trends in Figure 4-9 reflect competition between these two tendencies; trend *a* represents more rapid enrichment of mafic phases in iron than decrease in color index, while trend *b* results from more rapid decrease in the modal proportions of mafic phases compared with the change in their overall Fe/Mg ratio.

In the last few years, variation diagrams have become regarded with increasing caution and scepticism. Do they really solve petrogenetic problems or do they actually obscure the data? This is a topic that cannot be discussed until after Chapter 7, which describes the mechanisms by which magmas vary in composition. Variation diagrams are treated in illuminating detail by Wilcox (1979) and Cox and others (1979, especially Chapters 2 and 6). Specific variation diagrams will be used in later chapters of this book.

4.5 TRACE ELEMENTS

Elements with concentrations less than 0.1 wt % in igneous rocks are called *trace elements*, to distinguish them from the *major elements*. This arbitrary limit of 0.1 wt % is not an efficient discriminant because some elements, namely

Mn, P, and Ba, hover near that value in many igneous rock analyses. Ti, Cl, F, C, and even Mg can also be major elements in some igneous rocks and trace elements in others. To compound the confusion, some researchers use the term "minor elements" for those constituting between 5 and 0.1% of a rock; this third category is not needed.

Langmuir and Hanson (1980) propose another, more rational, classification, but one that requires more sophisticated information on the chemistry of the modal phases as well as bulk composition of a rock. Their threefold classification defines trace elements as those that are present in such low abundances that they do not affect the stability of phases; rubidium and strontium are examples. At the other extreme, essential structural constituents are elements that fully occupy one crystallographic site in a phase; silicon and aluminum are examples. Intermediate elements are those causing the significant substitutions in solid solutions; they do not completely occupy one site in a phase but are sufficiently abundant to change its stability; examples are magnesium and ferrous iron in olivines, and calcium and sodium in plagioclases. Although the Langmuir–Hanson classification has much to recommend it, this book will perpetuate the arbitrary distinction between major and trace elements.

Trace elements are customarily reported in parts per million (ppm); 1000 ppm equals 0.1 wt %. The development of new analytical techniques, especially since 1950, has permitted rapid and precise measurement of trace element contents. Specific trace elements will be mentioned throughout this book, but a few general principles need summary here.

In Chapter 2, valence charge and ionic radius were shown to be important controls on whether an ion would readily enter the crystalline structure of a mineral. Table 4-7 compares some trace elements with major elements in terms of valence charges and ionic radii for octahedral coordination. After examining this table we can postulate, correctly, that germanium can substitute for silicon (but is much less abundant); that chromium and vanadium can proxy for ferric iron; that nickel, cobalt, copper, zinc, and manganese can fill the same sites as ferrous iron; that strontium should play a geochemical role similar to that of calcium; and that rubidium should concentrate in potassium-bearing minerals. On the other hand, there are many ions, including those of tantalum, niobium, hafnium, zirconium, scandium, yttrium, the rare earth elements, barium, uranium, and thorium, with ionic radii that fall outside the ranges for major elements with the same valence charges. These *"large-ion" elements* therefore do not normally enter solid solutions as proxies for major elements. Instead, they form minerals in which they are major constituents (such as zircon, monazite, and uraninite), or they are concentrated in residual liquid or gas phases as magmatic crystallization progresses. The tendency for these large ions to remain in a fluid phase, rather than to enter crystals, has led to the labels *residual elements* and, misleadingly, *incompatible elements*.

TABLE 4-7 IONIC RADII AND CHARGES OF MAJOR ELEMENTS AND SOME TRACE ELEMENTS

Ionic charge	Ionic radius[a]	
	Major elements	Trace elements
5+	P, 0.35	V, 0.62; Sb, 0.69; Ta, 0.72; Nb, 0.72
4+	Si, 0.48; Ti, 0.69	C, 0.16; Ge, 0.62; Sn, 0.77; Hf, 0.79; Zr, 0.80; U, 0.97; Th, 1.08
3+	Al, 0.61; Fe, 0.73	Ga, 0.70; Cr, 0.70; V, 0.72; Sc, 0.83; Y, 0.98; rare earths (La–Lu), 1.13–0.94
2+	Mg, 0.80; Fe, 0.86; Ca, 1.08	Be, 0.35; Ni, 0.77; Co, 0.83; Cu, 0.81; Zn, 0.83; Mn, 0.91; Sn, 0.93; Sr, 1.21; Pb, 1.26; Ba, 1.44
1+	Na, 1.10; K, 1.46	Li, 0.82; Cu, 0.96; Ag, 1.23; Rb, 1.57; Cs, 1.78

Source: Data from Whittaker and Muntus, 1970, except for Be, C, Cu^+, P, Sn^{2+}, and U, which are from Mason, 1966, p. 297–299.
[a]The ionic radius in angstroms is for sixfold coordination.

In contrast, some ions, such as nickel, although present in low abundance, show strong tendencies to enter certain sites in crystals; olivine scavenges nickel from a silicate liquid efficiently, as Table 4-7 predicts.

To express these tendencies quantitatively requires definition of the *distribution coefficient*, $K^{x/f} = c^x/c^f$, where c^x is the concentration of an element in a crystalline phase and c^f is the concentration of the same element in a fluid (liquid or gas) that is in equilibrium with the crystalline phase. An element that is neither "preferred" nor "rejected" by a growing crystal will have $K^{x/f} = 1$. An element that tends to remain in the fluid will have a distribution coefficient significantly less than unity, and an element that is preferentially incorporated in the crystalline phase will have a distribution coefficient greater than unity.

The distribution coefficient for a specific ion will vary not only with the kind of crystalline phase but with temperature, to a smaller extent pressure, and significantly with the composition of the fluid. Takahashi (1978) has experimentally demonstrated that distribution coefficients for crystal–liquid equilibria are strongly influenced by the composition of the liquid, and many more data are needed before the great promise of trace element chemistry in

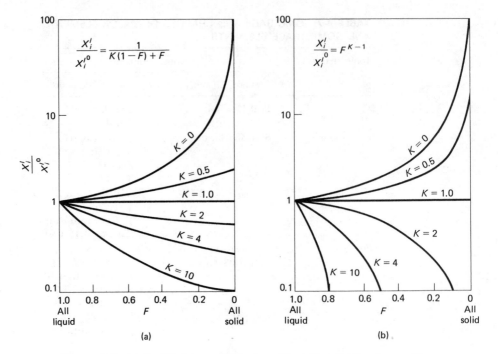

Figure 4-11. (a) Equilibrium crystallization; (b) perfect fractional crystallization. The vertical axis in both diagrams is the ratio of a trace element in
residual liquid (X_i^l) to the concentration of the same trace element in the
initial liquid before crystallization began (X_i^{l0}); note that the scale is
logarithmic. The horizontal scale is F, the fraction of the mass of the system
that is liquid. Heavy curves trace values of X_i^l/X_i^{l0} with varying F for
different values of the distribution coefficient K, for equilibrium crystallization, and for perfect fractional crystallization. (After Arth, 1976.)

solving magmatic problems can be fulfilled. An element that is excluded from
crystalline phases can provide a good index of the degree of crystallization or
melting. The theoretical treatment of relations among distribution coefficients, bulk composition of the system, and proportions of crystalline and liquid phases has been summarized by Arth (1976). Figure 4-11 shows the effects
of varying $K^{x/l}$ and the fraction of liquid on trace element concentrations for
two extreme models of crystallization. Just as we found for major elements
and for proportions of phases in Chapter 3, the fractionation of trace elements for equilibrium fusion is the reverse of that for equilibrium crystallization, but fractional fusion and crystallization do not necessarily yield the same
end results. Allègre and Minster (1978) have comprehensively summarized the
ways in which trace elements might be used to decipher stages in magmatic
processes.

4.6 SUMMARY

This chapter has been an overview of the bulk compositions of igneous rocks as combinations of phases or of components. Both means of expressing compositions are valid and necessary. There are severe problems in expressing compositional variation, both numerically and graphically, and some of the conventional ways of presenting data have been introduced. The mode, the CIPW norm, and variation diagrams make up a working vocabulary with which rocks can be described and compared in terms of composition. Trace elements promise to provide much additional insight when more data become available on their partition between crystals and fluids.

5

Classification
of Igneous Rocks

5.1 PURPOSE OF CLASSIFICATION

Names are applied to rocks for two reasons. The first is to emphasize differences between rocks in the same area, and the second is to emphasize similarities of a rock in one area to a rock in another area. These two aims may be in conflict. The second tends to produce a less finely structured classification than the first.

Any classification must be designed for a specific use. Probably the first classification of rocks was into two categories, those that can be chipped to produce a cutting edge or piercing point, and those that cannot. Excavation contractors are most interested in whether a rock mass must be broken with explosives in order to be removed, and whether the fragments will be large or small. Sculptors would classify rocks according to color, ability to be carved and polished, and resistance to weathering. These classificatory schemes are equally valid, for their specific purposes, and there is no single classification that can satisfy the architect, sculptor, contractor, and geologist.

As geologists, we require a classification system that conveys as much information as possible about texture, composition, style of occurrence, and significance of a rock. We are thus faced immediately with a choice between descriptive and genetic classifications. To classify rocks genetically is of course the ultimate goal, but a descriptive classification, according to what we see in a rock, is more honest and less subject to debate. We have already made an assumption concerning rock genesis when we decide to apply an igneous classification, rather than sedimentary or metamorphic nomenclature.

Even when we restrict ourselves to purely descriptive considerations, we find too many significant variables to allow a simple classification. During the great surge of microscopic petrography from 1870 to about 1920, geologists were so impressed with the details of texture and mineralogy made visible with the microscope that they emphasized differences between rock samples, rather than similarities. As a result, more than 1200 igneous rock names were introduced, many honoring exotic localities (jacupirangite, fasibitikite, llanite, kakortokite). Some of the distinctions between rock types are so trivial and restrictive that the entire mass of rock that fits the specific definition could be carried away by a small child.

This chaos in nomenclature has retarded the growth of igneous petrology by confusing its practitioners and by repelling the people (chemists and physicists) who must be recruited to solve petrologic problems. In this book, many rock names are deliberately excluded in order to discourage their continued use. Definitions of the less frequently used terms can be found in the following references, none of which is complete: Bates and Jackson (1980); Holmes (1920, reprinted 1972); Hatch and others (1973); Johannsen (1939); and Sørensen (1974, p. 558–577).

5.2 ATTRIBUTES TO BE CONSIDERED IN A DESCRIPTIVE CLASSIFICATION

Field relations and style of occurrence are vital information for describing rocks and eventually for understanding how they formed. However, with the exception of the pyroclastic rocks (Chapter 8), the type of occurrence does not influence the rock name.

Age was once considered an important attribute, to the extent that mafic intrusive rocks were given one set of names if they were Mesozoic or older, and another set if they were Cenozoic. Such distinctions are abhorrent to a uniformitarian view, and have been abandoned.

There remain texture and composition as important characteristics, and it is upon these that igneous rock classifications are based. The most important textural feature is grain size; a different set of names is required for rocks with average grain diameters of 1 mm or larger than is used for fine-grained and glassy rocks. Other aspects of texture (Chapter 6) can be significant, but are best expressed as prefatory terms modifying a rock name without changing it.

Rock composition, as discussed in Chapter 4, can be expressed in terms of components or phases. A chemical analysis is essential for deciphering igneous rock genesis, and is useful in description. However, chemical composition alone cannot serve as the basis of a practical classification for two reasons. In the first place, the geologist needs a classification that can be applied in the field, then modified in the laboratory if necessary. In spite of great advances in instrumentation, there is still no portable unit that will provide a reasonably complete analysis of a rock sample in the field. Rock analysis remains an expensive and time-consuming procedure. Second, two rocks may have the same chemical composition, but very different textures and mineral assemblages, and therefore deserve different names (consider an obsidian and a coarse-grained granite, for example). For the completely glassy obsidian, chemical data are essential for its classification. The composition of the granite, in terms of oxide components, could be estimated from the mode.

Nevertheless, igneous rock classification based on proportions of components has been historically important, and has provided broad categories of classification that remain useful. Therefore, the influence of chemical composition (in terms of components) will be discussed before modal classification (in terms of phases).

5.3 CHEMICAL CLASSIFICATION

The CIPW norm calculation procedure (Chapter 4) was originally intended as a basis for classification of igneous rocks. The norm is still with us, but the classification is not. The CIPW classification of course was restricted to samples that have been chemically analyzed, and this limitation is one reason for its not being adopted. In addition, such a logical, detailed classification, based on proportions of normative constituents, necessitated an entirely new terminology, one based on mnemonic syllables combined to make words. For example, the terms "perfelperquafelpersalone" and "domiric permagnesic belcherose," products of this logical system, might convey a lot of chemical information to the initiated, but geologists almost unanimously and immediately refused to speak or write such terms.

A less detailed chemical classification, with less bizarre terminology, developed before and after the CIPW normative scheme and survives. This is based on the proportions of a few, relatively abundant, oxide components. Of these, SiO_2, the most abundant, is accorded the greatest importance. The concept of *silica saturation* (Chapter 4) reflects the presence or absence of normative analogues of quartz and feldspathoids. As such, it is an ideal basis for classification, involving a yes-or-no decision; either quartz is present in the norm or it is not. This concept of silica saturation is carried over into the modal classification.

Another widely used set of terms is based on the weight percentage of SiO_2, unfortunately applying archaic terminology; a rock can be called *ultrabasic, basic, intermediate,* or *acidic,* depending on whether the SiO_2 weight percentage is respectively less than 45, between 45 and 52, between 52 and 66, or greater than 66 (Williams and others, 1954, p. 27). The "acidic" and "basic" terms are relics from a time when it was thought that silica formed an acid in magmas, and magnesium, iron, and calcium acted as bases.

TABLE 5-1 ALUMINA SATURATION IN IGNEOUS ROCKS

Term	Molecular proportions	Diagnostic normative constituents	Characteristic modal phases
Peraluminous	$Al_2O_3 > (CaO + Na_2O + K_2O)$	c	Corundum, muscovite, biotite, topaz, tourmaline, almandine–spessartite garnet
Metaluminous	$Al_2O_3 < (CaO + Na_2O + K_2O)$ but $Al_2O_3 > (Na_2O + K_2O)$		Melilite, olivine, amphibole, biotite
Peralkaline	$Al_2O_3 < (Na_2O + K_2O)$	ac, ns (no an or c)	Aenigmatite, sodic amphiboles and clinopyroxenes, olivine, rare Zr and Ti silicates

Source: After Shand, 1951.

A logical extension of silica saturation leads to *alumina saturation,* portraying variations in what is ordinarily the second most abundant oxide component. Shand (1951) divided igneous rocks according to the molecular proportions of Al_2O_3, CaO, and $(Na_2O + K_2O)$; his divisions are summarized in Table 5-1. Shand's concept is useful, but only the term *peralkaline* has been widely adopted. Note that it is quite possible to find a rock with SiO_2 exceeding 66 wt % but with molecular proportions of $(Na_2O + K_2O)$ exceeding that of Al_2O_3; such a rock is both "acidic" and "peralkaline," demonstrating that these terms, referring to silica content and alumina undersaturation, do not have their conventional chemical meanings.

The relative proportions of CaO to $(Na_2O + K_2O)$ are another variable that is significant in a strictly chemical classification. Peacock (1931) devised an *alkali-lime index* for classification of *rock suites* (large numbers of samples that are genetically related). On a variation diagram in which weight percentages of other oxide components are plotted against silica, CaO generally declines with increasing SiO_2 and is eventually exceeded by the sum of Na_2O and K_2O (Figure 5-1). The silica content at which CaO equals $(Na_2O + K_2O)$ is estimated by interpolation and is defined as the alkali-lime index. Rock suites in which the total of alkali oxides exceeds that of CaO at a silica content less than 51 wt % are called *alkalic* (not to be confused with peralkaline!). Those with an alkali-lime index between 51 and 56 are *alkali-calcic*, those between 56 and 61 are *calcalkalic*, and those in which CaO still exceeds $(Na_2O + K_2O)$ at silica percentages over 61 are *calcic*.

The alkali-lime index is useful when applied to a suite of rocks, but it requires a large group of analyzed samples that span a considerable range of silica content. Peacock's terms, particularly alkalic and calcalkalic, are very

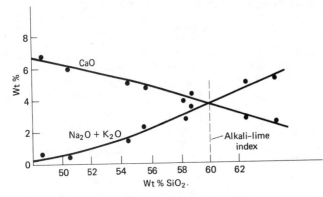

Figure 5-1. Peacock's (1931) alkali-lime index on a silica variation diagram. Each dot represents a rock analysis. In this example, the suite is calcalkalic. The solid curves are smoothed through the data points, and the alkali-lime index (approximately 59.7 wt % SiO_2) is estimated by interpolation of the crossover point for the CaO and (Na_2O + K_2O) curves.

widely used, often without regard for the original definition, and are wrongly applied to single samples rather than suites.

Silica content, silica saturation, alumina saturation, and alkali-lime index would not be sufficient by themselves for a complete classification of igneous rocks, even if the required chemical data were easy to obtain. These chemical indices are most useful when applied to groups of rocks comprising *magmatic provinces* (defined as geographic areas where, in a specified span of geologic time, magmatic rocks of restricted range in chemistry were produced). A few examples may clarify the definition. The silica-oversaturated, metaluminous, alkalic Mesozoic intrusions of the White Mountain magma series, New Hampshire, make up a classic magmatic province. The silica-undersaturated, ultrabasic to intermediate, alkalic Cenozoic shallow intrusions and volcanic rocks of the Navajo country, Arizona and New Mexico, make up another, and the silica-oversaturated, basic to acidic, calcalkalic Pleistocene and Holocene volcanoes of the Cascades in Washington and Oregon represent a third. As these examples show, the terminology is cumbersome, and conveys nothing concerning the specific rocks of each province. To communicate this information, we need a classification that applies to an individual sample rather than a suite of rocks.

5.4 MODAL CLASSIFICATION

Classification of igneous rocks according to the modal proportions of phases depends on arbitrarily defined boundaries between classes. The arbitrary nature of these boundaries is illustrated by Figure 5-2, which compares the definitions of "granite" as used by petrologists of two nations; the two fields do not even touch, let alone overlap.

There are really two separate problems involved in defining the boundaries. The first, and one that is obvious in Figure 5-2, is that international agreement should be reached so that a name means the same thing all over

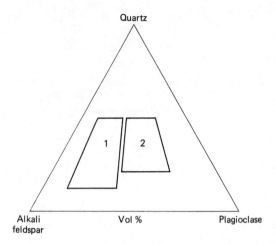

Figure 5-2. Modal proportions of quartz, alkali feldspar, and plagioclase permitted in definitions of "normal granite" used by (1) West German and (2) Soviet petrologists. There has been no agreement among North American petrologists, so their definition covers both areas and more. (Modified from Streckeisen, 1967, Fig. 1, with permission of E. Schweizerbart'sche Verlagsbuchhandlung.)

the world. Quite apart from this problem, which reflects the diversity and complexity of people, not of rocks, is the task of placing boundaries where the fewest rocks plot, to minimize "fence-straddling." With the dual aim of establishing an international accord on nomenclature and fixing the classification boundaries where there are the fewest samples, A. L. Streckeisen compiled data on the modes of igneous rocks and conducted an international poll among petrologists. The results (Streckeisen, 1967; IUGS, 1973) form the basis for a workable classification of igneous rocks with grain size greater than 1 mm.

There is general agreement that the significant modal phases for classification purposes can be grouped in five categories:

Q   (quartz; the high-temperature polymorphs tridymite and cristobalite are not found in the coarse-grained rocks under consideration)

A   (alkali feldspars, including albite if the anorthite content does not exceed 5%)

P   (plagioclase with 5% or more anorthite)

F   (feldspathoids)

M   (mafic minerals—in effect, all those not included in Q, A, P, and F)

The first step is to determine the modal percentage of M, the color index. If the color index is 90 or greater, a rock is classified according to the kinds and proportions of mafic minerals. If the color index is less than 90, the mafic minerals need not be told apart for purposes of classification, because the proportions of A, P, and Q or F determine the name of the rock. Q and F are mutually exclusive (quartz and feldspathoids cannot coexist in equilibrium), so modal compositions can be projected into the double triangle Q-A-

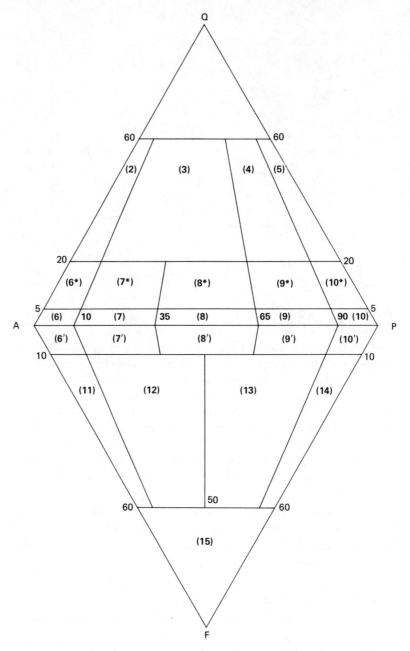

Figure 5-3. Modal classification fields for igneous rocks with color index less than 90. Q, quartz/(quartz + total feldspars); F, feldspathoids/(feldspathoids + total feldspars); A, alkali feldspar/total feldspars; P, plagioclase/total feldspars, all in volume percentages. Plagioclase is defined for purposes of this classification as feldspar containing more than 5% anorthite endmember. The numbered fields correspond to numbers in brackets in Tables 5-2 and 5-3, listing the root names to be applied using this classification. Other numbers indicate the Q, F, and P limits of the fields. (Modified from IUGS, 1973, p. 26, with permission of the American Geological Institute.)

P-F (Figure 5-3) by neglecting the mafic constituents (if they total less than 90% of the rock) and recalculating $Q + A + P$ or $F + A + P$ to sum 100%.

Between Streckeisen's publication of a preliminary report in 1967 and the 1973 report by the IUGS Subcommission (of which Streckeisen was chairman), much debate centered on the names to be used; there was much more general agreement on the positions of the boundary lines in the classification. The eventual result is tabulated in Figure 5-3 and Table 5-2. The IUGS Subcommission admitted (1973, p. 27) that "although some systems are better than others, a 'best' way to classify rocks may not exist."

The "root names" listed in Table 5-2 merely provide the bare bones of a classification, and must be fleshed out with modifiers to convey significant information about color index (which may vary from 0 to 90 with no change in the root name), texture, and mineralogy. On occasion, it will be necessary to include the color or the weathering characteristics, in order to distinguish one rock mass from another in the same area. Before appending any modifier to the root name, ask yourself "Is this modifier necessary?"

Significant variations in color index are pointed out by the modifiers *leucocratic* (color index 0 to 35), *mesocratic* (35 to 65), and *melanocratic* (65 to 90) before the root name. Textural differences can be emphasized by inserting terms (defined in Chapter 6) between the color-index modifier (if used) and the root name. Examples are "leucocratic porphyritic diorite" and "melanocratic coarse equigranular syenite." Two rock masses in an area may have the same root name and texture, and nearly the same color index, yet deserve distinction. They can be discriminated by mineral modifiers, listed in order of increasing amounts; for example, "biotite-hornblende granite" contains more hornblende than biotite. Phases that do not affect the root name of a rock are called *accessory minerals*, to distinguish them from *essential minerals* whose presence or proportions determine the root name. Rarely, it will be necessary to mention a mineral that is a unique characteristic of one type, although present in trace amounts. Such a modifier should be placed as far from the root name as possible, as in "monazite-bearing mesocratic coarse porphyritic quartz syenite." In this example, quartz is part of the root name.

Granted, these terms are cumbersome, but it seems better to sacrifice brevity for clarity, and these root names, even when loaded with a long string of modifiers, convey much more information than "lusitanite" or "ngurumanite," whose only real virtue is their compactness.

A few additional comments are needed concerning some of the root names in Table 5-2. The expressions "foid" (a contraction of feldspathoid), "foid-bearing," and "foidolite" are only employed to make the table neater. In practice, they are always replaced with the specific feldspathoid, as in "nepheline syenite," "analcime-bearing gabbro," and "leucitite."

Root names containing "diorite" and "gabbro" are determined by average plagioclase composition (less than and greater than 50% anorthite, respec-

TABLE 5-2 MODAL CLASSIFICATION OF IGNEOUS ROCKS
WITH COLOR INDEX LESS THAN 90 AND AVERAGE GRAIN
DIAMETER AT LEAST 1 MM

Modal values	Classification
[1] Q > 60	Not igneous
[2] Q = 20–60, P < 10	Alkali-feldspar granite
[3] Q = 20–60, P = 10–65	Granite
[4] Q = 20–60, P = 65–90	Granodiorite
[5] Q = 20–60, P > 90	Tonalite
[6*] Q = 5–20, P < 10	Alkali-feldspar quartz syenite
[7*] Q = 5–20, P = 10–35	Quartz syenite
[8*] Q = 5–20, P = 35–65	Quartz monzonite
[9*] Q = 5–20, P = 65–90	Quartz monzodiorite (An < 50)
	Quartz monzogabbro (An > 50)
[10*] Q = 5–20, P > 90	Quartz diorite (An < 50)
	Quartz gabbro (An > 50)
	Quartz anorthosite (M < 10)
[6] Q = 0–5, P < 10	Alkali-feldspar syenite
[7] Q = 0–5, P = 10–35	Syenite
[8] Q = 0–5, P = 35–65	Monzonite
[9] Q = 0–5, P = 65–90	Monzodiorite (An < 50)
	Monzogabbro (An > 50)
[10] Q = 0–5, P > 90	Diorite (An < 50)
	Gabbro (An > 50)
	Anorthosite (M < 10)
[6'] F = 0–10, P < 10	Foid-bearing alkali-feldspar syenite
[7'] F = 0–10, P = 10–35	Foid-bearing syenite
[8'] F = 0–10, P = 35–65	Foid-bearing monzonite
[9'] F = 0–10, P = 65–90	Foid-bearing monzodiorite (An < 50)
	Foid-bearing monzogabbro (An > 50)
[10'] F = 0–10, P > 90	Foid-bearing diorite (An < 50)
	Foid-bearing gabbro (An > 50)
[11] F = 10–60, P < 10	Foid syenite
[12] F = 10–60, P = 10–50	Foid monzosyenite
[13] F = 10–60, P = 50–90	Foid monzodiorite (An < 50)
	Foid monzogabbro (An > 50)
[14] F = 10–60, P > 90	Foid diorite (An < 50)
	Foid gabbro (An > 50)
[15] F > 60	Foidolites

Source: Modified from IUGS, 1973, p. 30, with permission of the American Geological Institute.

Note: Numbers in brackets refer to fields in Figure 5-3. Q, quartz/(quartz + alkali feldspar + plagioclase); F, feldspathoids/(feldspathoids + alkali feldspar + plagioclase); P, plagioclase/(plagioclase + alkali feldspar); M, color index; An, % anorthite in plagioclase.

tively). Streckeisen (1976) provides other comments on this IUGS modal classification, including justifications and alternatives.

Classification of ultramafic rocks (those with color index above 90) will be considered in Chapter 10. Subdivision of field 15 ("foidolites," Figure 5-3), examples of which are rare, need not concern us. Finer subdivision of field 10 will be needed in Chapter 11.

There remains the severe problem of classifying glassy and fine-grained rocks. Lava flows, pyroclastic deposits, and shallow intrusive bodies are exposed over a major portion of the earth's surface, and these rocks are equally deserving of classification as the coarser rocks. However, the modal proportions of minerals are more difficult to estimate as the grain size becomes smaller, and the mode is indeterminate if magmatic liquid has quenched to glass.

Suggestions that these rocks be classified according to the minerals large enough to identify have not been widely adopted, because of awareness that the few large crystals may be completely out of equilibrium with the rest of the rock. For example, a rock might contain 20 vol % plagioclase crystals (with, say, 30% anorthite; written An_{30}) in a completely glassy groundmass; in a modal classification, the sample would plot at the P corner of Figure 5-3. On the other hand, the glass might contain abundant normative q, or, and ab, so that, if the rock had completely crystallized, its modal composition might fall near the center of the Q-A-P triangle. A modal root name applied to a rock with more than about 20% glass or irresolvable groundmass is almost certain to differ from the root name that would be applied if all the constituents could be identified and tallied in the mode.

This seems to provide one of the few opportunities in this book for an intentionally dogmatic statement: glassy or very fine-grained rocks that are not represented by chemically analyzed samples should not be formally classified. Such a declaration clarifies the problem, but does not solve it. Granted that we need a chemical analysis in order to classify a glassy or fine-grained rock, how do we use the chemical data?

An obvious step is to calculate the CIPW norm, then equate the normative constituents to the modal groups Q, A, P, F, and M. This is easier said than done. Hydrous mafic silicates such as micas and amphiboles do not appear in the norm but contribute Na, K, Al, and Si to the salic normative constituents. Furthermore, and of greater significance, ab in the norm represents a component in both modal alkali feldspar (A) and modal plagioclase (P). We return to this problem, and a suggested solution, after considering the IUGS classification.

After publishing the classification for coarser-grained rocks, the IUGS Subcommission deliberated for several years before proposing a classification for finer-grained rocks (Streckeisen, 1980). The decision was to leave the boundaries in the Q-A-P-F double triangle as they were for the coarser-grained rocks, so that the classification fields for volcanic rocks are congruent

TABLE 5-3 MODAL CLASSIFICATION OF IGNEOUS ROCKS WITH COLOR INDEX LESS THAN 90 AND AVERAGE GRAIN DIAMETER LESS THAN 1 MM

Modal values	Classification
[1] Q > 60	Not igneous
[2] Q = 20–60, P < 10	Alkali-feldspar rhyolite
[3] Q = 20–60, P = 10–65	Rhyolite
[4] Q = 20–60, P = 65–90	Dacite
[5] Q = 20–60, P > 90	Dacite
[6*] Q = 5–20, P < 10	Alkali-feldspar quartz trachyte
[7*] Q = 5–20, P = 10–35	Quartz trachyte
[8*] Q = 5–20, P = 35–65	Quartz latite
[9*] Q = 5–20, P = 65–90	In all six fields, the names andesite
[10*] Q = 5–20, P > 90	and basalt are applied; basalt is
[9] Q = 0–5, P = 65–90	used if SiO_2 < 52 wt % after H_2O
[10] Q = 0–5, P > 90	and CO_2 are deleted and the
[9'] F = 0–10, P = 65–90	analysis recalculated to
[10'] F = 0–10, P > 90	sum 100%
[6] Q = 0–5, P < 10	Alkali-feldspar trachyte
[7] Q = 0–5, P = 10–35	Trachyte
[8] Q = 0–5, P = 35–65	Latite
[6'] F = 0–10, P < 10	Foid-bearing alkali-feldspar trachyte
[7'] F = 0–10, P = 10–35	Foid-bearing trachyte
[8'] F = 0–10, P = 35–65	Foid-bearing latite
[11] F = 10–60, P < 10	Phonolite
[12] F = 10–60, P = 10–50	Tephritic phonolite
[13] F = 10–60, P = 50–90	Phonolitic tephrite
[14] F = 10–60, P > 90	Tephrite (modal olivine < 10%)
	Basanite (modal olivine > 10%)
[15] F > 60	Foidite

Source: Modified from Streckeisen, 1978, Figure 1, with permission of E. Schweizerbart'sche Verlagsbuchhandlung.

Note: Numbers in brackets refer to fields in Figure 5-3. Abbreviations are the same as in Table 5-2.

with those for deep intrusive rocks, and to set up a parallel system of root names (Table 5-3). Modifiers, as needed, are appended to these roots as they are in naming coarser-grained rocks.

Some petrologists have objected to calling all fine-grained rocks of field 3 "rhyolite" without regard for their intrusive or extrusive nature, and advocate the term "microgranite" for fine-grained intrusive rocks of granite composition. Similarly, the prefix can be applied to other root names, as in "microsyenite" for intrusive trachyte. This terminology is not used in this book, to avoid proliferation of root names. It seems preferable to say "intrusive latite" rather than "micromonzonite," to cite an example from field 8.

Comments made concerning Table 5-2 generally apply to Table 5-3; the specific feldspathoid is substituted for "foid," for example. Note that the same name is applied to fields 4 and 5, and that two options ("basalt" for low SiO_2, "andesite" for higher) cover fields 9, 10, 9*, 10*, 9′, and 10′. Unfortunately, the majority of volcanic rocks fall in these six fields, and further subdivision does not follow the boundaries in Figure 5-3. The haphazard evolution of terminology for these rocks is described in Chapters 11 and 12.

The IUGS Subcommission thus provided a classification scheme for fine-grained rocks based on modal proportions. Concerning the grave problem of translating normative to modal constituents, the Subcommission made no recommendation. The most difficult part of that problem, the distribution of normative *ab* among modal alkali feldspar and plagioclase, has been addressed by Streckeisen and LeMaitre (1979). Figure 5-4 summarizes their classification model. Although the orientation of the diagram remains the same as the Q-A-P-F double triangle, Streckeisen and LeMaitre chose an orthogonal plot to emphasize that this is a normative, not a modal, classification. Analogous to Q and F of the modal scheme are Q′ [equal to 100 times $q/(q + or + ab + an)$] and F′ [equal to 100 times $(ne + lc + kp)/(ne + lc + kp + or + ab + an)$]. The horizontal axis, analogous to P/(A + P) in the modal diagram, is 100 times $an/(or + an)$. Normative *ab* is thus ignored. The numbered fields correspond to those of Figure 5-3, and the boundaries were found empirically by plotting the normative ratios for large numbers of rocks that had already been classified modally. The concordance between normative and modal classification fields for the same samples is not perfect, as Streckeisen and LeMaitre point out. Nevertheless, the normative plot is a valiant and necessary attempt, so it is included here although it may soon be superseded.

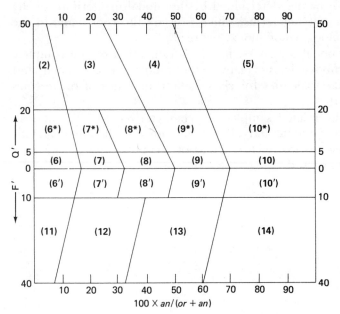

Figure 5-4. Normative classification of igneous rocks; numbered fields correspond in name to those of Tables 5-2 and 5-3. Field boundaries are best fits that separate chemically analyzed rocks named according to the modal scheme. (After Streckeisen and LeMaitre, 1979, Fig. 1b, with permission of E. Schweizerbart'sche Verlagsbuchhandlung.)

In the meantime, the most widely accepted chemical classification of volcanic rocks is that of Irvine and Baragar (1971). Their method, based on a combination of normative calculations and ratios of oxide components, is too involved to be summarized here. The original publication should be consulted by anyone wishing to classify analyzed fine-grained or glassy rocks. The Irvine and Baragar classification is particularly useful for plagioclase-rich rocks.

Many other classification methods are in use. Some petrologists discriminate between "basalt" and "andesite" solely on the basis of weight percent SiO_2. Whether one is classifying fine-grained or coarse-grained rocks, the specific classification method being used should always be explicitly stated.

5.5 SUMMARY

The least important attribute of a rock is its name, but names are necessary for communication among geologists. In the classification and nomenclature of igneous rocks, anarchy persists, and the geologist is obligated to specify the classification method being employed.

Individual rock samples are classified according to grain size and composition. Although classification by composition can with equal validity be based on components or phases, modal classification (in terms of phases) is more practical and informative. One modal classification scheme that is gaining international acceptance uses the volume proportions of quartz (or feldspathoids), alkali feldspar, and plagioclase, unless the mafic minerals exceed 90% of the sample. Root names determined by the modal proportions of the major phases require modifying terms, which precede the root name in the following order: mineralogy, color index, texture.

For fine-grained and glassy rocks, in which estimation of the complete mode is difficult or impossible, normative constituents are calculated and translated to their modal analogues for classification, or ratios of oxide components are used directly.

Suites of genetically related samples are characterized by silica and alumina saturation, silica content, and alkali-lime index to define magmatic provinces. These are regions in which magmas of a common parentage were emplaced in a restricted episode of geologic time.

6

Crystallization and Textures

6.1 CRYSTALLIZATION FROM A MAGMATIC LIQUID

The transition from liquid to crystalline state in magmas is not as abrupt as one might expect, because considerable order exists in silicate systems even at temperatures above the liquidus. Silicate melts contain SiO_4^{4-} tetrahedra that can become linked or polymerized by sharing oxygens (Hess, 1980). *Polymerization,* the degree to which tetrahedra are linked, increases as the temperature drops. Other components are present as ions (for example, Mg^{2+}, Fe^{2+}, Ca^{2+}, Fe^{3+}, Al^{3+}, Na^+, K^+) and as radicals and complexes (OH^-, PO_4^{3-}, AlO_4^{5-}, CO_3^{2-}, CO_2, $KAlO_2$, and $NaFeO_2$ are examples).

The analytical difficulties are formidable in determining which chemical species (single ions, radicals, or complexes) are present in hot silicate liquids. Even in such a cool and accessible material as sea water, it is difficult to determine the proportions of magnesium present as the Mg^{2+} ion and as the $MgSO_4$ complex, although it is easy to measure the total magnesium content.

Nevertheless, considerable progress has recently been made toward understanding the structure of silicate liquids. Small, highly charged ions that are in fourfold coordination with oxygen, including Ti^{4+} and P^{5+} in addition to Si^{4+}, are called *network formers*; they occupy the central sites in easily polymerized tetrahedra. *Network modifiers,* such as Mg^{2+} and Fe^{2+}, are octahedrally coordinated with oxygen. The sixfold polyhedra share edges with each other and with tetrahedra, and tend to break the oxygen-sharing links between tetrahedra, thereby loosening the polymerized structure. Other cations (Al^{3+}, Fe^{3+}, Ca^{2+}, Na^+, and K^+) can apparently act as network formers or modifiers, depending on the composition of the liquid. When Al^{3+} and Fe^{3+} enter tetrahedral coordination, other cations must closely associate with them to balance the charge deficiency that is set up when a trivalent cation, rather than silicon, occupies a tetrahedron. As examples, Na^+, K^+, or Ca^{2+} can associate with Al^{3+} or Fe^{3+} to make network formers with hypothetical formulas $NaAlO_2$, $KFeO_2$, or $CaAl_2O_4$ (Nielsen and Drake, 1979; Virgo and others, 1980; Wood and Hess, 1980).

X-ray diffraction studies of silicate glasses (Taylor and Brown, 1979) indicate that liquids rich in *q, or, ab,* or *ne* have an almost completely polymerized structure consisting of interconnected rings, each ring being made of six linked tetrahedra. Na and K ions can fit into large voids within the rings. In contrast, glasses rich in normative *an* and *di* contain rings made up of four tetrahedra (Hochella and others, 1978) without abundant large holes for alkali cations. Other glasses corresponding to mafic liquids rich in *ol* and *hy* contain isolated tetrahedra and short chains and thus are much less polymerized (Figure 6-1).

Ryerson and Hess (1980, p. 611) make the important point that the structure of a silicate liquid, dependent on composition, temperature, and pressure, can vary continuously over a broad range, whereas the structure of a crystalline phase can vary only through a narrow range before abruptly changing at a polymorphic inversion. "As a result, variations in mineral-melt equilibria caused by changes in either composition or external parameters may very well reflect the change in properties of the melt phase to a greater extent than those of the crystalline silicates." The curvature of a liquidus in a temperature–composition phase diagram, and variations in trace element distribution coefficients with changes in liquid composition, probably reflect the greater sensitivity of liquid structure compared to crystal structure.

The structural study of glasses is important as a means of understanding the structure of silicate liquids, but a glass is not merely a liquid with the partially ordered structure frozen in by rapid cooling; Carmichael (1979) has summarized the evidence for definite structural rearrangement taking place at a transition from true liquid to glass. However, the available evidence (Taylor and others, 1980) indicates that the structure of albite glass, at least, does indeed mirror that of the higher-temperature liquid, and that, even well above the liquidus, silicate glasses can possess considerable order and polymerization. Among this evidence is *viscosity* (essentially, resistance to stirring).

Viscosity is extremely high for nearly pure SiO_2 liquids, and decreases markedly as other components (network modifiers) are added. Therefore, viscosity is a function of both composition and temperature. At lower temperatures (but still above the liquidus for the composition being considered), the tetrahedra show increased polymerization, and viscosity increases. At constant temperature, increases in Al and Si contents cause an increase in viscosity; dissolved H_2O, in contrast, is a very efficient network modifier, as described in Chapter 9, and reduces viscosity markedly.

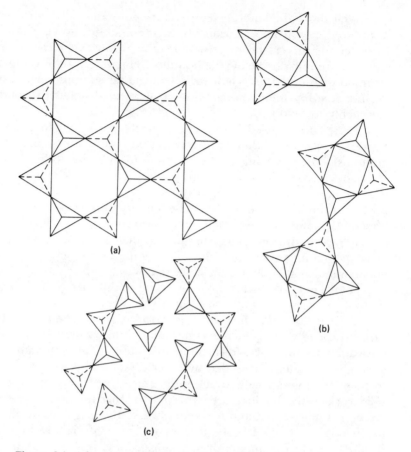

Figure 6-1. Schematic two-dimensional projections of inferred structures in silicate liquids; only the fourfold coordination polyhedra are shown. (a) *ab, or, q,* and *ne*-rich liquids; (b) *an* and *di*-rich liquids; (c) *ol* and *hy*-rich liquids.

Natural silicate liquids (observed as lavas) have viscosities that range from that of olive oil at 20°C [1 poise (P)] up to those of hot asphalt or cold honey (10^5 to 10^8 P). Until recently, the effect of pressure on viscosity was not anticipated. Kushiro (1976) experimentally showed that the viscosity of dry $NaAlSi_2O_6$ liquid at 1350°C decreases from 6.8×10^4 P at 1 atm pressure to 5.3×10^3 P at 2.4×10^9 Pa. For one sample of basalt (Kushiro and others, 1976), the experimentally measured viscosity decreased from 107 P at 1250°C and 1 atm to 8 ± 2 P (that of glycerin at 20°C) at 1500°C and 3×10^9 Pa. For comparison, the viscosity of liquid water at 20°C and 1 atm is 10^{-2} P. According to Mysen and others (1980), the viscosity of a highly polymerized melt decreases as pressure increases, at constant temperature, but the viscosity of less polymerized and hydrous melts increases.

Well above the liquidus temperature, ions and polyhedra begin to cluster as ordered crystal nuclei, probably each about 10^{-5} to 10^{-4} mm in diameter. In order for the crystal to grow, the system must be *undercooled* (that is, the temperature must be below the liquidus). For water, undercooling to $-20°C$ is commonly observed, as when turbulent glacial streams remain liquid far below the freezing temperature, and small water droplets in clouds can persist metastably to $-40°C$.

In an undercooled liquid, crystal nuclei form but, in order to survive at a given temperature, they must attain a minimum size (accumulate a quorum of atoms). The surviving nuclei continue to grow by adding successive layers of particles from the surrounding liquid. To enter a crystalline structure, a particle (ion, radical, or complex) must give up some of the energy that it possessed while in the liquid. The energy difference between liquid and crystalline states is the *enthalpy of fusion* (or of crystallization), colloquially called "heat of fusion." For example, to melt ice at $0°C$, 335 J must be provided per gram of ice (enthalpy of fusion). Conversely, to crystallize 1 g of liquid water at $0°C$, 335 J must be removed (enthalpy of crystallization).

Two kinds of nucleation must be distinguished; one is *homogeneous nucleation,* occurring spontaneously in a liquid through random encounters between particles, and the other is *heterogeneous nucleation,* occurring upon earlier formed crystals of the same mineral or another phase. Heterogeneous nucleation is much less difficult to accomplish, and probably most crystals growing in magma are seeded by other crystals. However, the first nuclei had to form by homogeneous nucleation unless they were fragments torn from solid rock as the liquid passed through it. In either case, heterogeneous nucleation is more rapid and tends to obscure the original nucleus by surrounding it with a much larger volume of new material. Dowty (1980) gives a detailed but clear summary of mechanisms of crystal nucleation and growth in magmas.

At a given temperature, once a nucleus is established, crystal growth is controlled by four factors:

1. The supply of essential constituents in the liquid.
2. The rate at which these constituents can diffuse through the liquid to reach the surface of the growing crystal.
3. Competition of other crystals of the same phase, and of other phases, for the same ingredient.
4. The rate at which heat, surrendered by particles joining the crystal structure, can diffuse away from the crystal (if heat does not flow away rapidly enough, the temperature at the surface of the growing crystal will climb back up to the liquidus temperature, and growth will stop). Heat diffuses more rapidly than matter, but the transport of heat and of matter is slower in cooler, more viscous liquids.

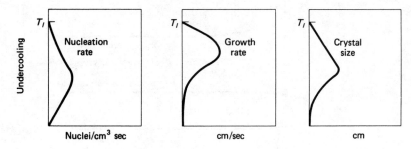

Figure 6-2. Influence of temperature upon nucleation rate, crystal growth rate, and resultant crystal size. T_l is the liquidus temperature.

In crystallization, there are two distinct processes to consider, each with its own rate. The *nucleation rate* (expressed as the number of crystal nuclei formed per cubic centimeter per second) is independent of the *growth rate* (expressed as increase in radius in centimeters per second). Both rates reach maximum values at specific degrees of undercooling for a specific composition, and there is no necessity for both maxima to occur at the same temperature (Figure 6-2). Generally, the homogeneous nucleation rate reaches its maximum at greater undercooling than the growth rate (comparable data are lacking for heterogeneous nucleation). A high nucleation rate coupled with a low growth rate will produce many small crystals. A low nucleation rate combined with a high growth rate will produce a few large crystals. The most common crystal size, as indicated in Figure 6-2, is a compromise between the maxima of nucleation and growth rates.

Note, in Figure 6-2, that if the temperature drops rapidly, both nucleation rate and crystal growth rate drop to zero, as they also do at the liquidus, and that the disordered although polymerized structure of the liquid persists metastably in glass. With less rapid cooling, the maximum in the growth rate is passed before the maximum in nucleation rate is achieved, leading to small crystals. Much slower cooling allows rapid growth on few nuclei, producing large crystals. This is illustrated in Figure 6-3, where each field contains the same percentage of crystals, so their sizes are inversely related to their numbers.

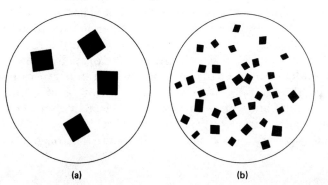

Figure 6-3. Consequences of (a) rapid growth on a few nuclei and (b) slow growth on many nuclei.

(a)   (b)   **103**

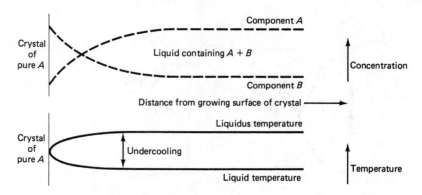

Figure 6-4. Changes in liquid composition and in undercooling in the vicinity of a crystal that is accepting component *A* and rejecting component *B* as it grows in a liquid containing both *A* and *B*. (With permission, after R. J. Kirkpatrick, *American Mineralogist, 60,* 798–814, Fig. 17, 1975, copyrighted by the Mineralogical Society of America.)

As a crystal grows, it rejects components in the melt that are not needed (or permitted) to build its structure, and gives off heat that must diffuse away from the crystal surface. These actions result in a precarious balance between crystal growth and dissolution (Figure 6-4). "Impurities" such as a higher concentration of rejected component *B* in the liquid near the crystal will cause a local reduction of the liquidus temperature. At the same time, flow of heat from the crystal causes a local rise in liquid temperature; undercooling (the difference between the liquidus temperature and the liquid temperature) increases away from the crystal–liquid interface.

The crystal growth rate is extremely sensitive to undercooling. For example, Kirkpatrick and others (1976), watching diopside grow from $CaMgSi_2O_6$ liquid, found that the growth rate parallel to the *c*-axis was 5×10^{-4} cm/sec (or about 0.4 m/day) at an undercooling of 10°C, and 2×10^{-2} cm/sec (more than 17 m/day!) at an undercooling of 115°C. If there are any irregularities or protuberances on the surface of the growing crystal, they will stick out into a more strongly undercooled region in the liquid, and will therefore grow faster. "When a crystal face becomes unstable, then protuberances develop and grow. Heat and uncrystallizable components flow away from the growing protuberance, not only perpendicular to the general trend of the interface, but parallel to it. Thus, additional protuberances will be inhibited from developing in the vicinity of one already formed because of the locally increased temperature and impurity content. A regular distribution of protuberances will develop with a periodicity depending upon the ratio of the growth rate and the diffusion coefficient in the melt" (Kirkpatrick, 1975, p. 810). With increased undercooling, diffusion slows, and protuberances become more closely spaced (Figure 6-5). With a combination of rapid growth but slow diffusion, the protuberances may survive on "skeletal" crystals (Figure 6-6). Experimental results support these conclusions; Lofgren (1974), Donaldson (1976), and Kirkpatrick and others (1979) found that plagioclase

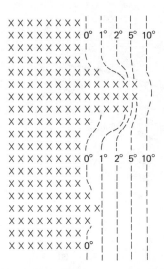

Figure 6-5. A growing crystal on the left (×'s represent ions) with protuberances projecting into more undercooled liquid. Degrees (Celsius) of undercooling are contoured.

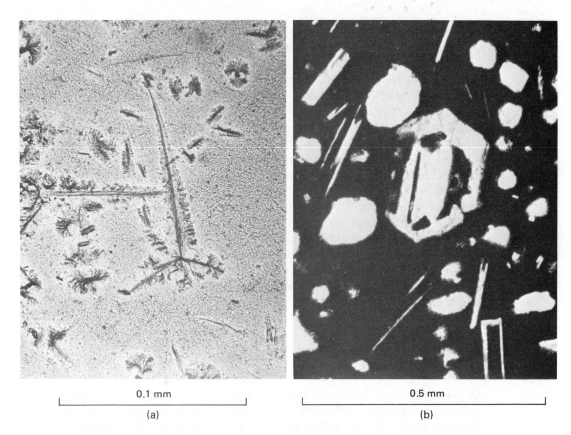

0.1 mm

(a)

0.5 mm

(b)

Figure 6-6. (a) Feathery or dendritic clinopyroxene in glass, Arran, Scotland. (b) Skeletal olivine (near center) and plagioclase in opaque basaltic glass, Pinacate volcanic field, Sonora, Mexico. Round light areas are gas bubbles (vesicles). Both photomicrographs were taken in plane-polarized light.

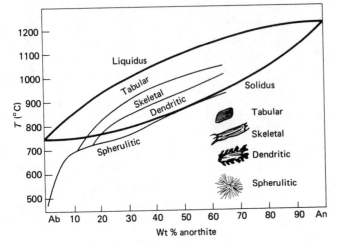

Figure 6-7. Variation in habit of plagioclase grown from synthetic liquids, as a function of composition and undercooling. (In part after Lofgren, 1974, Fig. 2, with permission of the American Journal of Science and G. Lofgren.)

and olivine crystals vary systematically in their shapes (also called *habits*) depending on the degree of undercooling (Figure 6-7).

Alkali feldspar liquids are much more viscous than those of diopside, for reasons discussed at the beginning of this chapter. Fenn (1977) found that the maximum growth rate of alkali feldspar from a hydrous melt was 5×10^{-6} cm/sec (about 4 mm/day). In Fenn's experiments, alkali feldspar crystallizing within 40°C of the liquidus was tabular, resembling the forms of plagioclase grown by Lofgren, but with greater undercooling mostly spherulitic alkali feldspar developed with only minor skeletal or dendritic forms. Taylor and others (1980, p. 116) account for the difficulty in crystallizing alkali feldspars in terms of structural contrasts between liquid and crystals. "The transition from the six-membered ring structure of the melt to the four-membered rings of the feldspar structure . . . involves breaking and reforming" the strong bonds between oxygens and tetrahedrally coordinated Si and Al. Such drastic rearrangement must control rates of nucleation and growth.

The sensitivity of crystal growth rate to undercooling is the foundation of a fundamental axiom in igneous petrology; glassy or fine-grained rocks cooled more rapidly than igneous rocks containing fewer, and larger, crystals per unit of volume.

6.2 ZONING IN CRYSTALS

As a crystal grows, the liquid remaining becomes enriched in the components that are not incorporated in the crystal, and depleted in those elements taken up by crystallizing phases. If diffusion of components through the liquid is rapid, the compositional inhomogeneity will tend to be erased, but if diffusion is slow relative to crystal growth, the liquid in the immediate neighborhood of

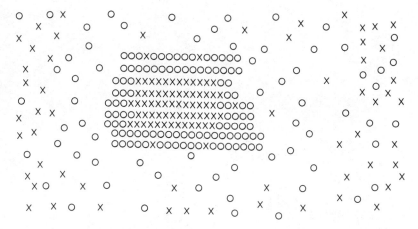

Figure 6-8. A crystal of solid solution series ×-O is growing; initially component × was preferred, but the local supply of × has become depleted, and now component O is being incorporated in the crystal (along with any ion of × that encounters the crystal surface).

a growing crystal will become depleted in some components more rapidly than those components can be replenished by diffusion. As a result, when the supply of some essential component becomes low, the crystal will stop growing until more is available or, if the crystallizing phase allows solid solution, another component may enter the crystal structure to substitute for the one that is in short supply. Figure 6-8 illustrates this situation in a schematic way. Because of differences in their radii and charges, different ions will diffuse through the same liquid at different rates.

Among the common igneous minerals, plagioclase most frequently shows compositional variation from center to periphery of a crystal. The changing ratio of albite to anorthite is indicated by different degrees of extinction in thin sections between crossed polarizers, because the optical orientation of plagioclase changes with composition. It is important to realize that the crystallographic orientation remains uniform from the center of the grain to the periphery. Such concentric zoning (Figure 6-9) may be simple and progressive, from anorthite-rich cores to albite-rich rims, or it may be oscillatory.

Haase and others (1980) have investigated oscillatory zoning in plagioclase. The wavelength of the oscillation varies from 10 to 100 μm, and the amplitude typically from 5 to 15% of the albite or anorthite component. The compositional variation during crystal growth reflects an exceedingly complex interplay between rates of crystal growth and of diffusion of components through liquid. Of the essential ingredients in plagioclase, sodium diffuses much more rapidly than calcium, calcium more rapidly than aluminum, and aluminum more rapidly than silicon–oxygen tetrahedra. There is a feedback mechanism between growth rate and composition of the surface layer being

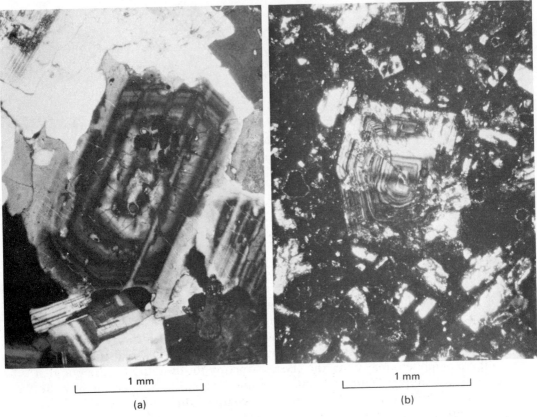

1 mm

(a)

1 mm

(b)

Figure 6-9. Concentric zoning in plagioclase (crossed polarizers): (a) quartz monzonite, Boulder Batholith, Montana; (b) andesite, Black Mountains, Nevada.

added to the crystal, such that addition of anorthite is favored by an already anorthite-rich surface, and addition of albite by an albite-rich surface. If anorthite-rich plagioclase is being deposited, aluminum is rapidly extracted from the liquid close to the crystal. Consequently, the growth rate of anorthite-rich plagioclase must decrease if the essential ingredient aluminum is depleted and cannot be replenished rapidly. Albite-rich plagioclase, requiring less aluminum, therefore begins to grow, at an accelerating pace. But now "the amount of Al [in the immediately surrounding liquid] can increase since less Al is being used up; but the crystal is still advancing, pushing the excess Al ahead of it" (Haase and others, 1980, p. 274). The liquid envelope therefore is enriched in Al by growth of more Si-rich plagioclase, and anorthite-rich plagioclase begins once again to precipitate, repeating the cycle.

Many other minerals show zoning, including olivine, pyroxenes, amphiboles, and micas, but it may not be obvious in thin section because optical properties in these minerals are not as sensitive to compositional variation as they are in plagioclase.

Concentric zoning is important for two reasons. In the first place, it records the crystallization history of a phase, and demonstrates that complete reequilibration, including homogenization of a phase, has not been achieved by reaction at subsolidus temperatures. Second, zones of a different composition protect the core of a crystal from reaction with the surrounding liquid. The consequences of this lack of communication between the interior of a crystal and the liquid are emphasized in Chapter 7.

Another type of zoning is less common, and is ineffective in sheltering early formed crystals from later reaction with liquid. This type, most prominently seen in clinopyroxenes in the more alkalic rocks, is called *sector zoning*. One or more components may be preferentially taken up in one crystallographic orientation within a growing crystal (Figure 6-10). In clinopyroxenes, Ti and Al are favored by the sectors perpendicular to (100) (Figure 6-11).

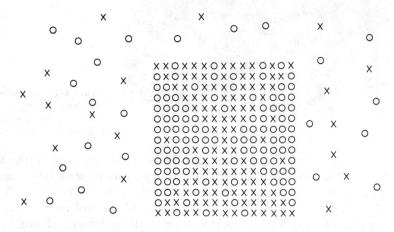

Figure 6-10. Schematic diagram of a sector-zoned crystal; component × is preferentially concentrated in crystal boundaries growing toward the top and bottom of the diagram, and component ○ is concentrated in boundaries advancing to the left and right.

6.3 GLASS AND ITS DEVITRIFICATION

Glass is metastable and tends to crystallize (*devitrify*). Reheating to temperatures at which nucleation and growth rates are higher (Figure 6-2) will have the same net effect as slow cooling. High temperature (but below the liquidus) therefore promotes devitrification, but water is the single most effective agent in speeding the crystallization of glass.

0.5 mm

Figure 6-11. Combined sector and concentric zoning in clinopyroxene (crossed polarizers), leucite tephrite, Kaiserstuhl, Germany.

Devitrification apparently proceeds in two stages; first, the glass becomes hydrated by introduction of water (probably as H^+ and OH^-), and then other ions, particularly Na^+ and K^+, promote the growth of crystals (Lofgren, 1970, 1971). Hydration leads to the "expansion of the polymeric SiO_4 network and the reaction with OH^- to break some of the polymeric chains. . . . The breaking of bridging bonds divides the SiO_4 polymeric chains into individual SiO_4 tetrahedra that are more readily acceptable into a crystalline silicate structure. The expanded glass network also allows diffusion of network modifiers such as Na, K, and Fe to proceed more rapidly and possibly facilitates reorientation of the SiO_4 tetrahedra" (Lofgren, 1970, p. 557–558). Crystallization is more rapid in the presence of alkali-rich solutions than with pure water.

The importance of aqueous solutions in devitrification is shown by the abundance in lunar samples of completely undevitrified glass more than 3×10^9 years old, surviving because of the lack of water on the moon. In contrast, on earth glasses older than 1×10^8 years are very rare.

Even if a sample of glass is completely isotropic (amorphous) in thin section, large quantities of water may have entered. The hydrated glass called *perlite* commonly contains 2 to 5% added water, and the volume change produced by this hydration causes peculiar curving cracks (Figure 6-12).

1 mm

Figure 6-12. Perlite (hydrated glass) with characteristic arcuate cracks accentuated by devitrification. Superstition Mountains, Arizona. Plane-polarized light.

Chemical analyses of natural glasses commonly are used to estimate compositions of magmatic liquids. However, there can be significant changes in glass compositions, including loss or gain of alkalis and oxidation of iron, during hydration and before devitrification (Stewart, 1979). Table 6-1 lists analyses showing chemical changes in a progression from black glass with sparse perlitic cracks, through light gray and thoroughly fractured perlite, to yellow-green and completely devitrified equivalent of the black glass. Note particularly the changes in Fe_2O_3, FeO, CaO, Na_2O, H_2O^+, and H_2O^-.

The devitrification products of a specific glass depend partly on the normative composition. For example, a rhyolite glass rich in *q, or,* and *ab* should produce a silica mineral (quartz, tridymite, or cristobalite) and one or two feldspars upon devitrification, but a basaltic glass rich in *an, di,* and *hy* should devitrify to plagioclase and pyroxenes. However, many of the phases produced by devitrification are hydrous (zeolites, clays, and chlorites, for example) and have no normative counterparts.

Devitrification commonly begins at a nucleation site provided by a crystal or a fracture, resulting in *spherulitic* growth (originating from a point source) or *axiolitic* growth (from a line source); see Figure 6-13.

111

TABLE 6-1 CHEMICAL ANALYSES OF RHYOLITE FROM THE ALLEN COMPLEX, PINTO CANYON, PRESIDIO COUNTY, TEXAS, SHOWING CHANGES ATTENDING HYDRATION AND DEVITRIFICATION

	Sample				Sample		
	1	2	3		1	2	3
SiO_2	73.09	72.50	71.60	q	30.29	32.49	42.69
TiO_2	0.29	0.17	0.26	or	30.61	28.83	29.54
Al_2O_3	12.15	13.15	10.80	ab	32.07	28.43	8.04
Fe_2O_3	1.12	0.51	0.74	an	0.85	3.78	8.66
FeO	0.75	0.36	0.30	c	—	0.96	0.65
MnO	0.08	0.09	0.02	di	0.60	—	—
MgO	0.07	0.07	0.29	wo	0.31	—	—
CaO	0.47	0.80	1.81	hy	—	0.30	0.88
Na_2O	3.79	3.36	0.95	mt	1.62	0.74	0.28
K_2O	5.18	4.88	5.00	il	0.55	0.32	0.49
H_2O^+	3.26	4.76	3.73	hm	—	—	0.55
H_2O^-	0.28	0.30	3.31	ap	—	0.05	0.02
P_2O_5	0.00	0.02	0.01	cc	—	0.02	0.09
CO_2	0.00	0.01	0.04	Water	3.54	5.06	7.04
Sum	100.44	100.88	98.86	Sum	100.44	100.98	98.93

Note: Analyses by G. K. Hoops. 1, black glass; 2, gray perlite; 3, yellow-green devitrified rock. Analyses and CIPW norms are in weight percent.

6.4 TEXTURES

The *texture* of a rock refers to the sizes, shapes, and arrangements of its constituent phases. Texture is influenced by the number of phases and by their sequence of formation, as well as by their sizes and shapes. Chapter 3 showed that equilibrium crystallization of phases, as predicted by a phase diagram, may be simultaneous or sequential. Recent experiments under conditions far from equilibrium, at varying cooling rates, show that phases can appear in a different sequence from that shown on an equilibrium phase diagram (Lofgren, 1979). For example, rapid cooling of a basaltic liquid caused crystallization in the order olivine–pyroxene–plagioclase, even though plagioclase was the liquidus phase at equilibrium. Such experiments tend to confirm inferences made on the basis of texture (discussed later in this chapter) that minerals tend to nucleate in order of the increasing complexity of their crystalline structures (Hawkes, 1967), with isometric oxides (spinel, magnetite) followed by silicates built up of isolated tetrahedra (olivine), then single chains (pyroxenes), and finally frameworks (plagioclase).

If crystallization is simultaneous, crystals of different phases may interfere with each other during growth. In sequential crystallization, one phase achieves its crystalline form before another begins to precipitate. A phase filling space between most other phases is said to be *interstitial*.

1 mm

(a)

1 mm

(b)

Figure 6-13. Devitrification structures in cross-polarized light: (a) spherulitic, in perlite, Superstition Mountains, Arizona; (b) axiolitic, in ash-flow tuff, Durango, Mexico. In both examples, elongated fibers of alkali feldspar intergrown with cristobalite are the devitrification products of these rhyolitic glasses.

To characterize igneous textures, five variables must be specified. These are the degree of *crystallinity* (the modal proportions of crystals to glass), grain size, grain size variation, grain shape, and the arrangement of nonequidimensional grains.

In decreasing order of crystallinity, the terms applied are *holocrystalline* (no glass), *hypocrystalline*, *hemihyaline* (roughly equal proportions of crystals and glass), *hypohyaline*, and *holohyaline* (all glass).

Two aspects of grain size are considered. *Aphanitic* rocks are those in which all grains are too small to be resolved with the unaided eye. *Phaneritic* rocks, in contrast, contain crystals that are large enough to be distinguished (not necessarily identified) without resort to a magnifying lens. In addition to this preliminary characterization, phaneritic rocks are further subdivided according to the average diameter of the constituent grains. *Fine-grained* igneous (and metamorphic) rocks are those with average grain diameters less than 1 mm (and, as outlined in Chapter 5, require a different set of names than coarser rocks); *medium-grained* rocks are those with average diameters between 1 and 5 mm; *coarse-grained* are 5 to 30 mm, and *very coarse-grained* exceed 30 mm (3 cm). Note that these terms do not apply to sedimentary rocks or unconsolidated materials.

Variations in grain size within a sample can be significant. *Equigranular* refers to one narrow size range, *seriate* to a wide and continuous range, and *porphyritic* to a bimodal size distribution. In porphyritic rocks, the larger crystals are called *phenocrysts*, and the finer-grained material makes up the *groundmass*. *Microphenocrysts* are larger than the constituents of the groundmass, but smaller than 1 mm.

Because a thin section is a planar slice through many crystals of differing orientation, estimating the true sizes of crystals can be difficult. The exact centers of very few crystals will happen to lie in the plane of the cut; commonly, the section will intersect a corner of a crystal, yielding a smaller apparent size. Kellerhals and others (1975) review the mathematics of the problem and propose a method for converting apparent dimensions as observed in thin section into true sizes and shapes. For practical purposes, such mathematical treatment is unnecessary, and adequate estimates of grain size can usually be obtained by examining both the rock sample and a thin section.

The shape of a crystalline particle can be described in two ways, emphasizing either the surface regularity or the equality of dimensions. Both ways convey important information. A threefold classification according to surface regularity uses the terms *euhedral* (entirely bounded by crystal faces), *subhedral* (partly bounded by crystal faces), and *anhedral* (showing no regular crystal faces). In the second way, if diameters are visualized running through the center of the grain and at right angles to each other, the shape of the grain can be termed *equant* (all three diameters are approximately the same length), *prismatic* (one diameter is longer than the others), *tabular*, *platey*, or *flakey* (one diameter is less than the others, becoming relatively shorter in the order named).

The fifth aspect of texture, the arrangement of nonequant grains, is covered by the self-explanatory terms parallel, subparallel, radiating, and unoriented.

Terms from each of the five categories can be combined in a textural description, and should be assembled in the same order as they were described: crystallinity, grain size, and grain size variation. Grain shape and arrangement refer to specific phases, and usually do not apply to the entire rock. Some examples of textural descriptions are: "holocrystalline, medium-grained seriate"; "hemihyaline, aphanitic, containing subparallel microphenocrysts of subhedral plagioclase"; "hypocrystalline fine-grained porphyritic, containing euhedral, unoriented phenocrysts of olivine and subhedral, subparallel tabular microphenocrysts of plagioclase in a groundmass of anhedral equant plagioclase and clinopyroxene with interstitial glass."

Because some textures recur widely, a specialized jargon has evolved as an abbreviated way of describing them without separately listing each aspect. The more useful of these terms are defined and illustrated here. *Glomeroporphyritic* means that some or all of the phenocrysts (of one or more phases) are clumped together (Figure 6-14). *Granitic* texture (by no means confined to granite) describes anhedral, roughly equant grains interlocking like the pieces of a jigsaw puzzle (Figure 6-15). *Mortar* texture is similar to porphyritic, but the grains have no crystal faces, and the texture arises by crushing (granulation or *cataclasis*) of a rock that was originally coarser-grained and more nearly equigranular (Figure 6-16). In *trachytic* texture (Figure 6-17), nonequant grains (either phenocrysts, groundmass constituents, or all) are parallel or subparallel. This texture commonly results from alignment of crystals in moving magma or lava.

Poikilitic texture means that one mineral surrounds another (Figure 6-18). There are varieties of poikilitic texture, distinguished according to the minerals involved. In *monzonitic* texture, alkali feldspar surrounds plagioclase. In *ophitic* texture, pyroxene surrounds plagioclase, and in the more common *subophitic* texture, pyroxene is smaller than plagioclase and does not completely enclose it.

A *vesicular* rock contains cavities (vesicles), millimeters to centimeters in diameter, that were gas bubbles trapped when the rock froze. Vesicles may be spherical or irregular. An *amygdaloidal* rock was formerly vesicular, but the vesicles have been partly or completely filled with minerals deposited from gas or solution.

Spherulitic and axiolitic textures, made up respectively of radiating and parallel needles (Figure 6-13), are common devitrification products. *Granophyric* texture, composed of quartz intergrown with feldspar (Figure 6-19), can result from devitrification or by direct crystallization from liquid.

Finally, four other textural terms require definition, but not illustration. *Aphyric* is a synonym for "nonporphyritic," referring to a rock lacking phenocrysts and microphenocrysts. *Vitrophyric* is the texture of a porphyry with a glassy groundmass. *Intergranular* means that the space between large grains in a porphyritic or seriate rock is filled with small, unoriented crystals, and *intersertal* means that spaces are occupied by glass. Additional textural terms, which tend to give confusion but little precision to petrology, are still used; their definitions can be found in the *AGI Glossary of Geology* (Bates and Jackson, 1980).

Textures of igneous rocks strongly depend on cooling history, and on the abundance of nuclei in the liquid when cooling started, and are much less dependent on magma composition. In the last decade, much better understanding of textures has come from experiments in which a bulk composition (with known equilibrium phase relations) was cooled at different rates, then quenched to permit examination of the progress of crystallization at known time intervals. Lofgren (1980) provides an excellent summary of this work. Only two such studies will be mentioned here.

1 mm

(a)

1 mm

(b)

Figure 6-14. Glomeroporphyritic textures in cross-polarized light: (a) cluster of quartz phenocrysts, all with the same optical orientation, in porphyritic rhyolite, Llano County, Texas; (b) aggregate of plagioclase crystals, and one smaller clinopyroxene left of center, in andesite dike, Tieton Volcano, Cascade Mountains, Washington.

1 mm

Figure 6-15. Granitic texture in intrusive rhyolite, Carn Chuinneag, Scotland. Cross-polarized light.

Figure 6-16. Mortar texture in pyroxenite, Webster, North Carolina. The large anhedral orthopyroxene grain in the center is surrounded by smaller grains broken from it; some retain the same optical orientation as the large grain. Cross-polarized light with first-order red accessory plate.

1 mm

1 mm

Figure 6-17. Trachytic texture in phonolite, Alamo Mountain, Otero County, New Mexico. Alkali feldspar crystals are subparallel; subhedral nepheline microphenocrysts are nearly equant. Cross-polarized light with first-order red accessory plate.

Figure 6-18. Poikilitic texture; amphibole encloses plagioclase and clinopyroxene in gabbro, Mount Rougemont, Quebec. Cross-polarized light.

1 mm

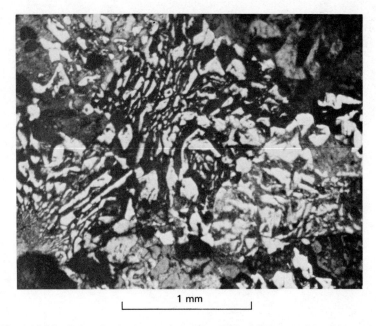

1 mm

Figure 6-19. Granophyric texture in cross-polarized light; quartz (white and gray) is intergrown with alkali feldspar (at extinction). Small quartz bodies share the same optical orientation over large domains. Skye, Scotland.

Swanson (1977) investigated the effects of varying nucleation and growth rates in a granitic melt. Figure 6-20 shows the results. At an undercooling of 50°C, the logarithms of growth rate and nucleation density (number of nuclei per cubic centimeter) for alkali feldspar are −8 and 1, respectively, while quartz and plagioclase have not begun to nucleate. At an undercooling of 100°C, the alkali feldspar growth rate is more than 300 times that of plagioclase, and quartz has still not nucleated. Nucleation densities for the two feldspar phases are the same at this degree of undercooling; alkali feldspar will form larger crystals than plagioclase because of its much higher growth rate. However, at an undercooling of 200°C, the growth rates for alkali feldspar, plagioclase, and quartz are all within 100 times each other, and nucleation densities are within a factor of 10 of each other. This combination of large and nearly equal nucleation densities and growth rates means that the texture produced at 200°C below the liquidus should be fine grained and more nearly equigranular.

Naney and Swanson (1980) added Fe and Mg to the system $NaAlSi_3O_8 - KAlSi_3O_8 - CaAl_2Si_2O_8 - SiO_2 - H_2O$, more closely approximating the compositions of natural magmas, and found that the added components, acting as network modifiers, drastically changed the order of appearance of crystalline phases upon cooling. Nucleation of feldspars and quartz was hindered, probably because the liquid was less highly polymerized, and pla-

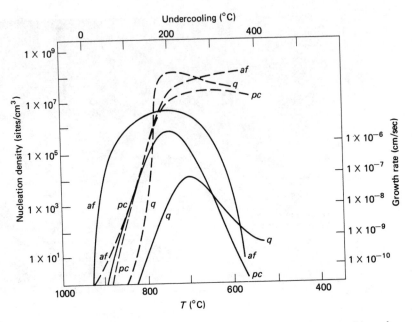

Figure 6-20. Crystal growth rates and nucleation densities in a granitic melt with 3.5 wt % H_2O at 2.5×10^8 Pa. Temperature is indicated on the lower horizontal scale, undercooling on the upper. The scales for nucleation density and growth rate are on the left and right. Growth rates are solid curves, nucleation density curves are dashed. *af,* alkali feldspar; *pc,* plagioclase; *q,* quartz. (With permission, after S. E. Swanson, *American Mineralogist, 62,* 966–978, Fig. 5a, 1977, copyrighted by the Mineralogical Society of America.)

gioclase appearance was suppressed to 250°C lower than its equilibrium liquidus temperature. In contrast, pyroxenes and biotite nucleated and grew above the equilibrium temperature ranges predicted by the phase diagram, probably because the Fe and Mg increased the nucleation of chains and sheet structures in the liquid, rather than three-dimensional framework structures of feldspars and quartz.

6.5 BOWEN'S REACTION SERIES

The nineteenth-century petrographers had observed that there are antipathies and mutual associations among the rock-forming minerals. Those pairs of phases least likely to coexist in the same rock sample are quartz and calcic plagioclase, muscovite and calcic plagioclase, muscovite and olivine, and K-feldspar and magnesian olivine. Pairs that do commonly coexist are quartz and alkali feldspar, pyroxene and calcic plagioclase, and hornblende and plagioclase of intermediate composition. N. L. Bowen (1928, p. 21) concluded that "those minerals that belong to the same general period of crystallization tend to be associated and those belonging to remote periods ordinarily fail of association. The controlling factors are thus analogous to those which determine that little girls ordinarily play 'London Bridge' with other little girls, occasionally with their mothers, seldom with their grandmothers, and never with their great-grandmothers."

Bowen summarized the mineral associations and antipathies in a simple scheme (Figure 6-21), which has since become known as *Bowen's reaction series*. The word "series" is plural in this usage; Bowen referred to a discontinuous series (the sequence on the left in the figure, relating the mafic silicates) and a continuous series (the plagioclase solid solution, on the right). The two series converge toward lower temperatures, and quartz, K-feldspar, and muscovite complete the sequence.

An additional line of evidence, supporting the scheme as erected on the basis of mineral antipathies and associations, was pointed out by Bowen (1928, p. 57); "A criterion of the reaction series, common to both the continuous and discontinuous type, and serving to show their fundamental likeness, is simply the tendency of one mineral to grow around another. . . . In the case of the continuous series this is commonly known as zoning . . . and, in the discontinuous series, as the formation of reaction rims. . . ." Bowen's generalization is a fruitful one, but it cannot rank as a fundamental law. In some peralkaline rocks, for example, mafic silicates can crystallize after alkali feldspar and quartz. Furthermore, iron-rich olivine can coexist with quartz, as do pyroxenes.

Note, in Figure 6-21, that the temperature scale is qualitative. Bowen's reaction series were not based on experimentally determined phase diagrams, although experiments do confirm the sequence as valid for many bulk compositions. The reaction series serve, not as explanations, but as a concise summary of observations that need explanations.

The discontinuous reaction series does not require, as commonly misinterpreted, that olivine be converted to orthopyroxene and this in turn to

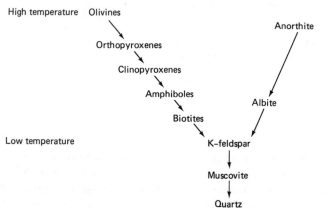

Figure 6-21. Bowen's reaction series.

clinopyroxene, and so on, as crystallization proceeds; instead, it gives the order in which successive mafic phases are likely to appear. When reaction rims do form, as at a peritectic, the tendency is for one phase to be surrounded by another that is lower in the series (olivine rimmed by orthopyroxene, for instance, or clinopyroxene by amphibole). Many magmas do not crystallize through the entire range of the reaction series, but yield assemblages of phases that fall in a relatively restricted temperature range in Figure 6-21.

6.6 SUBSOLIDUS MODIFICATION OF TEXTURES

The crystallization history of many igneous rocks does not end when the last drop of silicate liquid freezes. There are several processes that change the texture of a rock as it cools below the solidus, reacts with solutions that percolate through it, or deforms in response to a changing stress field.

Granulation or cataclasis, producing mortar texture (Figure 6-16), results during shear of solid rock, and may be caused by plastic movement of freshly crystallized rock (perhaps nudged by a fresh surge of magma from below) or, long after original crystallization, by faulting and folding.

Tuttle (1952), in a classic paper, outlined the textural changes that can occur through polymorphic inversion, exsolution, and recrystallization. One example of these changes is provided by Figure 6-22, which traces the partly hypothetical history of one phenocryst of a high-temperature solid solution of $(Na,K)AlSiO_4$.

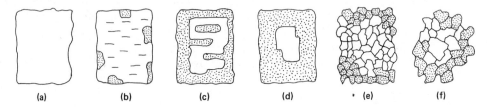

(a)   (b)   (c)   (d)   · (e)   (f)

Figure 6-22. Exsolution and recrystallization of one crystal of potassic nepheline, $(Na,K)AlSiO_4$. The stippled phase is Na-rich and the clear is K-rich. Stages (b), (c), and (d) show progressive exsolution, and stages (e) and (f) show recrystallization. (After Sahama, 1962, Fig. 4, *Norsk Geol. Tidsskr., 42,* 168–179, with permission of Universitetsforlaget.)

6.7 SUMMARY

Crystallization from a magmatic liquid requires undercooling and an adequate supply of components diffusing rapidly through the liquid to the surface of the growing crystal. Heat released during crystallization must be removed from the system. The size of a crystal is controlled by the growth rate, and the number of crystals per unit of volume reflects the nucleation rate. These two rates may reach their maxima at different temperatures below the liquidus. The average size of a large number of crystals, all competing for components, is therefore affected by both nucleation and growth rates. Zoning results when a solid solution takes up different ratios of components at different times or on different crystal faces at the same time.

Glass is metastable liquid with a partly disordered but polymerized structure frozen in by rapid cooling before crystals could grow. Devitrification of glass is enhanced by slow cooling or by reheating, and by infiltration of water. Glasses can tolerate significant changes in chemical composition before devitrifying.

The texture of a rock involves the ratio of crystals to glass, the grain size and size variation, and grain shapes and orientations. Because so many variables must be considered, the terminology applied to textures is complex, and textural interpretation is prone to ambiguity.

Bowen's reaction series describe the commonly observed associations, mutual exclusions, and textural relations of igneous minerals, by arranging the phases in approximate sequence according to the temperature ranges at which they begin to appear in magmatic liquids.

Igneous textures commonly are modified by subsolidus reactions, deformation, and reactions with fluids immediately after, or even long after, the last drop of liquid has frozen.

7

Generation and Evolution of Magma

7.1 PARTIAL FUSION

The essential phase, required by definition, in any magma is liquid. Seismic and other geophysical evidence strongly indicates that there is no permanent reservoir of mostly liquid magma in the earth's crust or upper mantle. Regions of anomalously low seismic velocities and high electrical conductivity, properties expected of rock with a small fraction of interstitial liquid (Mavko, 1980; Waff, 1980), have been recognized under oceanic lithosphere and under some volcanoes (for example, Etna; Sharp and others, 1980), but not under most continental regions. In the absence of a world-encircling and permanent source, magmatic liquid must be generated by melting previously solid rock. To initiate melting of a rock mass, it is not merely sufficient to raise its temperature above the lowest point on the solidus; additional energy (the enthalpy of fusion mentioned in Chapter 6) is needed to break bonds in the crystalline structures in order to produce liquid. Once the melting temperature has been reached, the extra heat needed to cause the change from solid to liquid can be as large a quantity as that required to raise the solid to that temperature.

Melting begins with the production of a small amount of liquid that has a composition determined by the shape of the solidus and by the relationship of the bulk composition to that solidus. In a simple eutectic system, for example, the first liquid will have the eutectic composition. If enough additional heat can be added, the amount of liquid will increase, and the liquid composition will migrate toward the bulk composition. For simplicity, and in accord

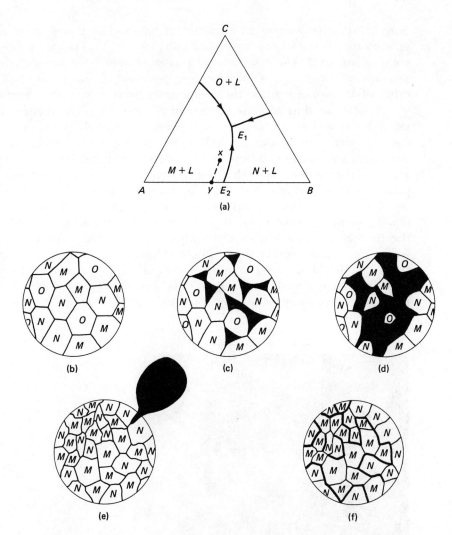

Figure 7-1. Hypothetical example of partial fusion. (a) Ternary system *A-B-C* with no solid solution; phases *M, N,* and *O* are pure *A, B,* and *C*. (b) Assemblage of bulk composition *x*, below the temperature of the ternary eutectic E_1. (c) Partial fusion has begun to produce liquid (black) with the composition E_1. (d) More advanced melting, still at E_1. (e) Liquid escapes after consuming all of phase *O*, leaving a solid residue of *M* + *N* at composition *y*. (f) Beginning of renewed partial fusion at the binary eutectic E_2.

with evidence mentioned later, we will assume that the melting process is perfect fractional fusion (Chapter 3). When a component essential to the liquid has all been consumed in the melt, there will be no further production of liquid, unless the temperature is increased so that melting can begin again in a simpler system lacking that component. Figure 7-1 illustrates these restrictions on the composition and amount of liquid. The liquid forms only at intersections of the three crystalline phases M, N, and O for eutectic E_1 composition, and at intersections of phases M and N for liquid with the composition of eutectic E_2. Other intersections, say of one grain of M with another of M, do not provide the components needed for either liquid E_1 or E_2.

Studies of thin sections of strongly heated, partially melted, rocks (for example, fragments of granite incorporated in much hotter mafic magma, then erupted and quickly cooled) show the expected distribution of glass, namely at intersections of grains of different minerals (Figure 7-2). Theory and experiment also show that liquid should form at intersections of unlike phases, and should not "wet" the flat faces between grains. Furthermore, at equilibrium, crystalline grains should not have sharp corners in contact with liquid (Bulau and others, 1979). Experiments (Waff and Bulau, 1979) confirm the theoretical inference that "tetrahedronlike liquid volumes occur at the four-grain corner intersections. These volumes are connected to similar volumes at neighboring corner intersections via channels occurring along crystalline edge intersections" (Bulau and others, 1979, p. 6106). These channels have shapes approximating triangular prisms.

1 mm

Figure 7-2. Photomicrograph (cross-polarized light with first-order red accessory plate) of part of a granitic block in mafic lava, Mount Meru, Tanzania. The dark patches are glass (now devitrified) formed by partial fusion at intersections of quartz, plagioclase, and K-feldspar. Compare with Figure 7-1c.

The network of volumes and channels can become continuously inter-connected when the amount of liquid exceeds some critical value (probably in the range 5 to 22%, according to Waff, 1980). Once a continuous network is established, liquid can percolate upward rapidly, driven by the density difference between it and the crystalline phases. This "buoyant percolation" apparently is inhibited at pressures exceeding 2.5 to 3×10^9 Pa (Walker and others, 1978) because liquids are sufficiently compressed above that pressure range so that their densities are too high. One factor limiting the rate at which liquid can escape is the rate at which solid grains in the residuum can flow or deform to fill the space being vacated by the liquid (Waff, 1980, p. 1815).

Gravitational instability, and the large amount of heat required to produce liquid, make it extremely unlikely that a given body of rock will melt completely. There is the additional constraint that heat cannot be focused into a specific space, because heat will only flow from a region of higher temperature to one of lower temperature. Addition of more heat will increase the total volume of rock involved in partial fusion, rather than markedly increasing the amount of liquid within a fixed volume.

The preceding paragraphs assumed that heat was added to the system from an external, unspecified source. Instead, we could postulate that the heat needed for fusion was generated within the rock during deformation. Feigenson and Spera (1980) have analyzed this alternative mathematically, assuming that partial fusion occurs in upper mantle rock deforming under constant shear stress. Because escaping liquid efficiently carries heat out of the system much faster than it can be removed by convection of the mantle rock, the volume fraction of liquid increases to only about 3 to 5% within 2 or 3 million years after deformation starts, and then remains constant (because heat is removed by liquid as fast as it is generated by deformation). As in the previous models, the volume fraction of liquid is therefore small.

Partial fusion in the upper mantle.   Circumstantial but strong evidence to be listed in Chapter 10 suggests that upper mantle rocks are predominantly assemblages of olivine + orthopyroxene + clinopyroxene + a highly aluminous phase (either spinel or garnet, depending on the depth). Much experimental work has been carried out on systems containing the essential components of these phases at pressures equivalent to those expected in the upper mantle, but much more needs to be done, and results so far are inconclusive. There is, however, general agreement that liquids generated by partial fusion in the upper mantle should be enriched, relative to the initial bulk composition, in SiO_2, Al_2O_3, FeO, CaO, Na_2O, K_2O, and H_2O.

Basaltic magma has been abundant and widespread both in space and time, and has erupted through continental as well as oceanic crust. The abundance and ubiquity of basalts in the geologic record make them a most likely product of partial fusion in the upper mantle. As we shall see in Chapter 11, however, there is considerable compositional variation among basalts, and the question therefore arises as to whether this diversity reflects differences in parent materials and conditions of fusion in the mantle, or whether it indicates modifications superimposed upon one originally uniform liquid as it moved from its source toward lower-pressure regions. This is a problem to which we must return several times in subsequent chapters.

Partial fusion in the continental crust.   The composition of the upper part of the earth's continental crust and the oceanic crust is reasonably well known (Table 2-1). Many salic igneous rocks in the continents have compositions that lie close to the plane representing the system $NaAlSiO_4-KAlSiO_4-SiO_2$ (Figure 7-3), called by Bowen (1937) the "residua system." The term arises because fractional crystallization tends to drive compositions of residual liquids toward this plane. As shown in Figure 4-4, the system embraces the compositions of the normative constituents *q, ab, or, lc, ne,* and *kp,* corresponding to the modal silica minerals, alkali feldspars, and feldspathoids.

The system $NaAlSiO_4-KAlSiO_4-SiO_2$ at 1 atm pressure (Figure 7-3a) shows two regions of low liquidus temperatures, and these reflect depressions in the underlying solidus. The "low-melting" regions persist, and become accentuated, when the system contains water at high pressures (Figure 7-3b). The incongruent melting of K-feldspar to leucite plus silica-rich liquid changes to congruent melting at high pressure, and ultimately, at still higher pressure, the stability field of leucite vanishes from the liquidus. At all water pressures so far investigated, the low-melting region in the quartz-normative

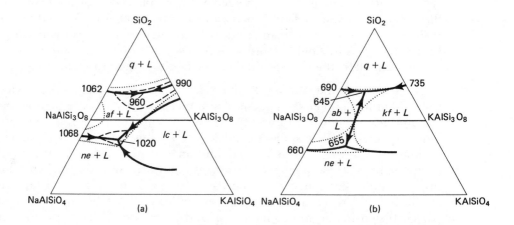

Figure 7-3. The system $NaAlSiO_4-KAlSiO_4-SiO_2$, in weight percent. (a) At 1 atm pressure. Numbers are liquidus temperatures. Dotted and dashed isotherms are respectively 1100 and 1050°C. (After Fig. 1 of Schairer, 1950, *Journal of Geology,* by permission of The University of Chicago Press.) (b) With excess water at 5×10^8 Pa. Dotted isotherm is 700°C. *L,* liquid; *q,* quartz; *af,* alkali feldspar; *lc,* leucite; *ab,* albite; *kf,* K-feldspar, *ne,* nepheline. (After Morse, 1969, Fig. 21, with permission of the Carnegie Institution of Washington and S. A. Morse.)

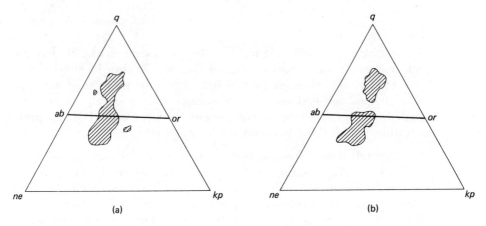

Figure 7-4. Normative compositions (in weight percent) of rocks with more than 80% *ab* + *or* + *q* or *ne* in Washington's (1917) compilation. (a) Extrusive rocks, 362 *q*-normative and 122 *ne*-normative. (b) Intrusive, 571 *q*-normative and 102 *ne*-normative. (From Tuttle and Bowen, 1958, Figs. 41 and 42, for *q*-normative, and from Hamilton and MacKenzie, 1965, Figs. 4 and 5, with permission of the Mineralogical Society of Great Britain, for *ne*-normative.)

portion of the system (above the alkali feldspar join in Figure 7-3) has liquidus and solidus temperatures below those of the analogous low-melting region in the nepheline-normative quadrilateral, at the same pressure.

In Figure 7-4 are plotted, for direct comparison with Figure 7-3, the normative compositions of salic extrusive and intrusive rocks. The ruled areas cover the compositions of all 1157 analyzed rocks in Washington's (1917) compilation that have a differentiation index of 80 or higher. Rock compositions clearly are concentrated in the regions with the lowest liquidus and solidus temperatures, that is, in those regions where the first liquid would be produced by partial fusion in this system and in which the last residual liquids would migrate during fractional crystallization. These concentrations strongly suggest that the rock compositions were controlled by interaction between liquid and crystalline phases, but cannot indicate whether the control was achieved by fusion or by crystallization. Partial fusion is more likely for the silica-oversaturated rocks than for those that contain normative feldspathoids, because the appropriate bulk compositions of parent materials to yield *ne*-normative liquids upon fusion are much less abundant in the earth's crust. On the other hand, the quartz-normative rocks shown in Figure 7-4 have bulk compositions that are expected of liquids generated by partial fusion of rocks known to be abundant in continental crust.

7.2 MAGMATIC DIFFERENTIATION

One of the fundamental concepts of igneous petrology is the premise that a range of liquid compositions can evolve from a single "parental" magma. Strong evidence for this is provided by compositional variations in lavas issuing from a single vent. Figure 4-6 illustrates compositional changes in lavas that erupted from Parícutin volcano, Mexico, during its short lifetime.

The set of processes by which a variety of rocks may derive from one initial magma is called *magmatic differentiation*. In general, the processes can be divided into fractionation of components between crystals and liquid, between or within liquids, and between liquid and gas.

Crystal–liquid fractionation.   The results of separating crystals from liquid to prevent continued equilibrium were discussed in Chapter 3. Now we examine the consequences in less general terms. Start with a two-component solid solution series (Figure 7-5) resembling olivine or plagioclase. Cool bulk composition M in this system slowly in four stages; when solidification is half completed in each stage, drain all of the remaining liquid off into a lower receptacle, leaving the crystals behind. Starting with 1000 g of composition M as an initial liquid, the first stage yields 500 g of crystals of composition N, plus 500 g of liquid of composition O. This liquid drains into a lower container, where crystallization is allowed to begin again, yielding 250 g of crystals of composition P by the time the process is interrupted by the removal of 250 g of liquid Q. The fractionation is repeated until 125 g of liquid with

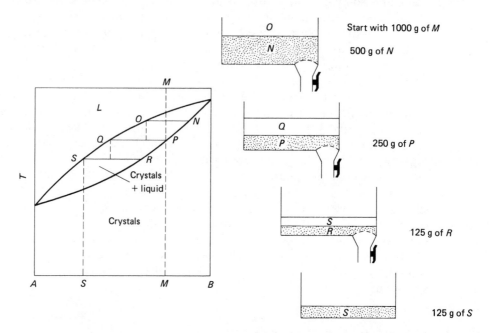

Figure 7-5. Four-stage fractional crystallization caricatured as an industrial process. Liquid is removed when crystallization is half completed in each of the first three stages. Note that the average composition of 500 g N + 250 g P + 125 g R + 125 g S is equal to the initial composition of 1000 g M.

composition *S* arrives in the bottom receptacle and crystallizes in a closed system, having nowhere else to go. This simple fractionation scheme yields four batches of crystals, with different compositions, from one initial liquid. The mechanism essential to such differentiation is the separation of crystals from liquid. There are four ways to achieve this separation.

Gravitational separation.   If crystals differ in density from the surrounding liquid, they may rise or sink, given sufficient time and a liquid of low viscosity and low yield stress that is not disturbed by rapid convection. The liquid will progressively be depleted in the components that are taken up by the crystalline phases, and a layered series of igneous rocks may form, in which not only the modal proportions but the compositions of phases may show progressive changes from bottom to top in the layered sequence (Chapter 8).

Liquid displacement by crystal growth.   If crystals grow on the floor, walls, or roof of a magma reservoir, they displace the remaining liquid, depleted in some components, toward the center of the magma body. Rapid burial of crystals by additional crystals squeezes interstitial liquid out of the pile, and prevents the more deeply buried crystals from reequilibrating with liquid. This mechanism of crystal–liquid fractionation is analogous to the concentric zoning illustrated in Figure 6-8, but inside out and on a much larger scale.

Filter pressing.   When a mixture of solids and liquid is squeezed as in a winepress, the liquid will escape, given the opportunity. In effect, this is the process discussed at the beginning of this chapter (Figure 7-1). The problem is in visualizing situations in which the liquid has an escape route that the crystals cannot follow. At high confining pressure, the buoyancy and low strength of liquid relative to solid phases combine to permit its separation from the system. At lower pressures, two possibilities seem likely to operate on a small scale; if a void opens by fracturing of the surrounding rock, liquid may drain from a mush, and if magma, flowing through a channel, encounters a constriction in which crystals become jammed, the liquid may continue to move.

Observations on slowly cooling basaltic lava lakes in Hawaii, obtained by drilling through the solidified crust to the still molten interior, indicate that a crystal–liquid mush begins to act as a brittle solid (and to sustain tensional fractures) when the crystalline portion exceeds about 20 vol % (Wright and others, 1968). Thus, there may be a large proportion of liquid available when tensional rifting of mush occurs, but there appear to be no estimates of the efficiency with which liquids can be separated from crystals by filter pressing. Propach (1976) has comprehensively reviewed the possible mechanisms of filter pressing, and concluded that the process is not very effective in magmas with a high ratio of liquid to solid and at low confining pressure.

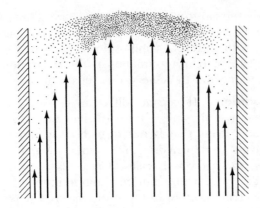

Figure 7-6. Cross section of a vertical magma conduit showing flow differentiation. Velocity vectors (arrows) indicate distance travelled in one unit of time. Friction and cooling near the walls reduce the velocity and increase the viscosity, causing suspended particles to migrate toward the center.

Flow differentiation.   As magma moves through a conduit, friction establishes a velocity gradient such that the central portion of the magma flows faster than the marginal parts. Increased viscosity near the walls, caused by lower temperature due to heat loss into the wall rock, will accentuate this velocity gradient (Figure 7-6). When solid particles (crystals or rock fragments) make up more than about 5 vol % of the magma, they tend to interfere with each other's movement. In effect, the streaming of highly incompressible liquid between two nearby crystals will cause them to repel each other in directions perpendicular to the direction of flow. This *grain dispersive pressure* (Komar, 1976) is a function of grain concentration, velocity gradient across the conduit, and viscosity of the liquid between the grains. Because the grain dispersive pressure must be constant across the entire width of the conduit, the concentration of suspended grains must decrease rapidly near the walls, where both the viscosity and the velocity gradient increase (Figure 7-6). In a vertical pipe, suspended particles will therefore be concentrated toward the center (Figure 7-7a), where they will tend to be larger as well as more abundant. In a horizontal conduit, this central concentration will be displaced downward by gravity (if the particles are more dense than the liquid), but will still reach its maximum some distance above the floor, because of the velocity gradient (Figure 7-7b).

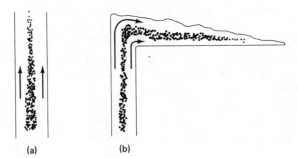

(a)   (b)

Figure 7-7. Flow differentiation (a) in magma moving in a vertical conduit; (b) modified by gravity in a horizontal conduit. (After Simkin, 1967, Fig. 3.10, with permission of John Wiley & Sons, Inc.)

132

Because it depends on local differences in flow velocity, flow differentiation is most likely to be effective only in relatively small igneous bodies. Barrière (1976) calculated that it will work only in bodies with widths or diameters of less than 100 m. To the extent that narrow conduits may feed larger bodies, however, the bulk composition of a large mass of magma may be changed as a crystal–liquid suspension passes through a flow-differentiation "filter."

Gravitational separation and flow differentiation both require a minimum liquid content of approximately 30 vol % in the magma (Arndt, 1977). At lower proportions of liquid, interference among crystals inhibits both processes. The effect of the solid-to-liquid ratio on mechanical properties of magma and lava is discussed in more detail in Chapter 8; for the moment, the significant point is that crystal–liquid fractionation by gravitational separation or flow differentiation requires a high percentage of liquid. When attributing compositional variation in a suite of igneous rocks to these mechanisms, we must assume that only those crystalline phases on or near the liquidus could be removed from the parental magma. It makes no sense to postulate fractional removal of a mineral that only begins to crystallize after 90% of the magma has solidified, or that occurs only in the groundmass, not as phenocrysts, in a porphyritic rock.

Liquid fractionation. There are two ways in which magmatic differentiation might be achieved in systems that are entirely liquid. One is by liquid immiscibility and the other is by diffusion along concentration gradients produced by temperature differences.

Liquid immiscibility. Coexistence of two liquid phases of different compositions at the same temperature has been recognized in some systems. Given enough time, even very viscous immiscible liquids will tend to separate into two layers (excepting the remote possibility that they have exactly the same density). Hence such a process, of cooling an initially homogeneous liquid down into a temperature interval where it unmixes into two liquids of markedly different compositions and densities, appears to offer an efficient mechanism of magmatic differentiation. However, serious objections have been raised against liquid immiscibility as an important magmatic process. In the examples recognized in the first experimentally investigated systems, immiscibility occurred at extremely high temperatures, and the liquid compositions did not correspond to those of any common igneous rocks. Among geologically significant systems, liquid immiscibility was known at reasonable temperatures only in the system $Fe_2SiO_4 - KAlSiO_4 - SiO_2$ (Roedder, 1956).

Reinforcing the lack of experimental evidence has been the absence of petrographic evidence. The most conclusive argument for liquid immiscibility would be globules of glass enclosed by glass of contrasting composition. Such evidence would not form in deep intrusive settings, and could be destroyed by devitrification at shallower depths and on the earth's surface.

Bowen (1928, p. 19) evaluated the evidence and concluded: "If any petrologist should choose to assume that this condition obtains for basalt and rhyolite it would be very difficult to prove him wrong, but the assumption would have nothing else to recommend it." The lack of evidence, and Bowen's authority, thereafter discouraged much consideration of immiscibility as an important mechanism in the formation of silicate rocks. However, petrographic and experimental evidence continued to accumulate in favor of immiscibility between a silicate liquid and liquids of greatly contrasted compositions (sulfides, carbonates, oxides, phosphates, and halides). Among the investigations that showed such immiscible relations are those of Roedder and Coombs (1967), and Skinner and Peck (1969).

There remained some puzzling occurrences of silicate rocks that were intimately mixed and showed contrasting mafic and felsic compositions. Ferguson and Currie (1971) demonstrated, by combinations of field observations, analyses of rock and mineral compositions, and melting experiments, that some of these are indeed best explained by liquid immiscibility, and that the temperature interval over which immiscible liquids may coexist can be as small as 20°C.

In 1969 a new world for igneous petrologists was literally opened as the first lunar samples were returned. Glasses that had escaped devitrification because of the absence of water on the moon were a high-priority topic of investigation, and the classic evidence for silicate liquid immiscibility was found. Droplets of glass, rich in K, Al, and Si, are enclosed by, or enclose, another glass rich in Fe and Mg (Roedder and Weiblen, 1971). These authors also found the same evidence, in the form of extremely small globules of one glass within another, in the groundmass of terrestrial lavas.

Textural and compositional features that imply immiscibility have now been widely recognized. A. R. Philpotts (1978) concluded, on the basis of petrographic and experimental evidence, that during fractional crystallization of some basalts, when the magma or lava is a little more than half crystallized, the residual liquid can separate into two liquids, one rich in iron, the other in silica. The more viscous silica-rich liquid generally quenches to a glass, but the small droplets of iron-rich liquid commonly crystallize to single spherical crystals of pyroxene. There still remains the vital question of whether the two liquids could ever separate sufficiently to produce two rock types. Roedder (1979a) reviewed the entire range of immiscibility in magmas and concluded that the process of liquid unmixing may be important even in the early stages of magmatic crystallization, and in rocks that are not particularly rich in alkalis or carbonates.

Liquid immiscibility illustrates the ways in which textural, analytical, and experimental data must be integrated if petrologic problems are to be solved, and shows that, in igneous petrology as in other fields, a concept can pass through stages of favor and disrepute. Specific examples of liquid immiscibility, and of other differentiation mechanisms, are deferred to later chapters.

Diffusion.   In his attempt to understand "our adversaries, the rocks" (Bowen, 1928, p. i), Bowen also considered diffusion within a homogeneous liquid phase as a possible differentiation mechanism: "A gradation of composition, in a completely liquid mass, resulting from a gradation of temperature in the mass (Soret effect) is among the possibilities considered, but there is every reason to believe that the greatest theoretical magnitude of this effect would be very small and that even this small effect would never be attained" (p. 5).

Although there is a tendency in aqueous solutions for certain components to diffuse from a region of high temperature to another of lower temperature, or vice versa, the rates of diffusion in dry viscous silicate liquids are so low that very large temperature differences would be required to drive this diffusion significantly. Bowen pointed out that such a steep temperature gradient almost certainly requires that crystallization begin, and then differentiation is not occurring in a homogeneous liquid phase.

With increasing appreciation of the effect of network modifiers on the structure of silicate melts, especially the effect of water, and with the advent of better data on diffusion rates (Hofmann and Magaritz, 1977), there has been renewed interest in diffusion as a differentiation mechanism. There is strong evidence that some large masses of nearly crystal-free magmatic liquid within a few kilometers of the earth's surface were stratified according to composition and density, and that successive eruptions tapped these reservoirs at different depths (Hildreth, 1979; 1981). Crystal–liquid fractionation high in the crust cannot account for the observed variations, and a combination of diffusion and convection may be required.

Liquid-gas fractionation.   If a separate gas phase nucleates as bubbles within a magmatic liquid, it is very likely that some components, particularly alkali metal ions, will be more soluble in the gas than other components. Thus entrained within the bubbles, such components will be carried upward, depleting the liquid in those constituents. A vertical composition gradient could be set up in a magma column by such *gaseous transfer* of components from bottom to top.

The gas phase, rather than redissolving in liquid, is more likely to escape from the magma into the overlying rock, water, or air. Minerals may precipitate from this gas phase, but it is unlikely that a sequence of magmatic liquids of different compositions would be produced. As we will see later, *vesiculation* (formation of gas bubbles) can be extremely important in transferring heat and volatile components out of magma, and in changing its mechanical properties, but its efficiency as a mechanism of magmatic differentiation has probably been overrated.

Nevertheless, gas transfer does deserve more experimental and theoretical investigation. Sakuyama and Kushiro (1979) have experimentally produced upward enrichment of alkalis in a vesiculating andesitic melt. By slowly reducing the confining pressure on a hydrous liquid at constant temperature, they produced gas bubbles that migrated upward, growing larger and more numerous toward the top of the liquid in the sealed capsule. Na_2O was depleted in the lower portion and enriched in the upper portion of the capsule; K_2O showed the same trend, less strongly. However, in Sakuyama and Kushiro's experiments, most of the alkalis added to the upper portion apparently remained in the gas bubbles, and did not return to the liquid. The analyses could not discriminate between alkalis in glass and in trapped gas bubbles.

Of all the mechanisms of magmatic differentiation, crystal–liquid fractionation by liquid displacement and gravitational separation, and liquid fractionation by diffusion–convection and possibly by immiscibility, seem the ones most capable of changing the composition of large masses of magma. There are, however, other ways to modify magma compositions through interaction with wall rock and with other, independently evolved magmas.

7.3 MAGMATIC ASSIMILATION

Hot magma may react with or dissolve wall rock and thereby change its composition. By shattering and hydraulic wedging, the magma may detach fragments of its walls and roof, which become incorporated in the magma. Such inclusions commonly are small and have relatively large ratios of surface area to mass. The large ratio, coupled with the high temperature of the magma, favors chemical reactions between magma and inclusions. The composition of the final igneous rock yielded when this contaminated magma crystallizes will depend not only on the compositions of the original magma and the inclusions but on the proportion of material assimilated by the magma.

It is unlikely that the inclusions will simply melt to mingle with the liquid of the original magma. As mentioned at the beginning of this chapter, considerable heat is required to melt a rock even after it has been heated to its solidus. This heat must be supplied by the magma, which must lose heat and partly crystallize during assimilation. Magma, unless greatly *superheated* (the opposite of undercooled; the temperature is above the liquidus), therefore is limited in the amount of rock that it can melt. Much melting can be expected only if the invaded rock is already very hot, if it is largely glass, or if its solidus temperature is considerably below that of the magma.

Assume that the heat capacity of wall rock is 1 joule per gram per degree Celsius. Then heating the rock from 200 to 1200°C requires 1000 J/g. Now assume that the enthalpy of fusion of the wall rock is 250 J/g, and that the rock melts at 1200°C. Therefore, the heat energy required for raising the temperature of 1 g of the wall rock from 200°C to its solidus of 1200°C and melting it by contact with magma is 1250 J. The magma supplies this heat by crystallizing. Further assume that the magma has the same enthalpy of crystallization as the enthalpy of fusion of the wall rock, 250 J/g (this is not only a convenient assumption, but a realistic one). To supply 1250 J for heating and melting 1 g of wall rock, the magma must crystallize 5 g of solid phases. Because additional heat is lost from the magma through conduction and groundwater convection, considerably more than 5 g must crystallize from the magma for every gram of wall rock that melts.

Instead of melting, inclusions are much more likely to react with the magma, converting the minerals in the solid inclusion into phases identical with those already crystallizing from the magma. Bowen (1928, p. 175–223) summarized the possible reactions in terms of his reaction series. "If the foreign material belongs to an earlier stage of the reaction series the tendency is to make it over into those phases with which the magma is saturated and to precipitate a further amount of these phases from the magma itself. If the foreign material belongs to a later stage of the reaction series it tends to become part of the liquid by precipitating phases with which the magma is saturated" (p. 215). "The material that can, by this reactive process, become a part of the liquid must consist of a later member of the reaction series, that is, must be material toward which the liquid could pass spontaneously by fractional crystallization. The net effect upon the liquid is, then, to push it onward upon its normal course" (p. 221). Some examples are shown in Figure 7-8.

Assimilation may therefore merely hasten the crystallization of magma. A large proportion of assimilated wall rock would be required to cause a perceptible change in the proportions or compositions of phases precipitated from the magma. McBirney (1979) provides an admirable review of the field, petrographic, and geochemical evidence for assimilation.

Magma of intermediate composition (andesite, dacite, quartz diorite, and other plagioclase-rich rocks of fields 4, 5, 9*, and 10* in Figure 5-3) has been postulated to form as mafic magma coming up through felsic continental crust assimilates some of the felsic material. The isotope chemistry of strontium testifies against this hypothesis as a general explanation for all intermediate magmas. The radioactive isotope of rubidium, ^{87}Rb, decays to ^{87}Sr with a half-life of 49×10^9 years. Rubidium is an alkali metal behaving like potassium. By analogy with the demonstrable concentration of potassium in the earth's crust relative to the mantle, we are certain that rubidium is similarly fractionated into the crust, leaving relatively little ^{87}Rb in the mantle to produce ^{87}Sr there. Strontium behaves like calcium, and is not as strongly fractionated between crust and mantle as is rubidium. Complicating the situation, ^{87}Sr is an abundant isotope of strontium and a large portion of it is primordial, not produced by decay of ^{87}Rb. Assuming that the overall ratio of two

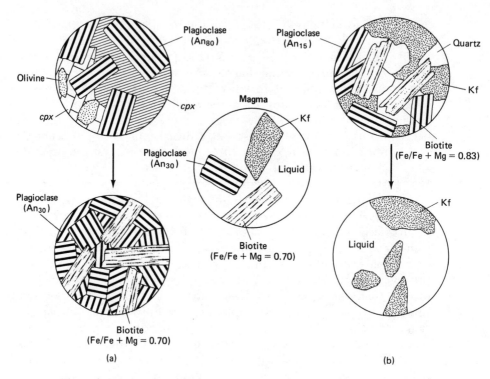

Figure 7-8. Two schematic examples of assimilation of different rocks in the same magma. The magma is precipitating the phases indicated in the central circle. Inclusion (a) contains phases higher in Bowen's reaction series than those crystallizing from the magma, and inclusion (b) contains phases that are lower in the series or, as in the case of K-feldspar, identical with those crystallizing from the magma. Inclusion (a) is made over in the solid state, and inclusion (b) melts and disaggregates.

strontium isotopes, ^{87}Sr and ^{86}Sr, was uniform just after formation of the earth, we can consider any change in the ratio $^{87}Sr/^{86}Sr$ as a function of time and original ^{87}Rb content (Figure 7-9). If the $^{87}Sr/^{86}Sr$ and $^{87}Rb/^{86}Sr$ ratios are measured for two or more phases in the same rock, or for two or more rocks known from other evidence to have formed from the same magma at the same time, the data should fall on a straight line, the slope of which is proportional to the age. Extrapolation back to $^{87}Rb/^{86}Sr$ equal to zero gives the *initial ratio* of $^{87}Sr/^{86}Sr$, correcting for that amount of ^{87}Sr produced by decay of ^{87}Rb since the magma crystallized.

It is this initial strontium ratio that has proved so important in modifying our estimate of the role of magmatic assimilation. The "primordial" value for $^{87}Sr/^{86}Sr$, estimated from measurements on meteorites that contain effectively no rubidium, was about 0.699. The ratio in samples from the earth's upper mantle is about 0.702 to 0.705. In the upper continental crust, where rubidium has accumulated and decayed through geologic time, the ratio is

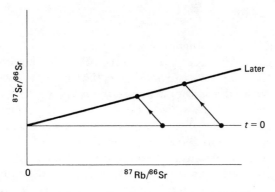

Figure 7-9. Decay of ^{87}Rb to ^{87}Sr increases the ratio of ^{87}Sr to ^{86}Sr while decreasing the ratio of ^{87}Rb to ^{86}Sr. At a time $t = 0$, before any rubidium has decayed, the strontium isotope ratio (vertical axis) is uniform and independent of the rubidium content, as shown by the horizontal line. As time passes, a rock with no rubidium (left edge of diagram) retains its original ratio of ^{87}Sr to ^{86}Sr, but rocks with more rubidium develop higher ^{87}Sr/^{86}Sr, and lower ^{87}Rb/^{86}Sr, as more ^{87}Sr is produced at the expense of ^{87}Rb (inclined line marked "later"). The changing ratios for two systems, which may be different rocks from the same magma or different minerals in the same rock, are shown by the arrows.

higher and more variable, exceeding 0.750 in old, rubidium-rich rocks. The strontium isotope ratio can be measured with precision to the fifth decimal place, and is used as a tracer in the following manner.

If a lava attained its intermediate composition by assimilation of crustal rock in mantle-derived mafic magma, the strontium isotope ratio should lie between those of the mantle and the crust. Processes involving reaction between crystalline phases and magmatic liquid do not fractionate strontium isotopes; the ^{87}Sr/^{86}Sr ratio cannot be changed by fusion or crystallization, but only by assimilation of material with a different isotopic ratio, or by addition of rubidium followed by its slow decay. The strontium isotope ratio of a magma generated by partial fusion in the mantle should therefore be identical to that of the parental mantle rock, and the ratio should remain unchanged if the magma differentiates.

Because the heat requirements are less stringent for assimilation of glassy rocks than of holocrystalline rocks, assimilation of volcanic debris at high levels within a volcanic pile may be effective in producing rocks of intermediate composition. Eichelberger (1974) has presented evidence for such contamination of magma by glass of a different composition. Two parts of the evidence are the presence in the "hybrid" lava of phenocrysts that have compositions that are not in equilibrium with each other or with the liquid represented by the groundmass, and the presence of undigested lumps of the contaminant. Assimilation of glass does not provide an answer to the larger question of why a volcano would erupt liquid of one composition, followed by liquid of a different composition.

7.4 MIXING OF MAGMAS

The remaining mechanism by which bulk compositions of magmas could vary is by mixing, in different proportions, magmatic liquids of contrasting compositions. This scheme was proposed in 1851 by Bunsen to explain the existence of intermediate Icelandic lavas by the mingling of basaltic and rhyolitic liquids. The absence of any world-encircling and permanent reservoirs of these two liquids, and the observation in many places that intermediate rocks are more abundant than those of extreme compositions, rule out the two-liquid model as a general cause of compositional variation.

Locally, particularly within the complex plumbing system at high levels within a volcano, mixing of magmas can be convincingly demonstrated (Wright and Fiske, 1971). At Kilauea, magma may differentiate, then sit, thermally insulated and remaining as a hot liquid within the volcanic pile, later to mix with a less differentiated magma supplied from below.

Anderson (1976) has evaluated the petrographic evidence for mixing of magmas. In porphyritic lavas containing abundant glass in the groundmass, the phenocrysts may contain glassy inclusions representing the liquid from which they grew. These glassy inclusions may have different compositions in different phenocryst minerals and all may in turn differ from the composition of the groundmass glass, if magmas have mixed. A hypothetical example based on data in Anderson (1976) follows. The liquid quenched as groundmass glass has 1.1 wt % K_2O. Olivine phenocrysts in the same sample contain glassy inclusions with 0.5 wt % K_2O, representing a mafic, potassium-poor liquid in which the olivine grew. In contrast, pyroxene phenocrysts in the same sample contain much more potassic glass inclusions arranged in concentric zones; the glass becomes richer in potassium (1.5 to 2.3 wt % K_2O) toward the outside of each pyroxene crystal, indicating that these phenocrysts grew in liquid that was becoming enriched in potassium. The composition of the groundmass glass is uniform and intermediate, with respect to potassium content, between the compositions of the two liquids from which the olivine and pyroxene grew. The glass compositions indicate that two liquids, one carrying olivine and the other pyroxene, mingled to produce the liquid that quenched as groundmass glass. Although only one component is considered in this example, the same relations should be found for other constituents in the three kinds of glass. Anderson (1976, p. 30) concluded that "inclusions of glass in crystals from mixed magma . . . are tiny windows looking into the magma reservoirs."

As McBirney (1980) points out, the mixing of two magmas, each liquid in equilibrium with a different crystalline phase, should result in the complete resorption of one or both of the minerals if equilibrium is maintained (for a simple example, consider the mixing of two liquids on opposite sides of a eutectic such as that in Figure 3-6). The persistence of phases crystallized from different liquids in the mixed magma testifies that equilibrium is not maintained during or after mixing.

Eichelberger and Gooley (1977) emphasize the importance of magma mixing, using much larger scale evidence than that used by Anderson (1976). Their model, based on time and space relations of mafic and felsic magmas in several regions, follows. Mafic liquid from the mantle invades lower continental crust, supplying heat for partial fusion of crustal rocks to produce a felsic liquid. The felsic liquid accumulates, and subsequent injections of mafic magma keep it hot and mix with it to produce intermediate liquids. The mixing varies in thoroughness because mafic magma is too dense to rise through felsic magma in spite of its higher temperature, and tends to form globular masses chilled against the felsic liquid. Convection and successive injections may produce a homogeneous intermediate magma or a stratified magma body.

Others (including McBirney, 1980; Huppert and Sparks, 1980) doubt that mixing of the two liquids can be effective before crystallization is far advanced. The mafic magma is more likely to spread as a higher-density pool beneath the felsic liquid, and exchange of components across the interface between the two will be very much slower than heat exchange. Both magmas will probably convect vigorously but independently.

7.5 MAGMATIC EVOLUTION IN OPEN SYSTEMS

This summary of magmatic evolution is incomplete without consideration of a major complication that affects all of the processes by which magma changes composition; such evolution almost inevitably takes place in an open system (O'Hara, 1977), with lava being erupted from the top of the magma reservoir, new magma entering from below, and some components being exchanged with the surrounding rock, groundwater, and atmosphere.

The most thoroughly audited volcanoes, from the viewpoint of magmatic income and expenditure, are Etna (Wadge, 1977) and Kilauea (Wright and Fiske, 1971). In both examples, magma is fed into shallow reservoirs where it accumulates for a few years, causing distension of the volcanic pile before eruption from the summit or flanks.

Pearce and Flower (1977) suggest that magma chambers may even reach a steady-state condition in which the effects of fractionation are balanced, over long time intervals, by mixing of fresh magma with differentiated magmas. The frequency of communication between magma source in the mantle and magma reservoir in the crust can impose cycles or long-term trends on the compositions of lava erupted during the activity span of a volcano. Newhall (1979) analyzed 51 sequential lava flows from Mayon volcano in the Philippines, and found cyclical variation in major elements and in proportions of phases. These variations he attributed to periodic injection of mafic magma into a shallow reservoir where it fractionated before injection of another batch of mafic magma which mixed with the fractionated residue. Rose and others (1977a) concluded that there is a relationship between the height of Santa Maria volcano in Guatemala and the composition of the lava emitted. Study of 26 successive outpourings of lava indicated that, as the cone grew higher, the younger lavas were enriched in Si, Na, and K but depleted in Ca, Mg, and Ti through crystal fractionation during longer periods of magmatic stagnation between eruptions. The phases richer in Ca, Mg, and Ti settled, thereby increasing the density of the magma and further inhibiting eruption, which became less frequent as the volcano evolved. These observations, and similar ones at other volcanoes, suggest that much of the chemical evolution of magma may take place at shallow depths, in the uppermost parts of generally stagnant but occasionally active magma columns.

7.6 SUMMARY

This chapter introduced many possible ways by which magma can vary in composition; specific examples of these processes appear in later chapters. Magmas are generated by partial fusion, and their compositions may be subsequently modified by differentiation, assimilation, and mixing. Displacement of liquid by growing crystals, and gravitational separation, are the dominant means of crystal–liquid fractionation, with filter pressing and flow differentiation affecting smaller volumes of magma. Liquid fractionation by immiscibility and by a combination of diffusion and convection may subsequently prove to be important. Liquid–gas fractionation and assimilation are locally effective, especially in magmas that have already reached unusual compositions through other mechanisms. Equilibrium crystallization in a closed system, as assumed in the necessary models provided by phase diagrams, is probably an extreme oversimplification of real magmatic processes.

8

Forms of Igneous Rock Bodies

Extrusive and intrusive rocks form masses that display a wide variety of sizes and shapes. For lavas, the underlying topography and the mechanical properties of the crystal−liquid mixture determine the dimensions of flows. The forms of intrusive rocks, on the other hand, reflect the interplay of several influences, some of which are controlled by the depth at which the magma comes to rest.

8.1 MECHANICAL PROPERTIES OF MAGMA AND LAVA

In Chapter 6 we considered viscosity of a liquid as it influences crystal growth. However, magmas that carry crystals, rock fragments, or gas bubbles tend to behave as *pseudoviscous* materials; a yield stress must be exceeded before flow occurs. In contrast, for a *purely viscous* or *Newtonian* fluid, even an infinitesimally small shear stress will initiate flow. Mutual interference among suspended particles increases both the viscosity and the yield stress.

Recent observations and calculations of yield stress and viscosity for lavas include those of Hulme (1974) and Pinkerton and Sparks (1978). It is likely that polymerization of silicate liquids close to the solidus will also cause yield stress and viscosity to increase, and that yield stress will drop to zero (more nearly Newtonian behavior will appear) as temperature climbs toward the liquidus. As a further complication, the rate at which stress is applied can also be important. Like the silicone fluid commonly sold under the trade name Silly Putty, lava can behave as a brittle solid, breaking when stress is rapidly imposed, or as a pseudoviscous liquid, flowing when stress is applied more slowly (Walker, 1969).

Another mechanical property of magmatic liquids, influential in controlling small-scale features of lava flows and in the formation of gas bubbles, is *surface tension*. Surface tension is a force acting on the interface between two liquids, or between a liquid and a gas. Liquids with high surface tension tend to form high-domed droplets rather than thin films. Murase and McBirney (1973) have measured the surface tension of a few lavas near the liquidus in an argon atmosphere; the values range from four times that of water at room temperature in air to slightly less than that of mercury, seven times that of water at the same conditions.

8.2 OTHER FACTORS CONTROLLING EMPLACEMENT OF MAGMA AND LAVA

In addition to viscosity, yield stress, and surface tension, several other variables control the dimensions of igneous bodies, especially of magmatic intrusions.

Important for intrusive bodies, but not for extrusive, is the stress field in that local volume of the earth's crust at the time of intrusion. The strength of rock is different for compressional, tensional, and shear stresses. Compared to wood or steel, rocks have very low tensile strength, but are strong in compression and shear. Deep in the crust, higher temperature reduces the compressional and shear strengths, and the stresses tend to become equal in all directions. This condition, comparable to the *hydrostatic* stress to which bodies immersed in a liquid are subjected, is called *lithostatic* stress. Closer to the earth's surface, greater compressional and shear strengths allow rock to sustain significant differences in the magnitudes of stresses in different directions. This state, in which the stress varies with orientation, is called *differential stress*.

The mechanical behavior of the *wall rock* being invaded by magma (country rock, to use an old mining term) is complexly related to the stress field and to the inherent properties of the rock. There may be many closely spaced planes of weakness, such as sedimentary bedding surfaces, to act as guides for incipient fracture or slip. These surfaces may be oriented at any angle to the directions of maximum stress, and the rock may behave in a brittle, plastic, or highly complex manner in response to stress. A large body of magma may encounter several rock types among its country rocks, each with a different style of deformation (Figure 8-1).

Two methods of magmatic emplacement are *forcible intrusion* (magma lifts the country rock or shoulders it aside) and *stoping* (magma breaks off fragments of country rock). *Passive emplacement*, in which magma fills a void opened for it by stresses not generated in the magma, is a third possibility but seems to be less common, with the important exception of constructive plate boundaries under ocean ridges.

Figure 8-1. Hypothetical cross section of unspecified scale, showing magma (black) intruding massive quartzite (stippled), thin-bedded shale (striped), and thick-bedded limestone (brick-wall pattern). The quartzite sustained a vertical fracture; as the magma-filling fracture dilated, the quartzite walls moved apart as rigid bodies. The overlying shale deformed plastically, and the limestone failed in a brittle manner, as blocks were stoped into the magma.

The stress field and the mechanical behavior of the country rock are complex functions of depth, but are also influenced by other factors. Less clearly related to depth, but partly affected by it, is the density contrast between magma and the surrounding and overlying rocks. Densities of some holocrystalline igneous rocks, glasses, and typical country rocks are listed in Table 8-1. Fusion of natural rock samples, followed by quenching at atmospheric pressure, yields glasses with densities 4 to 10% less than those of the corresponding holocrystalline rock. Upon heating to the range of magmatic temperatures (900 to 1300°C), glass densities decrease another 10%, approximately (Carmichael and others, 1974, p. 8). These figures take into account only the effects of thermal expansion, and neglect any density decrease caused by dissolved water or other gases under high pressure, and also ignore any density increase due to compression at high pressure. As a reasonable guess, magmas at sufficient depth to retain their dissolved gases, and containing mostly liquid with few suspended solid particles, probably have densities about 75 or 80% of those of the corresponding holocrystalline rocks. If the chemical composition of the liquid is known, its density at high temperature can be estimated (Nelson and Carmichael, 1979).

It is the buoyancy of magma, due to this density contrast, that propels it toward the surface, displacing denser rock downward. With decreasing depth, the density of magmatic liquid decreases but the viscosity generally increases. Because wall rock densities are less above the Moho than below, magma ascent velocity may decrease rather abruptly at the crust−mantle boundary, and "magma chambers should tend to form near the Moho discontinuity" (Kushiro, 1980, p. 117), and in the top few kilometers of crust, where the density contrast becomes small if the wall rocks are porous and poorly compacted.

Under some conditions, especially at higher levels in the crust, magma moves upward through fractures. The mechanisms of fracture formation and magma transport are reviewed by H. R. Shaw (1980) and Spera (1980b). The

TABLE 8-1 RANGES OF DENSITY FOR COMMON ROCKS AND GLASSES UNDER SURFACE PRESSURE

	Density (g/cm³)
Holocrystalline igneous rocks	
Granite	2.52–2.81
Quartz diorite	2.68–2.96
Diorite	2.72–2.96
Anorthosite	2.64–2.92
Gabbro	2.85–3.12
Peridotite	3.15–3.28
Dunite	3.20–3.31
Natural glasses	
Rhyolite	2.33–2.41
Andesite	2.40–2.57
Basalt	2.70–2.85
Sedimentary rocks	
Sandstones	2.17–2.70
Shales	2.06–2.66
Limestones	2.23–2.66
Dolomites	2.77–2.80
Metamorphic rocks	
Slates	2.72–2.84
Schists	2.70–2.96
Gneisses	2.70–3.06
Amphibolites	2.79–3.14
Eclogites	3.34–3.45

Source: Daly and others, 1966, with permission of the Geological Society of America.

rate at which magma is injected must influence the ultimate shape of an intrusive body. Pollard and others (1975) list three manifestations of mechanical work involved in intrusion: the flow of the magma itself, dilation of the wall rock to make room for the magma, and deformation of the wall rock by faulting and folding. The response of the wall rock through dilation and deformation may be rate dependent, because rapidly applied stress can cause brittle failure in the same material that may flow in response to more slowly applied stress. Furthermore, the flow behavior of the magma may also be rate dependent, as already mentioned. One example of interdependence between rate of emplacement and flow properties is that given by O. L. Anderson (1979) concerning propagation of a magma-filled crack: "A symbiotic relationship exists between the crack and the fluid. The crack tip cannot accelerate faster than the fluid within it can flow in the channel provided by the crack, and the speed of the fluid is limited by its own viscosity." The presence of a minimum yield stress (pseudoviscous behavior) and loss of heat to the wall rock combine to impose a minimum dimension (width of a crack, or radius of a cylindrical pipe) for the conduit. Magma cannot reach the surface of the earth unless the minimum dimension is at least 0.2 m for basaltic magma (Wilson and Head, 1981, p. 2973).

In addition to movement through cracks or pipes, there are two other means by which magma can ascend: diapirism and zone melting. A *diapir* is a buoyant mass, either connected to its parent layer or detached, that rises through more dense surroundings as a plastic solid (for example, a salt dome) or as a largely crystalline mass (for example, a magma at a temperature above its solidus but well below its liquidus). In addition to buoyancy, high heat content is a requirement for rise of an igneous diapir, because heat must be expended to soften "a thin rind of wall rock, causing it to flow past the magma" (Spera, 1980b, p. 291). As a diapir continues to rise, the volume percentage of liquid increases because the solidus temperature decreases as the pressure is lowered. Diapiric rise may be too rapid for liquid to escape (Anderson, 1981).

According to the *zone melting* hypothesis, magma invades the crust and transmits its thermal energy, but little of its mass, upward. A mafic magma, generated by partial fusion, moves through the uppermost mantle (through fractures or diapirs) and enters continental crust. Here its density may exceed that of the wall rock, so its ascent slows and crystallization accelerates due to increased heat loss. Crystals of mafic silicates accumulate at the bottom of the magma pool (Figure 8-2) while the heat released by crystallization moves upward through the roof. The crustal rock at the roof has a considerably lower solidus temperature than does the mafic magma below, so the roof begins to melt. Melting is aided by water expelled by the mafic magma as its crystallization proceeds. Slowly the base and roof of the magma body are displaced upward by crystal accumulation and by melting, respectively. Ultimately, the process comes to a halt because the heat absorbed by roof melting is always less than the heat liberated by crystallization below (heat is lost to the sides by conduction). Zone melting in the crust should leave a vertical column of igneous rock that becomes progressively less mafic upward and is enriched toward its top in those elements that are rejected by the accumulating crystals and concentrated in the residual liquid.

The efficiency of zone melting is difficult to estimate. Probably this mechanism has operated at the roofs of large mafic bodies to produce more felsic caps, as in the Bushveld Complex, South Africa, but it seems unlikely that the entire thickness of the crust can be pierced by a single pool of magma constantly changing composition and "burning its way" upward in this fashion.

By whatever mechanism magma is emplaced, the ratio between rate of emplacement and rate of heat loss must be crucial. When heat is lost rapidly, magma will either crystallize or quench to glass, in either case forming a more rigid sheath at the contact with wall rock and inhibiting emplacement of more magma. When enough heat has been lost that the temperature throughout

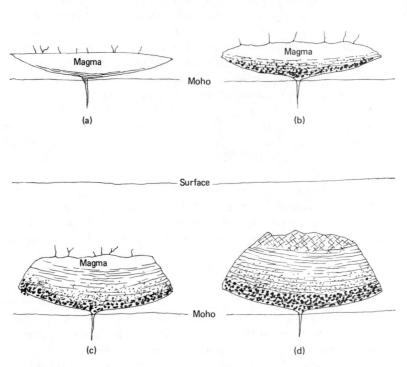

Figure 8-2. Model of zone melting in continental crust. (a) Mafic magma with high solidus temperature invades continental crust approximately 35 km thick. (b) Crystals accumulate on the floor of the magma pool and the roof rocks begin to melt. (c) A progressively smaller mass of magma of increasingly felsic composition is displaced upward by accumulation below and melting above. (d) The process closes as hydrous granitic magma crystallizes to form a cap (× pattern).

the magma reaches the solidus, the igneous mass can only move slowly as a plastic solid. Heat is lost from magma by convection within the magma body carrying heat to the contact, by conduction across the contact, by conduction and convection in the wall rock and fluids percolating within it, and, most efficiently, by escape of water and other volatile components (Chapter 9) from the magma. Heat loss is of course most rapid on and near the earth's surface, where the temperature differences between magma and its surroundings are greatest and where a steep pressure gradient permits rapid loss of dissolved volatile components. Norton and Cathles (1979) calculate that all of the heat in an intrusive body could be carried away by convection by 0.3 g of water for each gram of rock; much of that water would be supplied by the wall rock, rather than the magma, and could be recycled through long-lived convection cells. The importance of this circulation of heated groundwater in energy and mineral resources is emphasized in Chapter 16.

Heat loss is not only important in controlling the shapes of igneous bodies in the upper crust but in determining whether the magma can climb up to those shallow levels visible to us or must solidify in the mantle or lower crust. Marsh (1978, 1979a) has analyzed the relations between shapes, sizes, and rates of ascent of magma bodies. There are three extreme possibilities for style of ascent. Magma may move upward as approximately spherical globules (detached diapirs) or through fractures or pipes. Because "the rate of cooling is proportional to the product of the surface area of the magma and the temperature difference between it and the surrounding wall rock" (Marsh, 1979a, p. 169), magma filling a planar fracture must ascend 10^4 times as fast as a sphere of equivalent volume to reach the surface at the same temperature as the sphere. This is true not only because of the higher surface area-to-mass ratio of the crack but because magma in the crack is "continually thrust into contact with a medium at a lower temperature" (Marsh, 1978, p. 618). In contrast, a sphere, less prodigal with its heat, moves upward by heating, and thereby softening, the surrounding wall rock so that it can weaken and flow around the buoyant magma sphere. The problem with the third model, that of a vertical pipe connecting the magma source to the site of intrusion or extrusion, is that the ratio of surface area to mass is highly unfavorable for heat conservation and, if the pipe is continuous, the pressure differential between liquid at the bottom and that at the top will be so large that flow rates should exceed anything observed in volcanic eruptions. As a consequence, the spherical model seems most reasonable for magma traversing the upper mantle and perhaps the lower crust. We do find abundant examples near the surface of magma transport in pipes and fractures.

Figure 8-3 compares temperature-versus-depth trajectories for a sphere of magma, initially entirely liquid, leaving its site of generation in the mantle and ascending at three different rates. For the assumed geotherm, liquidus, solidus, and sphere radius of this example, magma climbing at less than 300 m/yr will reach its solidus before it reaches the earth's surface. More rapidly ascending spheres can, over a considerable portion of their paths, be at temperatures exceeding the assumed solidus of the surrounding mantle, and therefore may generate additional magma as in the zone-melting model. Because the cooling curves as calculated are concave to the assumed solidus and liquidus, magmas are implied to be superheated over much of their paths, precluding crystal–liquid fractionation except at shallower levels during their ascent.

Mathematical treatment of heat flow from magma to wall rock is formidable; some recent papers using reasonable geologic conditions are by Jaeger (1967), Whitney (1975), Norton and Knight (1977), and Spera (1980a). Figure 8-4 indicates, in a highly simplified way, a few of the complexities. The

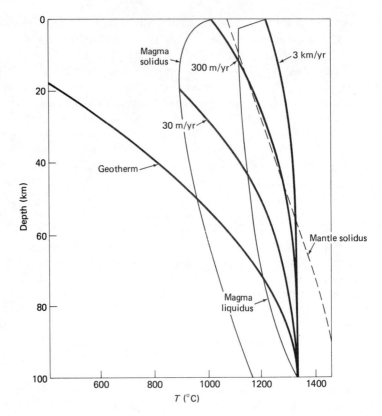

Figure 8-3. Three cooling curves for a sphere of magma 0.8 km in radius ascending at 30, 300, and 3000 m/yr after leaving its site of generation at 1350°C and 100 km depth. The geotherm, solidus of the mantle, and the liquidus and solidus of the magma are also estimated in this diagram. (After Marsh, 1979a, Fig. 7, with permission of the Society of Sigma Xi, The Scientific Research Society, Inc., and B. D. Marsh.)

temperature at the contact rapidly reaches a value that is midway between the initial (and assumed uniform) temperatures of the magma and the wall rock. By conduction the magma loses heat down this steep thermal gradient to the wall rock; cooling of magma and heating of wall rock are most rapid closest to the contact. The "waves" of temperature change move away from the contact, the distance covered in any time interval being proportional to the square root of time. Figure 8-4 assumes that all heat is lost by conduction, that the magma is emplaced instantaneously and at a uniform temperature, and that heat liberated by crystallization can be neglected. Actually, magma would not be emplaced in an instant (so the wall rock close to the contact would be preheated) and would already show temperature variations. Furthermore, convection and escape of volatile components would be more efficient processes of heat loss than conduction alone. Finally, the heat given off by crystallization (varying from 200 to 800 J/g for some common igneous minerals;

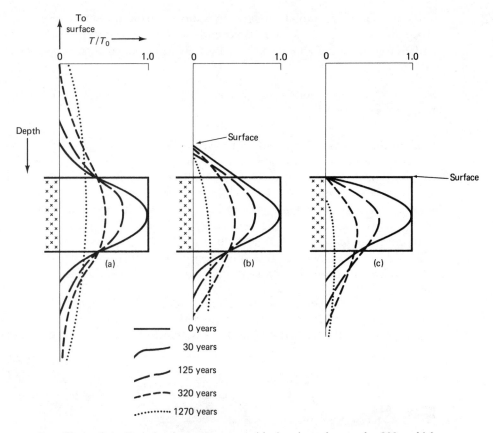

To surface
$T/T_0 \longrightarrow$

0   1.0   0   1.0   0   1.0

Depth

Surface

Surface

(a)   (b)   (c)

——————— 0 years

————— 30 years

— — — 125 years

– – – 320 years

············· 1270 years

Figure 8-4. Variation in temperature with time in and around a 200-m-thick layer of magma (assumed to extend for infinite distances to the left and right). Curves labelled in elapsed years show the ratio of temperature T to initial magma temperature T_0, where T/T_0 at the surface, and in the surrounding rock before magma emplacement, is assumed to be zero. (a) Intrusive sheet far below the earth's surface. (b) Intrusive sheet near the surface. (c) Extrusive sheet on the surface. (Modified from Jaeger, 1967, Fig. 2, with permission of John Wiley & Sons, Inc.)

Yoder, 1976, p. 91) would tend to counteract the temperature drop due to conduction, until all liquid had crystallized. These complications, plus varying absorption or liberation of heat energy during metamorphism of wall rock, will tend to make the temperature-versus-depth curves of Figure 8-4 become more highly asymmetric. The locally steep temperature gradients thus produced will have great effect on viscosity and on nucleation and crystal growth rates.

On a much larger scale, heat loss or retention can influence the form and internal complexity of an igneous body; later magma pulses will tend to follow the still-warm paths taken by earlier pulses. Magma that deviates from a prewarmed path will be more likely to freeze before climbing very far. The tendency for more mobile magmas to move through hotter, more restricted channels, where they need not expend so much energy in heating, dilating, and deforming the wall rock, will lead to multiple intrusion and to magma mixing.

8.3 INTRUSIVE FORMS

Classification of the shapes of intrusive bodies is based, first, on the relations between the intrusive contacts and structural surfaces (stratification, foliation, or other directional features) of the wall rock. If the intrusive contacts cut across the structural surfaces (or if no such surfaces exist in the wall rock), the body is said to be *discordant*. If the intrusive contacts are generally parallel to the structural surfaces in the wall rock, the body is *concordant*. Some concordant and discordant bodies are illustrated in Figure 8-5. Considerable leniency must be allowed in the definition of a concordant intrusion, because if the contact nowhere cut across the structural surfaces, the body could not be intrusive!

Among concordant bodies, *sills* are tabular (the contacts with the largest areas are approximately parallel), *laccoliths* have planar floors and domed roofs, and *lopoliths* have floors and roofs that are concave upward. Lopoliths

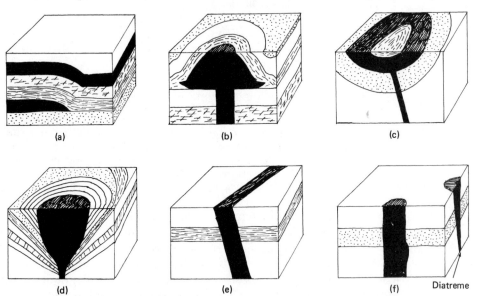

Figure 8-5. Generalized forms of simple intrusive bodies (black): (a) sills; (b) laccolith; (c) lopolith; (d) funnel; (e) dike; (f) neck and diatreme. Sills, dikes, necks, and diatremes range in scale from 1 m or less up to 1 km in thickness or diameter. Floor diameters of laccoliths average 1 or 2 km; lopoliths are larger, extending for tens of kilometers in outcrop on eroded surfaces.

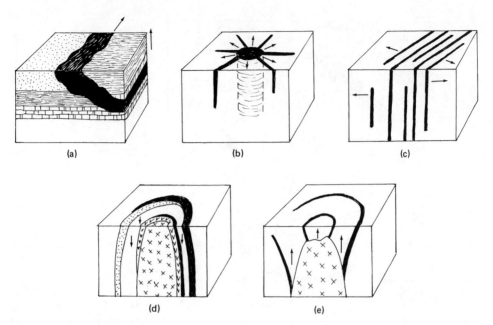

Figure 8-6. Forms of some discordant intrusive bodies: (a) discordant sheets; (b) radial dikes; (c) parallel dike swarm; (d) ring dikes; (e) cone sheets. Arrows indicate directions of displacement of the wall rock. Front edge of each block is from 1 to 10 km long.

grade into more discordant *funnels*. Lopoliths and funnels are generally filled with dense mafic rocks, whereas laccoliths usually contain less dense felsic rocks, suggesting that density contrast between magma and wall rock is an important control on shapes of concordant bodies.

Dikes are tabular and discordant; *necks* or *pipes* are crudely cylindrical, exhumed feeder conduits of volcanoes. *Diatremes* are downward-tapering pipes, commonly choked with rock fragments indicating explosive emplacement. *Discordant sheets* ("trapdoor intrusions") cut across the stratification of the wall rocks at low angles and lift the overlying rocks (Figure 8-6).

The local stress field at the time of emplacement can strongly influence the orientation of dikes (Figure 8-6). Outward horizontal push by magma may set up a radial pattern of vertical cracks that fill with magma, forming *radial dikes*. Flexing or distension of crust, with magma injected along faults or joints, produces *parallel dike swarms*. Subsidence or withdrawal of a magma column, or collapse of the roof over a magma body, forms vertical to outwardly dipping fractures, arcuate in plan view, along which *ring dikes* are injected. Upward push of magma produces arcuate fractures that, in contrast to ring dikes, dip inward and become more nearly vertical at greater depth. These fill with magma and become *cone sheets*. Ring dikes and cone sheets rarely form closed loops, and are usually short segments of arcs when traced on the ground surface.

● Phonolite	▢ Volcanic rocks	
▨ Nepheline syenite	▨ Wall rocks	
▨ Granite, quartz syenite, syenite	⊀35 Strike and dip in volcanic rocks	

Figure 8-7. Simplified geologic map of the Pilanesberg ring complex, South Africa. (Generalized from Mathias, 1974, Fig. 4, with permission of John Wiley & Sons, Inc.)

Ring complexes contain any combination and sequence of ring dikes and cone sheets, with or without radial dikes or a central intrusive body (Figure 8-7). They record a complex history of changing stress fields during intrusion.

On a larger scale, ring dike emplacement can lead to *cauldron collapse*, in which a block of overlying rock, several kilometers in diameter, founders into an underlying magma pool (Figure 8-8). Such collapse can be catastrophic; pyroclastic debris issuing from the arcuate fractures is described later in this

Figure 8-8. Cauldron collapse in cross section. Crudely cylindrical block of crust, bounded by ring dike fractures, drops into the magma (black). The purely mechanical effect of displacement of magma by the sinking block is trivial compared to the effect of pressure release when the magma chamber suddenly vents to the earth's surface; the magma loses its dissolved gases rapidly as bubbles nucleate and the upper part of the magma expands into a froth.

1 km

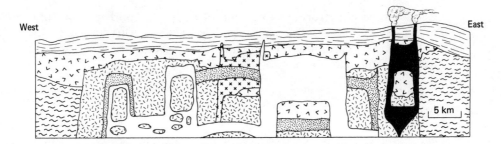

Figure 8-9. Schematic section across the Coastal Batholith of Peru. Younger pulses of magma were emplaced by cauldron collapse, cutting and dropping segments of older plutons. (Generalized from Myers, 1975, Fig. 10, with permission of the Geological Society of America and J. S. Myers.)

chapter. Some large igneous bodies apparently have been emplaced by successive episodes of cauldron collapse (Figure 8-9). According to the interpretation shown in the figure, cauldron collapse may result in venting to the surface (as in the farthest example to the right) or may involve only subterranean caving without eruption.

Pluton is a useful term for any large intrusive body, regardless of shape. A *stock* is a pluton with a surface area of exposure less than 100 km², and a *batholith* is a pluton of larger outcrop area. The distinction between stock and batholith is not significant and purely arbitrary. Many stocks are undoubtedly bumps or cupolas on underlying batholiths, and a little more erosion might increase the outcrop area enough to promote a stock to a batholith.

Many batholiths are enormous, up to hundreds of kilometers wide and a few thousand kilometers long. Evidence from gravity and seismic investigations suggests that batholiths are floored at depths of 5 to 20 km. Such large masses were not emplaced by single magmatic pulses; all batholiths show abundant evidence of repeated intrusion and, like the one in Figure 8-9, are best viewed as a cluster of separate plutons, each with a volume of a few cubic kilometers, intruded over an interval of several millions of years (Pitcher, 1978; Bateman and others, 1963).

Intrusive bodies can be classified in two other ways besides their shapes and sizes. One is according to the apparent depth of emplacement (Buddington, 1959). Shallow (*epizonal*) intrusions, within a few kilometers of the earth's surface when they crystallized, commonly have smaller grain size, cavities marking former gas bubbles, and perhaps a smaller content of hydrous silicates. The term *hypabyssal* has also been applied to such shallow intrusions, without any sharply defined limit on depth of emplacement. Those that crystallized deeper are called *plutonic* (subdivided into *mesozonal* and *catazonal*). The distinction is a difficult one, because it depends on the water content of the magma (Chapter 9), the temperature of the wall rock, and probably on other factors, and may be misleading. Another classification, not sufficient by itself, considers time relations between magmatism and regional deformation.

Rocks emplaced before, during, and after folding and metamorphism can be called *prekinematic*, *synkinematic*, and *postkinematic*. Directional features in the wall rock may continue into prekinematic bodies, but end abruptly at contacts with postkinematic ones.

8.4 EXTRUSIVE FORMS

Sizes and shapes of lava flows are controlled by viscosity, yield stress, and topography. Although magmatic liquids are much more viscous than water, lava does flow downhill, concentrates in channels, and becomes ponded behind barriers to form an approximately horizontal top surface. That lava is not a perfectly viscous fluid, but has a definite yield stress, is shown by the observation (Hulme, 1974) that lava generally stops advancing when eruption ceases at the source vent. A purely viscous liquid would continue to run downhill, like water, even after the supply was interrupted.

Lava issues from central vents ("point sources") and fissures ("line sources"); Figure 8-10 shows examples of both, and illustrates topographic control by valleys cut in a gentle southeasterly slope on the flank of Kilauea. Although continuous sustained emission of lava has been observed, most flows are fed in discrete spurts separated by intervals varying from minutes to days.

The effects of varying viscosity are encompassed by two compositional extremes, basalt and rhyolite. Basalt lavas can have viscosities a thousand times less than those of rhyolite lavas, and therefore tend to form thin but extensive flows (Figure 8-11a). Basalt flows vary in thickness from less than 1 m to more than 400 m, but average a few meters. Lengths can exceed 100 km. The largest flow in recorded human history, in Iceland in 1783, reached a volume of 12.5 km^3 (Thorarinsson, 1970a). Velocities on steep slopes can exceed 45 km/hr (Macdonald, 1967).

In contrast, highly viscous lavas such as rhyolites form thick short flows (Figure 8-11b), or pile up at the vent as *domes* or stick out as *spines*.

Surface features of lava flows, especially basalts, show variation. Lava is quickly cooled by contact with air and by radiative heat loss, forming a strong but pliable skin a few millimeters thick, which acts as a fairly efficient thermal insulator. Inside this skin, the lava can remain liquid for considerable time. Because the viscosity of the lava rapidly decreases inward (Fink and Fletcher, 1978), the plastic skin wrinkles as the underlying liquid moves, producing *pahoehoe* (Figure 8-12) with a smooth rolling or corrugated top. At the advancing front of a pahoehoe flow, the skin is breached and liquid gushes forward, quickly chilling to become sheathed with its own skin. Such protrusions, aptly named *pahoehoe toes*, are usually only 1 m or so wide; they advance for a few more meters before their skin becomes rigid and breaks, allowing a new toe to form.

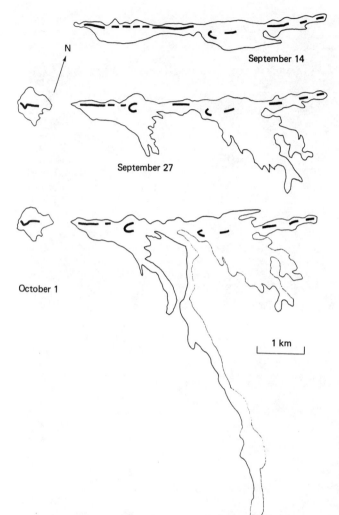

N

September 14

September 27

October 1

1 km

Figure 8-10. Distribution of pyroclastic vent deposits (solid black) and limits of lava flows at three stages during the 1977 eruption of Kilauea, Hawaii. (After Moore and others, 1980, Figs. 3, 6, and 9, with permission from Elsevier Scientific Publishing Company and R. B. Moore.)

Another common surface style on basalt, called *aa*, is rough, spiney, or fragmental (Figure 8-13). Lava has been seen to change from pahoehoe to aa downstream, particularly after cascading through a turbulent stretch where gas is most easily lost and cooling is most rapid. Peterson and Tilling (1980) conclude that the transition from pahoehoe to aa involves both viscosity and the rate of shear strain. If the viscosity is low, a high rate of shear strain is required to produce aa, but in lavas of high viscosity a low shear rate is sufficient. Pahoehoe forms and survives in lavas that stop moving before their viscosity exceeds a critical value. If, on the other hand, the flow keeps moving

(a)

(b)

Figure 8-11. Extremes in fluid behavior of lava. (a) Thin basalt flows, each becoming more vesicular toward the top, Portrillo volcanic field, New Mexico. Hammer indicates scale. (b) Rhyolite domes and thick flows, Mono Craters, California.

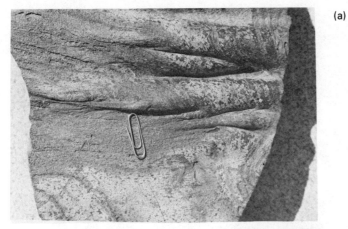

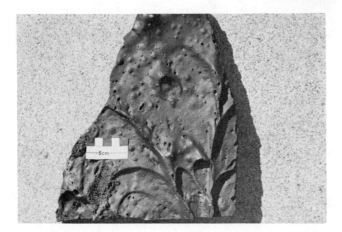

Figure 8-12. Three views of a pahoehoe slab from the Blue Dragon flow, Craters of the Moon National Monument, Idaho. (a) Top; (b) side; (c) bottom.

Figure 8-13. Aa lava, showing the rough, spiney, and contorted surface.

(perhaps over a locally steep gradient) at high viscosity, relatively rigid clots form where shear is most rapid, more fluid lava sticks to these, and the clots begin to rotate. In addition, pahoehoe crust is broken and stirred into the still-liquid interior of the flow. By both processes (growth and rotation of clots and disruption of pahoehoe crust), the aa flow acquires a surface that looks and feels like broken glass. It is.

Blocky lava (Figure 8-14) in turn is commonly less vesicular than aa. The surface, at least, is covered with smooth-faced but angular blocks, and the entire thickness of the flow may consist of loose fragments. Rapid movement breaks still-liquid but highly viscous lava into angular pieces; this brittle behavior at high rates of shear strain, combined with viscous behavior at lower rates, suggests a higher degree of polymerization in the liquid, which generally reflects higher silica content. Lavas more viscous than basalt rarely show pahoehoe surfaces, although some rhyolites are extruded as extremely viscous but smooth, toothpaste-like bodies rather than blocky flows.

More highly viscous lavas tend to move like the caterpillar treads on a tractor; the broken upper surface of the flow is carried forward upon the liquid interior, then rolls under (or slides off) the front of the flow to be deposited as a rubbly base over which the liquid continues to advance. The top and bottom of such a flow thus tend to be *brecciated*, while the interior cools to form a less fragmented rock. If the progress of a viscous lava flow is impeded, the more mobile interior may ride up and over a thick, solid basal portion, producing a *ramp structure* visible in cross section and a *pressure ridge*

160

Figure 8-14. Blocky lava; steep front of a rhyolite flow, Mono Craters, California.

on the flow surface (Figure 8-15). The brecciated and rigid portions on the sides of a flow may be pushed up into relatively high ridges or levees. The liquid interior may break through the toe of a flow and drain away, leaving *lava tubes* or tunnels that preserve the boundary between less viscous liquid and the rigid floor, walls, and roof.

Figure 8-15. Pressure ridges on a trachyte lava flow, Menengai Caldera, Kenya. View is from the caldera wall; the opposite wall is in the far right distance.

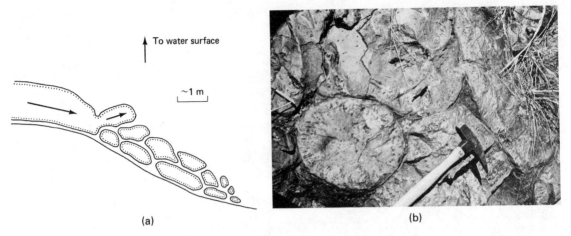

To water surface

~1 m

(a)   (b)

Figure 8-16. (a) Schematic cross section showing formation of lava pillows. Chilled skin of lava is stippled, with its thickness exaggerated for clarity. (b) Lava pillows, Kromberg Formation, Barberton Mountain Land, South Africa. Note the radial distribution of vesicles, well preserved after 3×10^9 years.

Lava of low viscosity may become ponded to form a lava lake; a solid crust forms rapidly at first, then continues to thicken more slowly, inhibiting heat loss from the liquid below. Much has been learned by observations on (and in) cooling lava lakes, which are the only large-scale yet accessible analogues to magma chambers (Wright and others, 1976).

When basaltic lava enters standing water, the interaction is less violent than one might expect. The hot liquid is so rapidly chilled by the water that it develops a skin similar to that on pahoehoe flows. An extremely thin envelope of steam separates the hot lava from the cool water, and rarely is there explosive expansion of steam bubbles. The lava continues to advance. As in the subaerial formation of pahoehoe toes, the chilled skin distends and breaks, allowing a new pod of lava to squeeze out and be chilled in turn. Unlike pahoehoe toes, these protrusions may break away from the main body of the flow, roll down slope, and accumulate while still hot and soft, forming *lava pillows*. Later-formed pillows cover them and tend to sag while still hot, adjusting their shapes to the substrate before solidifying completely (Figure 8-16). Underwater photography by geologists diving at the front of Hawaiian basalt flows that entered the sea has provided dramatic documentation of the way pillows form, oozing rapidly from cracks to achieve their full size in a matter of seconds.

Not all lavas, even basalts, flow this placidly into water. Some break into glassy granules upon contact with water. The greater surface-to-volume ratio of these hot particles, and the consequently greater opportunity for contact between lava and water, result in turbulent, even explosive, interaction.

8.5 PYROCLASTIC ERUPTIONS

Pyroclastic ("fire-broken") rocks form when igneous material (solid or liquid) is explosively disrupted into particles that then accumulate on the earth's surface under air or water. Such rocks therefore share igneous and sedimentary heritages, and it is not too cynical to remark that they may combine the most perplexing features of both. As magma approaches the earth's surface, the drop in confining pressure causes dissolved gas to separate as bubbles. The bubbles expand rapidly, and the magma may become a froth or be disrupted into particles when the gas reaches about 75 vol % of the mixture (Wilson and Head, 1981). Volcanic explosions usually result from liberation of dissolved gases from the magmatic liquid, but other causes are recognized. Interaction of viscous lava with standing water is one. More important is the violent expansion of groundwater into steam if magma is intruded into, or near, an aquifer. Explosions powered by release of dissolved gases are called *magmatic*; those powered by expansion of water external to the magma are called *phreatic*.

Sparks (1978) has summarized the processes of bubble nucleation and growth in magmas. After nucleation (which requires supersaturation of the liquid with respect to a dissolved gaseous component), a bubble grows by two processes; the first is the diffusion of the dissolved component through the liquid to reach the bubble, and the second is expansion caused by decreasing confining pressure. The pressure decrease can result from rise of the bubble through the liquid, rise of the liquid carrying the bubble with it, or removal of the top of the magma column by eruption. Bubbles stop growing for several interrelated reasons. Loss of dissolved gas to the bubbles increases the viscosity of the liquid (because water, a major gaseous component in many magmas, is a strong network modifier); thus more viscous liquid has to flow through more and more constricted spaces between closely packed bubbles, as the bubbles expand (Figure 8-17). Furthermore, adjacent bubbles are competing for gaseous components from the same thin wall of liquid between them, and tend to retard each other's growth. When the bubbles reach about 75 vol % in the magma, they are all in contact and stop growing. However, gas components can continue to diffuse into a bubble, even when the latter can no longer expand; the gas pressure within the bubble must increase. At the interface between the froth and overlying gas (Figure 8-17), the larger bubbles will burst, disrupting the magma and causing the interface to move downward through the froth. Bubbles in rhyolitic liquid are smaller (0.01 to 0.2 mm diameter, usually) than those in basaltic liquid (1 mm to 5 cm). Sparks (1978) attributes this difference not to viscosity or gas content, but to lower rates of diffusion of water through rhyolitic liquid and to higher rates of eruption of rhyolitic magma (the more rapid ascent leaves less time for bubbles to grow by diffusion, and growth by diffusion is more effective than growth by expansion, except at very low pressure).

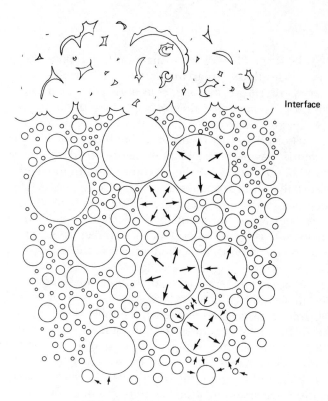

Interface

Figure 8-17. Growth and disruption of bubbles in magmatic liquid. Small dashed arrows show diffusion of dissolved gaseous components through liquid. Large arrows indicate directions of bubble movement. Except in liquids of very low viscosity, small bubbles do not coalesce to form large ones. The interface between froth and the overlying mixture of gas and broken bubble walls will move downward.

The formation of gas bubbles in magma has two effects: The volume of the liquid plus gas mixture is larger than the volume of the liquid alone before vesiculation, and the separation of a gas phase rapidly drains heat from the liquid. Consequently, magma ascending a narrow conduit must cool and accelerate as it approaches the surface if it contains an appreciable amount of gaseous components. Some magmas may even attain supersonic velocity within the last kilometer of their travel below the surface (McGetchin and Ullrich, 1973).

Two kinds of volcanic explosion can be envisioned at the interface shown in Figure 8-17 (McBirney, 1973). In the first, gas bubbles expand in low-viscosity liquid and rupture, forming a high-velocity but low-pressure stream carrying suspended particles. The velocity of a fragment will be the resultant of gas velocity (upward) and the terminal velocity (downward) that the fragment would have if it were dropping through stationary gas of the same density as the gas in the moving stream. In this type of explosive activity, which can be compared to the spray from an aerosol can, the size of fragments decreases with increasing distance from the vent, because big ones settle out first. Terminal velocities of pyroclastic particles (assumed to be smooth spheres with a density of 2.5 g/cm^3) are calculated to be 0.6 m/sec for a diameter of 0.01 cm, 25 m/sec for 1 cm, and 200 m/sec for 1 m (Walker and others, 1971).

In the second kind, gas expands in a high-viscosity liquid and pushes the fragments in a coherent mass in front of it. The fragments, acting like projectiles in a gun barrel, are accelerated by the high-pressure gas behind them and reach a velocity that is proportional to the square root of the gas pressure. In this style of explosion, large particles tend to be carried farther from the vent than small ones, because of their greater mass and consequently greater momentum. The horizontal distance traveled by an ejected fragment in one of these cannon-like explosions will depend on its initial velocity, ejection angle, mass, cross-sectional area, and atmospheric friction (Fudali and Melson, 1972). Initial velocities of these so-called *ballistic ejecta,* calculated from observed distance, mass, and shape of fragments, are up to 600 m/sec (Self and others, 1979).

Upon leaving the vent, however, the gas and suspended fragments rapidly decelerate, not only because of gravity and air friction but because of turbulent expansion and mixing with air. As the momentum of the lower portion of the eruptive cloud decreases, its density also decreases (partly from the fallout of suspended fragments, and partly from expansion of entrapped air being heated by the cooling fragments). As a result, the momentum-dominated or *gas-thrust* lower portion (generally forming only the bottom tenth of the total cloud height) merges with an upper, buoyancy-dominated or *convective-thrust* portion in which the bulk density of the gas plus entrapped air plus suspended particles is less than that of the adjacent atmosphere. The cloud height can theoretically reach 55 km on the earth (Wilson and others, 1978) and is controlled by the ratio of thermal to kinetic energy in the cloud and by the particle sizes (smaller particles are not only carried higher but give up more heat to their surrounding gas to drive convection, because they are more abundant, have higher surface area-to-mass ratios than larger particles, and stay up longer). Figure 8-18 summarizes the anatomy of an eruption cloud, and Figure 8-19 documents the growth of one convective-thrust cloud.

The cloud column may collapse at its base (Figure 8-18) if the density of the suspension exceeds that of the adjacent air when the upward gas-thrust motion ceases. Such high density is promoted by high eruption rates, discontinuous eruptions, and low water and high carbon dioxide contents in the magma (Sparks and Wilson, 1976). The consequences of column collapse are discussed later in this chapter.

Volcanic eruptions differ widely in style, and a complex terminology has evolved which reflects the relative violence of the eruption, the fluidity of the lava, the temperatures at which particles are ejected and at which they land, and the heights, shapes, and colors of the clouds. Bullard (1976), Macdonald (1972), and Williams and McBirney (1979) discuss this terminology. Walker (1973) has offered a more quantitative classification of explosive eruptions based on area of dispersal and degree of fragmentation of the pyroclastic material. A revised classification embracing all kinds of eruptions is needed, and should be based on properties of the preserved deposits, not on behavior that can only be recorded by eyewitnesses at the time of eruption.

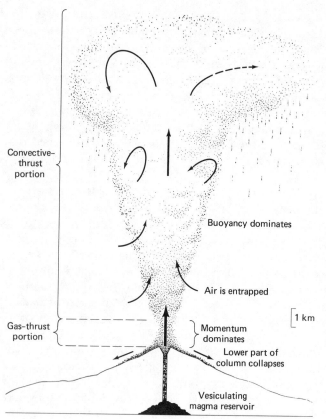

Figure 8-18. Cross section of an eruption column, showing regimes dominated by upward momentum (gas thrust) and by buoyancy (convective thrust). The column has started to collapse at the base, owing to the high density in the gas-thrust portion.

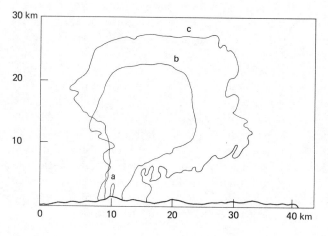

Figure 8-19. Growth of an eruption cloud from Hekla volcano, Iceland, March 29, 1947; a, cloud outline 6 ± 3 minutes after estimated instant of eruption; b, outline after 11 ± 3 minutes; c, outline after 18 ± 3 minutes. Note that the horizontal and vertical scales are equal. (After Thorarinsson, 1970, Fig. ix, with permission from Almenna Bokafelagid.)

8.6 FORMS OF VOLCANOES

Volcanic accumulations, involving varying proportions of lava flows and pyroclastic materials, build several kinds of edifices. *Shield volcanoes* have low flank slopes, gently convex-upward profiles, and commonly contain high ratios of lavas to pyroclastic debris. *Cones* have steeper slopes, are commonly concave upward in silhouette, and may be entirely lava or entirely pyroclastic. *Composite volcanoes* (also called *stratovolcanoes*) contain alternating sequences of lava and pyroclastic materials in subequal proportions.

One volcanic feature, the *caldera*, is as much a cause as an effect of explosive volcanism. Calderas are topographic and structural depressions, collapse features caused by foundering of roof into an underlying magma body (Figure 8-8). Calderas are therefore surface expressions of cauldron collapse and ring dikes, although some calderas are at least partly bounded by sagging, rather than fractures. Calderas commonly have roughly circular outlines, diameters of 2 to 30 km, and are partly filled with mixtures of lava flow rocks, pyroclastic deposits, and rubble that slumped from the caldera walls during and after collapse. They are also particularly susceptible to later intrusion, being hot and fractured areas overlying reservoirs that probably retain some magma for long times.

Caldera collapse results from withdrawal or lateral drainage of the magma column in basaltic volcanoes, or from sudden vesiculation and eruption of felsic magma. Although the first cause is not associated with pyroclastic rocks, this seems the best place to mention it. In 1968, the summit of the basaltic shield volcano comprising Isla Fernandina in the Galapagos Islands, previously marked by a caldera, underwent renewed collapse (Figure 8-20). The subsidence, approximating a volume of 1 or 2 km^3, dropped the floor of the older caldera by 300 m in "a series of short drops over a 12-day period" (Simkin and Howard, 1970). The collapse was probably initiated by withdrawal of the magma column or by near-surface lateral intrusion; associated flank eruptions were of insufficient volume to account for the subsidence.

Sigurdsson and Sparks (1978) have provided evidence that rapid and voluminous fissure eruptions in Iceland originated when faults intersected magma reservoirs (with volumes of 10 to 15 km^3) 3 to 7 km below the summits of basaltic volcanoes; magma drained off through the faults and migrated laterally for over 70 km within the crust, while the volcano summits collapsed into calderas. In 1977 the lava lake partly filling the caldera of Nyiragongo in central Africa reached its maximum level, then abruptly drained through flank fissures, releasing at least 20×10^6 m^3 of lava in less than 1 hour. The extremely low viscosity lava (high alkalis, and SiO_2 less than 44 wt %) reached velocities of 100 km/hr as it spread over an area of 20 km^2, leaving a coating only a few millimeters thick on the upper slopes (Tazieff, 1977). There were at least 60 fatalities.

Concerning felsic magmas, the cause−effect relationship between cal-

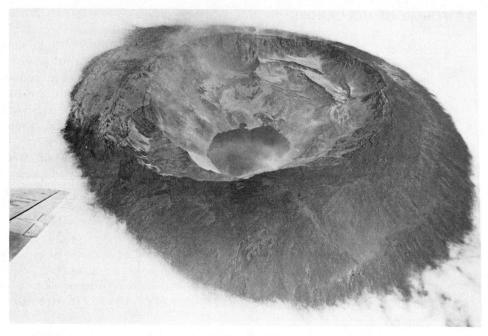

Figure 8-20. Summit caldera of Isla Fernandina, Galapagos Archipelago. The caldera is approximately 4 by 6.5 km. Clouds hide the lower flanks of the shield volcano. (From Simkin and Howard, 1970, *Science, 169,* 429–437. Copyright 1970 by the American Association for the Advancement of Science.)

dera formation and pyroclastic eruptions is clouded, because sudden removal of the underlying magma results in the collapse, but propagation of fractures to the surface causes the pressure release and vesiculation. After collapse and eruption, a central dome may grow in the caldera by intrusion of additional magma below the floor. The most likely explanation of these *resurgent domes* (Figure 8-21) appears to be that the magma column feeding the volcano rises in response to the removal of a large mass of magma from the top of the column during eruption (R. A. Bailey, 1976). Resurgent domes must not be confused with smaller lava domes in calderas, representing viscous, degassed magma squeezed up along the fractures after caldera collapse. Many calderas occur in clusters within volcanic fields (Figure 8-22), and may mark sites where plutons, representing pulses in the emplacement of a batholith, approached the earth's surface.

8.7 PYROCLASTIC DEPOSITS

Pyroclastic ejecta (also called *tephra*) are classified according to size and, among the larger particles, shape. Unconsolidated material with average grain diameter smaller than $1/16$ mm is called *dust*, that between $1/16$ and 2 mm is *ash*, that

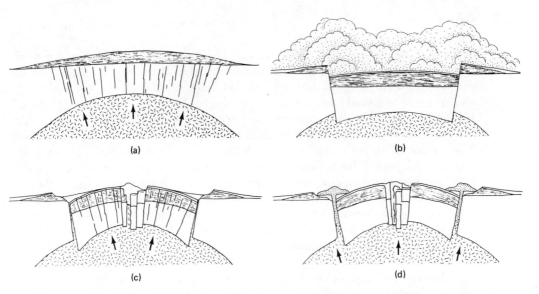

Figure 8-21. Development of a caldera: (a) approach of magma toward the surface, with preliminary volcanism; (b) caldera collapse with vesiculation and eruption of magma; (c) resurgent doming; (d) extrusion of late lava domes along the bounding ring fractures. The scale can vary; most calderas are a few kilometers in diameter. (After Smith and Bailey, 1968, Fig. 5, with permission of the Geological Society of America and R. L. Smith.)

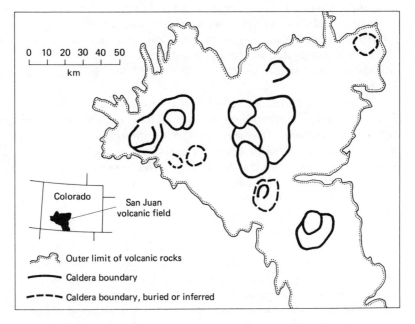

Figure 8-22. Generalized map of calderas in the San Juan volcanic field, Colorado. Note a tendency for younger calderas to be nested within older. (After Steven and Lipman, 1976, Fig. 1.)

between 2 and 64 mm *lapilli,* and those particles larger than 64 mm are *blocks* if angular, *bombs* if rounded. Upon consolidation (by compaction and/or cementation), dust becomes *dust tuff,* ash becomes *tuff,* lapilli become *lapilli tuff,* blocks become *pyroclastic breccia,* and bombs become *agglomerate* (Schmid, 1981). The constituents of pyroclastic particles are divided into three categories (glass, crystals, and rock fragments) for purposes of further classification. The breaking of vesiculated glass yields *shards,* segments of bubble walls. Shards and *pumice lumps* are the most abundant constituents of *vitric tuffs.* *Lithic* means that rock fragments dominate, and *crystal tuff* contains broken or whole crystals as the dominant particle type. These classificatory terms are combined, as in "crystal-lithic tuff" for one in which crystals are more abundant than rock fragments, or "lithic lapilli tuff," and using chemical and modal data as outlined in Chapter 5, these terms can be modified further by igneous root names, as in "crystal-vitric rhyolite tuff."

Heiken (1972) divided the glass particles of ash into two groups, depending on whether the disruption was caused by magmatic or phreatic explosion. Ash formed by magmatic gas expanding in liquid contains droplets or spheres of glass if the liquid was of low viscosity, and contains angular pumice (highly vesicular glass) fragments or shards if the liquid was highly viscous. In contrast, glass particles produced by phreatic interaction of magma with groundwater or surface water tend to be blocky or pyramidal, with few or no vesicles.

Cinder cones are local and rapid accumulations, around a vent, of particles that followed ballistic rather than airborne paths. They are the most common subaerial volcanic form on earth (Wood, 1980); each generally completes its growth in less than a year. Commonly, the volume of lava poured out as flows far exceeds that of the pyroclastic accumulation in the cinder cone from the same vent. Lava flows generally escape from the base of a cone, not from its apex.

Maars are broad and shallow explosion craters resulting from near-surface, commonly phreatic, explosions. They are usually surrounded by thin pyroclastic deposits. In maars and other manifestations of explosive volcanism, it is often useful to distinguish three kinds of fragments which may be mixed in any proportions. *Juvenile* particles represent "new" igneous rock solidified from magma during the eruption. *Accessory* particles represent earlier magmatism from the same volcano, but a previous eruption. They are, in effect, recycled. *Accidental* fragments are of sedimentary, metamorphic, or igneous rocks not genetically related to the volcano.

Wright and others (1980) formalize a widely used genetic classification of pyroclastic deposits into falls, flows, and surges. *Fall deposits* accumulate as material drops approximately vertically from an eruption cloud; Wright and others (1980, p. 317) characterize these as "maintaining a uniform thickness over restricted areas while draping all but the steepest topography.... They are generally well sorted and sometimes show internal stratification due to variations in eruptive column conditions." *Pyroclastic flow deposits* are emplaced by lateral movement of hot, high-density suspensions, and are "topographically controlled . . . poorly sorted . . . generally lack internal stratification" (p. 318). *Pyroclastic surge deposits* are often the distal equivalents of flow deposits, emplaced by turbulent lateral movement of low-density suspensions; they tend to drape over topography but are thickest in valleys, and thus behave in a way intermediate between fall and flow deposits. Surge deposits often show unidirectional features, including cross-stratification. Some stratification styles in pyroclastic deposits are illustrated in Figure 8-23.

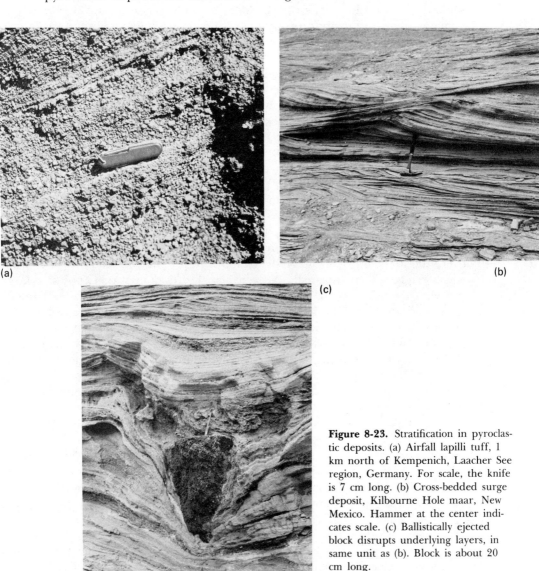

(a)

(b)

(c)

Figure 8-23. Stratification in pyroclastic deposits. (a) Airfall lapilli tuff, 1 km north of Kempenich, Laacher See region, Germany. For scale, the knife is 7 cm long. (b) Cross-bedded surge deposit, Kilbourne Hole maar, New Mexico. Hammer at the center indicates scale. (c) Ballistically ejected block disrupts underlying layers, in same unit as (b). Block is about 20 cm long.

Suspended particles, unless they travel so far before falling out that they are diluted with other detritus in sediments, form air-fall or water-laid ash. *Water-laid ash* accumulates when airborne particles fall into standing water. There may be a large or small admixture of detrital sediment, organic material, or cement. Vitric ash is particularly responsive to reaction with water, as unstable glass alters to clay or zeolite minerals. A common altered water-laid ash, called *bentonite*, contains the expanding clay, montmorillonite.

Air-fall ash can be distinguished from water-laid ash more easily by the sediments with which it is interbedded, and of course by any fossils that it contains, than by intrinsic properties. Air-fall ash is generally less well stratified, and tends to drape over the previously existing topography with higher initial dips. It is also more easily eroded and less likely to be preserved in the stratigraphic record unless buried soon after deposition. *Aerodynamic sorting* of suspended crystals in a volcanic cloud, according to size, shape, and density, can be efficient; olivine crystals may fall at one place in the same instant that biotite flakes are falling elsewhere from the same convective-thrust cloud.

Ash-flow tuffs, the most spectacular and voluminous of pyroclastic deposits, accumulate from high-density suspensions of hot glass shards (with varying amounts of pumice, rock fragments, and crystals), transported as *nuées ardentes* ("glowing clouds"). Most nuées ardentes probably originate by collapse of eruption columns, but some may represent frothing lava that spills over the rim of a crater or caldera. Ash-flow tuffs are the deposits most frequently associated with caldera collapse, although they have also issued from vents in lava domes and from linear fissures. One of the most fully documented, however, "occurred as a result of oversteepening of the [dacite lava] flow front which triggered an unusually large avalanche; that in turn exposed a large hot vesiculating lava surface, from which the nuée simultaneously erupted" (Rose and others, 1977). The destructive May 18, 1980, eruption of Mount St. Helens similarly began with a landslide from a flank oversteepened by shallow intrusion. The slide unloaded the confining pressure on a shallow reservoir of hydrous magma, causing immediate vesiculation and eruption as a laterally directed blast and pyroclastic flow. Subsequent ash flows at Mount St. Helens, however, formed by the collapse of dense eruption clouds, as abundantly documented on film and videotape.

A single ash-flow tuff unit may cover thousands of square kilometers, be hundreds of meters thick, and have a volume of thousands of cubic kilometers (Smith, 1960). In mass, such a unit may be equivalent to a large pluton. To cover a large area while still hot, the flow must be extraordinarily mobile. The devastating nuée ardente that came from a lava dome on Mont Pelée in 1902 traveled more than 100 km/hr, but stayed close to the ground and tended to follow the course of a valley rather than to climb over ridges. Other ash flows, however, have surmounted obstacles more than 200 m high that were in their paths some 20 to 40 km from the vents (Miller and Smith, 1977), and have been able to negotiate right-angle turns in glaciated valleys,

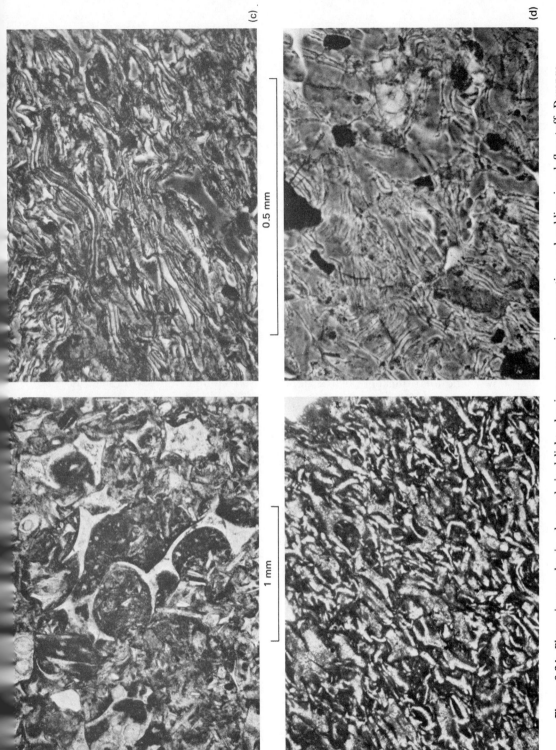

0.5 mm

1 mm

(c)

(d)

(a)

(b)

Figure 8-24. Photomicrographs, in plane-polarized light, showing progressive compaction and welding in ash-flow tuff, Durango, Mexico. (a) Shards are undeformed. (b) Shards are appressed and beginning to interlock. (c) Shards are strongly flattened and welded together at points. (d) Shards are merging into a continuous glassy mass.

demonstrating high density combined with high velocity. Nuées ardentes from Mayon Volcano in the Philippines were timed at 9 to 63 m/sec during a 1968 eruption (Moore and Melson, 1969).

Ash flows continually lose gas as they travel, gradually deflating and becoming more dense. Mobility is largely lost when the bulk density of the suspension increases to about 1 g/cm, because it is approximately at this density that individual particles come in contact and interfere with one another's motion. Many ash flows retain a high temperature until they stop moving. When the flow comes to rest, if the temperature is above approximately 550°C, glass shards will deform, interlock, and stick together as they are compacted, forming a *welded tuff* (Figure 8-24). Another common term for ash-flow tuffs, especially for welded ones, is *ignimbrite* (from the Latin, *ignis,* "fire," plus *imber,* "shower"). It is an unnecessary synonym.

Compaction is achieved not only by deformation and welding of shards but by a general decrease in porosity and by flattening of pumice lumps (Figure 8-25). The latter represent larger clots of vesiculated magma that were not as thoroughly disrupted as they left the vent. Glass may undergo further vesiculation after deposition in ash flows (Schmincke, 1974), and the entire mass may flow (but only for centimeters) during compaction, giving ash-flow tuffs some small-scale features that have been mistaken for those of lava flows. The final thickness of the deposit, after compaction is complete, is estimated to be 10% or less of the original thickness of the mobile suspension that passed that point. Because compaction will be greater in thicker sections, the top surface of an ash-flow deposit will tend to sag toward the center of a filled valley.

One ash flow may follow another within a few hours or days, so that the two may cool as one unit, or there may be several hundred thousand years between eruptions. There is a tendency for successive ash flows to be hotter, farther traveled, and more highly welded than their predecessors. Also, subtle chemical and mineralogical changes, expressed as vertical zoning, are observed in sequences of ash-flow tuffs or even within a single unit. The general rule in this compositional zoning is for more highly fractionated magma, richer in silica and alkalis, and with a higher ratio of liquid to crystals, to be erupted first. This tendency suggests that a vertically zoned magma chamber is tapped from the top down in successive eruptions (Hildreth, 1979, 1981; Smith, 1979).

Superimposed on this primary compositional zoning, and commonly obscuring it, is zoning formed after the flow has been deposited. The basal part of an ash flow may be less welded than that above if it lost heat rapidly to a cool or wet substrate. On the other hand, if the substrate were hot, or if the flow were unusually thick or hot, the shards might be totally welded into a *basal vitrophyre,* so that no shard outlines would be visible. Toward the tops of most ash flows, the welding (if present) diminishes, and the very tops are commonly not at all welded, reflecting rapid heat loss. Still other kinds of

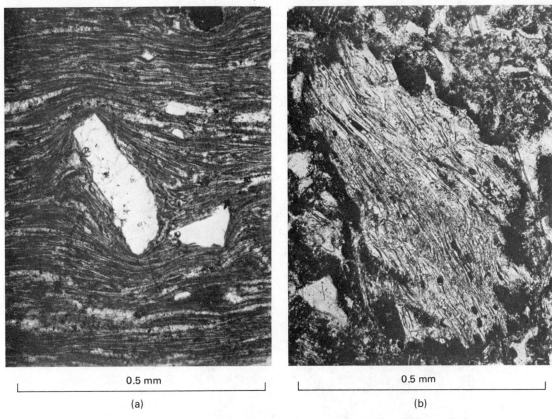

0.5 mm

(a)

0.5 mm

(b)

Figure 8-25. Compaction features in ash-flow tuffs. (a) Photomicrograph, in plane-polarized light, showing flattened shards wrapping around crystal fragments. (b) Photomicrograph, in plane-polarized light, of small collapsed pumice lump. Note the cuspate edges of the lump, and the elongation of the vesicles. (c) Flattened pumice lumps (dark) oriented with the shortest dimension perpendicular to the top and bottom surfaces of the deposit.

(c)

5cm

1 mm

Figure 8-26. Photomicrograph, in cross-polarized light, showing vapor-phase deposit of alkali feldspar (white) and tridymite (darker gray) as euhedral crystals partly filling a deformed vesicle in a pumice lump in an ash-flow tuff, Durango, Mexico.

zoning may obscure the zones formed by variations in magma composition and in degree of welding. The combination of trapped gases and slow cooling may encourage devitrification at some levels within the flow, and gas, migrating upward through the compacting mass, may precipitate minerals in voids near the top (Figure 8-26), or alter plagioclase and mafic crystals and rock fragments.

As in other types of pyroclastic deposits, the proportions of glass, crystals, and rock fragments vary widely. Individual glass shards and broken pumice fragments strongly tend to escape from the dense ash flow, into the less dense convective-thrust portion of the eruption cloud. Ash-flow tuffs are consequently enriched in crystals and rock fragments, relative to the original magma composition, and as much as half of the finer glass particles may be lost before deposition of the ash flow (Sparks and others, 1973). Rock fragments tend to increase in size downward in an ash flow, but pumice lumps tend to become larger toward the top of the flow, indicating that the bulk density of the transporting suspension was intermediate between the densities of the rock fragments and the pumice. Because of the winnowing out of glass relative to crystals and rock fragments, chemical analysis of a bulk sample of ash-flow tuff is not likely to correspond to the composition of the initial magma (Walker, 1972); undevitrified glass must be analyzed to reconstruct the magmatic liquid composition. Nearly all ash flows are felsic, but rare mafic examples are known.

Because ash flows are extensive and, on the scale of geologic time, are emplaced instantaneously, their bases are excellent stratigraphic markers, and the phenocrysts in ash flows are usually high-priority targets for isotopic age measurements.

Although not strictly pyroclastic, *lahars* (volcanic mudflows) must be mentioned. Volcanic slopes are rapidly built, chemically and mechanically unstable accumulations of lava and pyroclastic particles, susceptible to mass movement down the slopes. Heavy rains accompany many eruptions as clouds condense from the water vapor released from magma, or nucleate at higher atmospheric levels on dust. Water, falling on the unstable slopes, commonly mobilizes the loose debris and transports it as an unsorted dense slurry. Lahars (the term comes from Indonesia, where such mudflows have caused many fatalities) may be cold or hot. Displacement of water in a crater lake by an eruption or by growth of a lava dome is often cited as a cause of hot lahars. Volcanic mudflows can be distinguished from pyroclastic deposits by their lack of sorting, their sharp terminations, limited extent, and the common occurrence of mudcracks on their tops.

8.8 INTERNAL FEATURES OF IGNEOUS AND PYROCLASTIC ROCKS

Some smaller-scale features found inside the rock bodies described in this chapter give clues to emplacement mechanisms and postemplacement history.

Vesicles (cavities formerly filled by gas bubbles) occur in shallow intrusive bodies, lavas, and ash flows (and in volcanic rock fragments in other pyroclastic deposits). A bubble can form only when the gas pressure exceeds the confining pressure. We therefore expect vesicles to be smaller and less numerous at greater depths in intrusive and subaqueous extrusive rocks. This expectation is borne out by study of submarine lava pillows (Moore and Schilling, 1973); vesicles have not been found in lavas dredged from water depths exceeding 4 or 5 km. Vesicles can become elongated by upward migration of gas bubbles, or can be stretched and sheared by flow of the liquid, and so record the direction and magnitude of late movement.

Water-rich magmas, even those crystallizing at considerable depth, can liberate a vapor phase in the last stages of crystallization. The vapor-filled space may be preserved as a *miarolitic cavity*, distinguished from a vesicle by its more irregular shape (commonly lined with euhedral crystals jutting into the space) and by the smaller ratio of cavity diameter to average diameter of crystals in the surrounding rock.

If elongated or flat constituents (crystals, rock fragments, glass shards, pumice fragments, vesicles, miarolitic cavities) are present, their longest dimensions may be parallel in one direction (*lineation*) or in a plane (*foliation*). These two types of preferred orientation can be produced by flow or compaction. Careful mapping of lineation and foliation can help in deciphering the mechanism of emplacement of an intrusive body (Balk, 1937) and of lava, but are less useful in pyroclastic rocks because local perturbations are produced by compaction, late flow during adjustment to underlying topography, and edge effects. Figure 8-27 shows an example of flow foliation in lava.

Layers are produced in igneous rocks by varying rates of nucleation and crystal growth, causing phases to precipitate in different abundances and sizes; crystal settling and flotation; deposition and erosion by currents in magmatic liquid; and, at least in theory, by flow differentiation.

Jackson (1967) provided a vocabulary for *cumulates* (layered igneous rocks) that is condensed here with some modification. *Cumulus crystals* were originally defined as those that had grown elsewhere, before being deposited where we find them. They were thus thought to be analogous to detrital grains in sandstone. In the light of subsequent interpretations to be discussed later, this definition must be broadened to include all crystals that grew from a larger volume of magma than did the other two kinds of crystals in cumul-

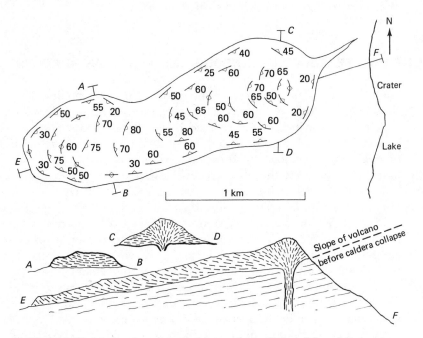

Figure 8-27. Map and sections showing strike and dip of flow foliation in a lava dome and flow, The Watchman, west shore of Crater Lake, Oregon. (After Williams, 1942, Fig. 9, with permission of the Carnegie Institution of Washington.)

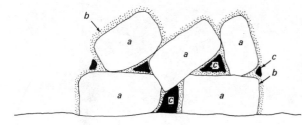

Figure 8-28. Idealized cumulate showing *a*, cumulus crystals; *b*, overgrowths; and *c*, intercumulus precipitate.

ates; many cumulus crystals are now known to have grown in place. The *postcumulus material* is of two kinds (Figure 8-28); *overgrowths* form on cumulus crystals, and *intercumulus precipitate* is analogous to pore-filling cement in sandstones. Both overgrowths and intercumulus precipitates crystallized from liquid trapped between the cumulus crystals, and this liquid generally had a very different composition from that of the larger volume of liquid that fed the cumulus phases.

Stratification in cumulates can be classified as horizons and layers. *Horizons* mark former surfaces of deposition or erosion, in other words, the interfaces, at specific times, between cumulus mush and liquid. They need be neither planar nor horizontal; many cumulate horizons show high initial dips, even vertical, and some cumulus sequences grew downward into liquid. Horizons can be marked by the appearance or disappearance of a cumulus phase, an abrupt change in relative abundances of cumulus phases, or an abrupt change in size or shape of a cumulus phase.

Layers are bounded above and below by horizons and are characterized by uniform or uniformly gradational properties, which include modal proportions, sizes, and compositions of the cumulus phases. Layers in cumulates imitate many features of subaqueous and subaerial stratification, including cross bedding, graded bedding, current lineation, channels, and slump structures. Examples are illustrated by Wager and Brown (1967) and Irvine (1974), among others. There can be little doubt that most of these features represent differential movement between a liquid and a cumulus mush. Many questions remain, such as how cyclic sedimentation can produce such rhythmic stratification as that in Figure 8-29.

Until recently, the favored explanation for cumulus layers in general was sinking or floating of crystals in liquid. The velocity of a single spherical particle in a stationary Newtonian liquid is given by Stokes's equation,

$$V_s = \frac{2gr^2D}{9n}$$

where V_s is the velocity (cm/sec), g the gravitational acceleration (cm/sec²), r the particle radius (cm), D the difference in densities of the particle and liquid (g/cm³), and n the viscosity of the liquid (poises). Departure from spherical shape will decrease the velocity, but this effect can be calculated (Komar and Reimers, 1978). If the crystal continues to grow, its velocity will increase.

179

Figure 8-29. Rhythmic stratification in cumulates, Ilimaussaq intrusion, Kangerdluarssuq, southwest Greenland. Darkest layers are approximately 2 to 10 m apart. Layers drape gently over a large stoped block in the center of the photograph.

There are many other complexities with which the simple model must be embellished. The velocity of each particle will diminish as the volume fraction of particles increases. An empirical equation for this effect is

$$V = V_s(1 - F)^{4.65}$$

(Walker and others, 1976), where V is the velocity, V_s the velocity given by Stokes's equation for a single particle, and F the volume fraction of particles. If $F = 0.01$ (that is, the magma contains 1 vol % crystals), $V = 0.95V_s$; if $F = 0.10$, $V = 0.61V_s$, and if $F = 0.50$, $V = 0.04V_s$. For silicate liquid below its liquidus, pseudoviscous behavior (with a yield stress to be overcome before a particle can move) will further decrease the velocity. In addition, convection is very likely whenever a magma body is more than about 15 m thick (Bartlett, 1969), and the upward component of velocity of a convecting liquid is quite likely to exceed the settling velocities of crystals.

In addition to theoretical and experimental problems with crystal sinking or flotation, there are observational problems. Plagioclase is a common cumulus phase, accompanying olivine or pyroxenes, yet at least at low pressures, plagioclase is less dense than the liquid and the mafic silicates are more dense (Irvine, 1980). It is possible that cumulus phases settle as clumps or chains, or that they are carried to the floor of a magma chamber by density currents (where admixed mafic phases impart a bulk density to the current that is higher than that of the liquid), and the plagioclase may remain on the floor, rather than buoyantly rising, if the magma is pseudoviscous.

Maaløe (1978) attributes rhythmic layering to variation in nucleation, and Campbell (1978) concludes that nucleation near the floor of a magma chamber is favored for two reasons: First, the increase in liquidus temperature with pressure is more rapid than the downward increase in temperature in a convecting magma, thus causing greater undercooling near the floor; second, crystals at the floor provide nucleation sites for more crystals. Crystals may therefore nucleate and grow most rapidly at a mush–liquid interface, without moving through a column of magma.

In an unsettling paper, McBirney and Noyes (1979) built a strong case against gravitational accumulation as the origin of cumulates, drawing much of their evidence from the Skaergaard intrusion of East Greenland (Chapter 11), which had previously served as a prime example of crystal settling (Wager and Brown, 1967). Among the evidence against gravitational separation in the Skaergaard intrusion are the following.

1. Plagioclase crystals, less dense than the liquid from which they are inferred to have crystallized, are concentrated in layers near the bottom of the magma body, whereas the much more dense olivine and clinopyroxene are concentrated near the roof. A solidification front must have advanced into the magma body from walls, roof, and floor more rapidly than the crystals could sink or float, trapping them in place.

2. There is textural evidence (Figure 8-30a) that crystals grew in the place we find them.

3. Layers are uniform in thickness and composition over horizontal distances of kilometers, right up to the walls of the magma body. Greater cooling along the walls, combined with descent of convection currents along the walls, should cause more rapid accumulation and thickening of layers near the walls, if the crystals there were gathered from a larger vertical thickness of magma. Furthermore, the thickness, and grain sizes and orientations, within layers are unaffected by such obstructions as blocks stoped from the roof (Figure 8-30b).

McBirney and Noyes therefore propose that crystallization takes place in a more or less static boundary layer between solid rock and convecting liquid. The boundary layer has a steep temperature gradient, and is therefore a likely site for crystallization. At times, the convecting liquid is perturbed, perhaps by roof stoping, and scours channels in the stagnant, crystal-rich boundary layer and forms cross bedding and graded bedding. Layering is attributed to rhythmic nucleation in the boundary layer, with the liquid being depleted in the components of one phase so that it stops precipitating while another phase, in which the liquid has been enriched in essential components, begins

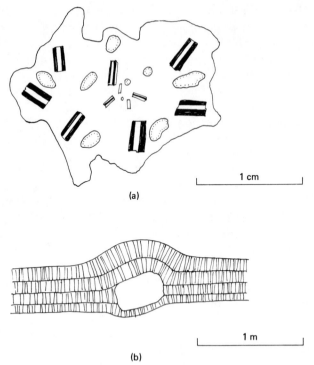

(a)

1 cm

(b)

1 m

Figure 8-30. Evidence for crystallization of cumulus phases in place. (a) Clinopyroxene poikilitically encloses grains of olivine and plagioclase that are progressively larger toward the margin of the pyroxene. Olivine and plagioclase were growing in place until their communication with liquid was cut off when they were engulfed by pyroxene; grains near the margin of the pyroxene had more time to grow larger. (b) Layers, with branching elongated crystals perpendicular to the layers, draping without disruption over a stoped block (compare with Figure 8-29).

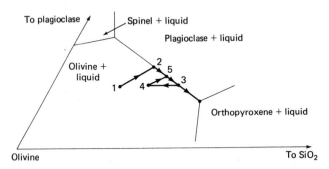

Figure 8-31. Schematic liquidus phase diagram showing combined effects of fractionation and magma mixing. Magma with composition 1 is emplaced and precipitates olivine; the remaining liquid reaches point 2 on the olivine + plagioclase + liquid cotectic, then migrates to point 3 as both olivine and plagioclase crystallize. Then liquid 3 mixes with an equal amount of unfractionated liquid (of composition 1) to form liquid 4, which precipitates olivine alone until it reaches point 5. Rhythmic layers form, of olivine alternating with olivine plus plagioclase. (After Usselman and Hodge, 1978, Fig. 8, with permission of Elsevier Scientific Publishing Company and T. M. Usselman.)

to crystallize in its stead. This heretical view, backed by many data, deserves critical testing in other cumulates. Crystal settling may be effective in mafic sills, but appears to be overrated as a means of producing layers in larger and more felsic intrusive bodies. Magma mixing (Figure 8-31) also can account for rhythmic layering. There is no shortage of possible explanations for the sequence observed in cumulates; the problems arise in selecting among them.

Repeated intrusion produces *internal contacts* separating the products of different magmatic pulses (Figure 8-32). Such multiple injection of magma into already partly solidified mush, on a scale varying from meters to kilometers, is particularly common in batholiths.

Inclusions of other rocks enclosed by an igneous rock are of two kinds. *Autoliths* are fragments of rock derived from the same magma as the enclosing rock, such as stoped blocks of a chilled margin, or clots of cumulates stirred up by a fresh surge of magma into the chamber. *Xenoliths* are fragments of wall rock, basement, or upper mantle, through which the magma has passed. Inclusions can also represent fragments of parent rock or unmelted residue from partial fusion, or incomplete mixing of magmas (Figure 8-33).

Joints are fractures along which no visible displacement has occurred. Brittle fractures can form at any time after a magma has become about 20% crystallized (H. R. Shaw, 1980, p. 258). Joints can therefore form during emplacement and cooling of an igneous rock or long afterward. A genetic classi-

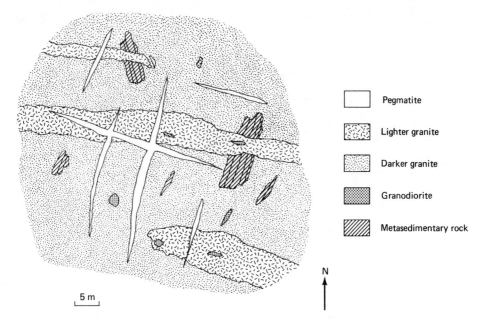

Pegmatite

Lighter granite

Darker granite

Granodiorite

Metasedimentary rock

N

5 m

Figure 8-32. Sketch map showing generalized relations of repeated intrusive pulses, Hallowell granite, Maine. Note the varying orientations of internal contacts and inclusions peculiar to each pulse.

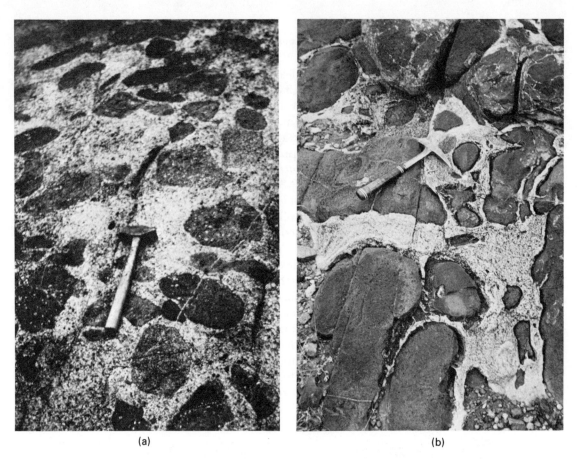

(a)   (b)

Figure 8-33. Inclusions in igneous rocks. (a) Abundant rounded xenoliths of
wall rock near contact, Saganaga pluton, Minnesota. (b) Intrusive "pillows" of
mafic rock in granite, Wichita Mountains, Oklahoma, interpreted as evidence
of incomplete magma mixing. The mafic rock has chilled margins against the
granite.

fication seems valid because the orientation of joints reflects quite accurately
the stress conditions under which they form. Four categories are recognized,
based on the origin of the stresses; emplacement, thermal contraction, regional deformation, and expansion due to unloading. Note that these are
listed in a time sequence.

During intrusion or extrusion of partly crystalline mush, tensional and
shear stresses can produce brittle fractures. These early formed joints may
merge outward into marginal thrust faults, as still-liquid magma pushes already solid rock out of its way. Joints produced during emplacement should
show regular orientation patterns relative to contacts and to foliation and lineation.

As heat flows outward from an intrusive body, the temperature of the
wall rock is raised and fluid, trapped in unconnected pores, expands much
more than the surrounding minerals, fracturing the rock (Knapp and Knight,
1977). Joints therefore propagate away from the pluton, continuing to

lengthen long after the magma has crystallized. These joints can promote stoping and otherwise aid intrusion by weakening the wall rock, and they also enhance the cooling of the pluton by permitting more ready escape of volatile components, which may circulate through the wall rock as hot solutions, precipitating mineral deposits (Chapter 16).

Volume decrease in igneous rocks during cooling can produce a distinctive stress pattern that is reflected in columnar jointing (Figure 8-34). The rock breaks into long polygonal prisms, usually oriented with their long axes parallel to the direction of greatest heat flow (Spry, 1962). The prisms can have three to eight sides, but five or six are most common; diameters in cross section range from a few centimeters to 5 m. Columnar jointing occurs in ash-flow tuffs and static lavas as well as in shallow intrusive bodies and may, in the latter, continue for short distances into the wall rock.

Regional folding or faulting, perhaps unrelated to magmatism and occurring long after it, tends to produce sets of parallel joints, commonly in three sets perpendicular to each other. The joints continue into the wall rock, unless its mechanical properties differ considerably from those of the igneous rock.

Figure 8-34. Columnar jointing in a basaltic lava flow, Mont Rodeix, Auvergne, France. The cliff is approximately 20 m high.

Deep-seated (plutonic) magma crystallizes under high confining pressure. When uplift and erosion eventually combine to bring the intrusive rock near the earth's surface, the pressure is drastically lowered. The rock responds by expanding, if and when it can. Blocks of granite cut from a quarry wall expand so that they cannot fit back into the hole they left. Commonly, the stress is released by detachment of slabs roughly parallel to the topographic surface and progressively thicker farther from the surface.

The patterns of brittle failure in igneous rocks are important not only to a geologist deciphering a portion of the earth's history but to anyone needing to understand a rock mass's excavation behavior and hydrologic properties. A contractor must know whether a rock will break easily into small blocks without blasting, and a groundwater hydrologist may find that joints will increase the permeability of a large granite mass many orders of magnitude above the permeability measured in the laboratory on a small sample.

8.9 SUMMARY

There is much more igneous rock outside the classroom and teaching laboratory than there is inside. This chapter combines the larger-scale field aspects with phase relations and with textural and compositional features visible in detached samples and thin sections.

Viscosity and yield stress are two fundamental properties governing intrusive and extrusive forms, and are controlled by degree of crystallinity and by polymerization. Other factors that determine intrusive forms are the local stress field, the density contrast between magma and wall rock, rates of intrusion and heat loss, and mechanical response of the wall rock. Forms of lava flows are controlled by topography, viscosity, yield stress, and rate of heat loss. Pyroclastic deposits take a variety of forms, depending on explosive style, liquid viscosity, particle size, horizontal component of velocity, and density of the particle–gas suspension.

Magma can move upward as roughly spherical blobs or diapirs, through cracks or pipes, or by zone melting, and can be intruded into its crystallization site by forcible intrusion, stoping, or passive emplacement.

Internal features (gas bubbles, inclusions, preferred orientation, internal contacts, and joints) testify concerning conditions during the last stages of consolidation of igneous and pyroclastic rocks. Layers in igneous rocks are subject to several interpretations; they record much of the crystallization histories of some magma bodies, and we are beginning to decipher their testimony.

9

The Effects
of Volatile Components

Throughout the preceding chapters, magma has been referred to as liquid with dissolved gases and suspended solids. *Volatile components* are those that tend to escape from liquid to concentrate in a gaseous phase; confining pressure inhibits, and may prevent, the liberation of a gaseous phase from a magmatic liquid. The terms "gas" and "vapor" will be used synonymously. Except very close to the earth's surface, pressure is high enough to exceed the critical pressure of most gas phases so that a gas can, through change in either temperature or pressure, change its density continuously to that of a liquid without passing through an abrupt transition. Such a gas is, strictly speaking, a supercritical fluid. However, we will consider only fluid phases with chemical compositions that are distinctly different from those of coexisting silicate liquids, where there is much less danger of confusion than there is in one-component systems.

9.1 PARTIAL PRESSURE, MOLE FRACTION, AND FUGACITY

A familiar analogy to the liquid and gas phases in a magmatic system is a sealed container of carbonated beverage. Of the two abundant volatile components in this system, CO_2 shows a much greater tendency to separate into the gas phase than does H_2O. Above the top surface of the liquid all the space is occupied by a gas phase, largely CO_2, at a pressure exceeding 1 atm. If the seal on the system is broken, some of the gas escapes, and the pressure on the gas phase suddenly drops to 1 atm. Bubbles of carbon dioxide rapidly nucleate in the liquid, as the distribution of this volatile component is no longer at equilibrium between gas and liquid phases. If we replace the seal on the system with a membrane that is permeable only to CO_2, not allowing H_2O and other components of the gas mixture to pass through, we can measure the pressure exerted solely by CO_2 in the gas mixture. The contribution to total pressure that is added by the CO_2 alone is called the *partial pressure* of CO_2, abbreviated as P_{CO_2}. If we replace the CO_2-permeable membrane with one that allows only water vapor to pass through, we can measure the partial pressure of H_2O, P_{H_2O}, and so on, for each component in the gas phase.

The total pressure exerted by the gas phase is the sum of the partial pressures of all components. For a gas phase containing only the two components CO_2 and H_2O, the total pressure is equal to $P_{CO_2} + P_{H_2O}$. It seems only reasonable that, if the proportion of CO_2 in the gas increases, P_{CO_2} approaches the value of P_{total} until, in a gas containing only CO_2, P_{CO_2} equals P_{total}. To quantify this relationship between partial pressure and composition, we must express the composition of the gas phase in terms of *mole fraction*. The mole fraction of a component in a phase is equal to the number of moles of that component divided by the sum of the number of moles of all components in the phase.

If the gas mixture consists of gaseous components that obey the ideal gas law ($PV = nRT$, where P is the pressure, V the volume, n the number of moles, R the gas constant, and T the temperature in degrees Kelvin), and if the gaseous components do not react with each other, the partial pressure of each component is equal to the mole fraction of that component times the total pressure: $P_{CO_2} = X_{CO_2} P_{total}$ and $P_{H_2O} = X_{H_2O} P_{total}$, where X stands for mole fraction.

Gases obey the ideal gas law at low pressures and high temperatures, but in many geologic settings pressure is high enough for gases to deviate significantly from ideal behavior. We could, by means beyond the scope of this book, take nonideal behavior into account, but we will merely define another function called *fugacity* (from Latin "fugare," to escape), which can be thought of as the partial pressure that a gaseous component would have if it obeyed the ideal gas law. Fugacity, symbolized by f, is expressed in the same units as pressure.

9.2 COMPOSITIONS OF MAGMATIC GASES

Observations of volcanic eruptions, and the presence of vesicles and miarolitic cavities in rocks, convince us that a gas phase does separate from magmatic liquid under some conditions. Estimating the composition of such a gas phase is extremely difficult. Samples of hot gas can be collected from volcanic vents and lava lakes, but these are already contaminated with groundwater and atmospheric components before collection. Estimates of the compositions of gas issuing from Kilauea lava lakes, after the contaminants have been estimated and subtracted from the analyses, range in mole fractions from 0.50 CO_2, 0.35 H_2O, 0.15 SO_2 to 0.10 CO_2, 0.70 H_2O, 0.20 SO_2 (Gerlach, 1980a), with much smaller proportions of CO, H_2, H_2S, S_2, COS, and HCl.

Such a gas mixture will react readily with rocks and with the atmosphere. Furthermore, the components in the mixture will react with one another, particularly those that contain carbon or sulfur in different valence states, under the rapidly changing conditions of temperature and total pressure as the gas mixture approaches the surface (Heald and others, 1963). Finally, Anderson (1975, p. 38) points out that "most data on volcanic gases pertain to the late quiet eruptive stage."

Other, less direct, methods of identifying gases in magmas are those involving liberation and analysis of volatiles trapped in fluid inclusions within minerals, and microscopic observation of the behavior of included fluids during heating and cooling. Murck and others (1978) found that fluid inclusions in ultramafic rocks from the mantle are largely CO_2, with small amounts of SO_2, H_2S, and COS. Felsic magmas, in contrast, contain phenocrysts with fluid inclusions that are rich in H_2O with subordinate CO_2 and CO (for example, Sommer, 1977).

9.3 SOLUBILITY OF WATER IN SILICATE LIQUIDS

The solubilities of a few volatile components in silicate liquids have been experimentally measured. At low pressures, water is the most soluble of the abundant volatile components. A liquid is said to be *water-saturated* if it has dissolved all the water that it can hold under the given temperature and pressure conditions, so that any excess water must form a vapor phase that coexists with the liquid. In the system $NaAlSi_3O_8$–H_2O (Burnham and Davis, 1974), albite liquid becomes water saturated at 1×10^8 Pa and 900°C with 4.4 wt % H_2O, at 3×10^8 Pa and 800°C with 7.8 wt %, and at 1×10^9 Pa and 700°C with 17.1 wt %. Solution of water in silicate liquids causes a drastic lowering of the solidus and liquidus. The low abundance of water in the deeper crust and upper mantle of the earth, however, makes it unlikely that much magma is even close to water saturation, except in the earliest stages of partial fusion or the latest stages of crystallization.

The mechanism by which water dissolves in silicate liquids can be modeled by the reaction (Burnham, 1975)

$$(Si_2O_7)^{6-} + H_2O = 2[SiO_3(OH)]^{3-}$$

In this equation, water reacts with an oxygen anion that was shared by two adjacent $(SiO_4)^{4-}$ tetrahedra, represented by the formula $(Si_2O_7)^{6-}$, to produce two hydroxyl, $(OH)^-$, anions, each of which occupies one apex of each of the two tetrahedra, which are no longer linked (Figure 9-1). Interaction of water

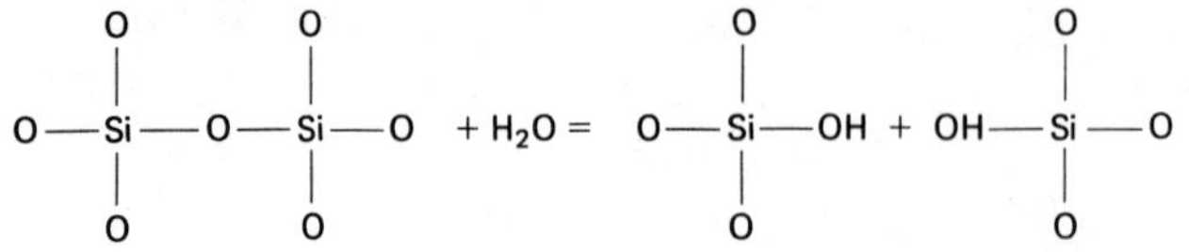

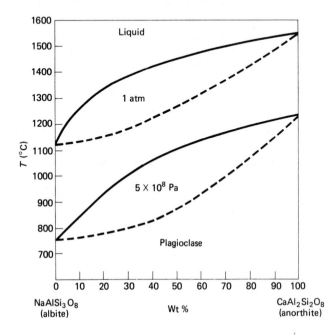

Figure 9-1. Model showing the depolymerization of silicon–oxygen tetrahedra by H_2O.

with $(AlO_4)^{5-}$ tetrahedra is more complicated, but also results in breaking bonds between tetrahedra. The resulting *depolymerization* of the liquid, breaking frameworks into sheets, sheets into chains, and chains ultimately into single tetrahedra, drastically reduces the viscosity. For a granitic liquid at 1000°C, changing the mole fraction of H_2O from 0.00 to 0.50 drops the viscosity from 10^{10} to $10^{4.2}$ P; under the same conditions, the electrical conductivity increases from $10^{-4.8}$ to 10^{-1} per ohm-cm, indicating a great increase in the "free" ions in the liquid (Burnham, 1975). Depolymerization pushes the liquidus and solidus to lower temperatures as more H_2O enters the liquid (Figure 9-2). The solidus temperature decreases most rapidly as P_{H_2O} (and X_{H_2O} in the liquid) initially depart from zero (Figure 9-3). Solidus curves, projected onto a P_{H_2O}–temperature plane, are therefore strongly concave toward higher temperature and pressure, but tend to straighten at very high pressure unless the solid phases invert to higher-density polymorphs.

The effect of pressure on a dry system is in distinct contrast to the effect on a water-bearing system. Without H_2O (or other strong network modifiers, which are far less abundant than water), silicate systems have solidus curves with positive slopes when projected onto pressure–temperature planes; in other words, solidus temperature increases as pressure increases (Figure 9-4).

Figure 9-2. Solidus and liquidus curves for the system $NaAlSi_3O_8$–$CaAl_2Si_2O_8$; the upper set of curves refers to the anhydrous system at 1 atm pressure, and the lower set shows the solidus and projected liquidus with enough water to saturate the liquid at 5×10^8 Pa. The liquid contains water, but only the albite and anorthite proportions can be shown on this diagram. (After Fig. 174 in *The Interpretation of Geological Phase Diagrams* by E.G. Ehlers. W. H. Freeman and Company. Copyright © 1972.)

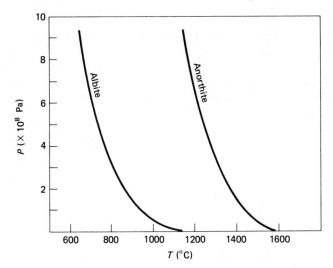

Figure 9-3. Melting curves for albite and anorthite, projected onto the temperature–P_{H_2O} plane. (After Yoder, 1958, Fig. 11, with permission of the Carnegie Institution of Washington.)

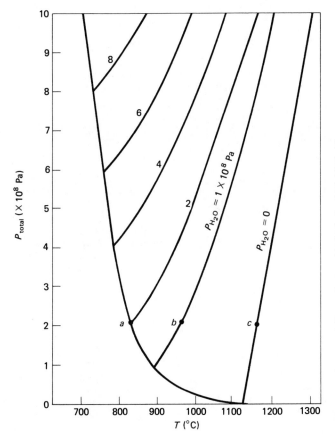

Figure 9-4. Pressure–temperature diagram for the system $NaAlSi_3O_8$–H_2O. The solidus curves upward to the left when the liquid remains water-saturated ($P_{H_2O} = P_{total}$), but increases at a constant rate to the right when no water is present ($P_{H_2O} = 0$). Other curves represent the solidus for values of P_{H_2O} (1, 2, 4, 6, and 8 $\times$ 10^8 Pa) less than that required for water saturation. For example, at a total pressure of 2 $\times$ 10^8 Pa, a water-saturated liquid would appear at approximately 820°C if the water pressure were 2 $\times$ 10^8 Pa (point a), at 960°C at 1 $\times$ 10^8 Pa (point b), and anhydrous liquid would not appear until approximately 1170°C at P_{total} = 2 $\times$ 10^8 Pa (point c). (After Burnham and Davis, 1974, Fig. 19, with permission from the American Journal of Science.)

If a system contains enough water to saturate the liquid at low pressure, but not enough to saturate it at higher pressure (where the liquid would dissolve more water if more were available), the solidus will follow a pressure–temperature path such as that outlined in Figure 9-5, first moving to lower temperature as the pressure increases and then, when all available water in the system has dissolved in the liquid, moving to higher temperature as the pressure continues to rise.

Figure 9-5 emphasizes the importance of water content in controlling the level to which a magma can rise before it reaches its solidus and completely crystallizes. A water-rich magma (such as A in the figure) will reach its solidus before it reaches the earth's surface (Cann, 1970), whereas a dry magma such as B can reach the surface and erupt as lava well above its solidus temperature. This simplified model does agree with observation, in that lavas tend to be more mafic, on the average, than deep-seated intrusive rocks, and mafic liquids carry less dissolved H_2O, on the average, than felsic liquid.

Whether a liquid is water saturated depends on the amount of H_2O in the system, the temperature, the total pressure, the composition of the liquid, and the previous crystallization history. Figure 9-6 provides an example. As-

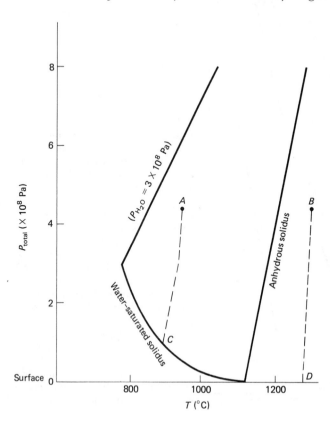

Figure 9-5. Pressure–temperature diagram for the system $NaAlSi_3O_8$–H_2O showing the contrasted behavior of two magmas. A is water-rich (P_{H_2O} = 3 × 10^8 Pa, P_{total} = 4 × 10^8 Pa, and T is 100°C above the solidus). B is anhydrous (P_{H_2O} = 0, P_{total} = 4 × 10^8 Pa, and T is 100°C above the solidus). Both magmas ascend toward the earth's surface; B reaches the surface at D, well above its solidus, and erupts as lava, but A exsolves a vapor phase at 3 × 10^8 Pa total pressure and cools due to water vapor escape, encountering its solidus at C to solidify at depth.

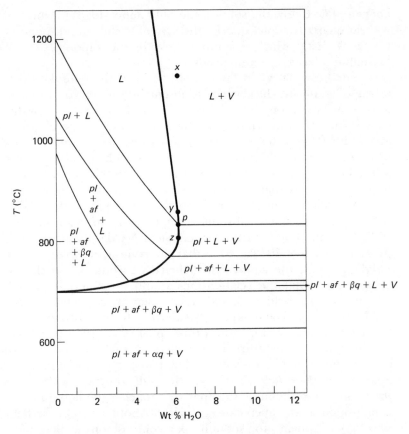

Figure 9-6. Effects of adding water to a synthetic granite at a total pressure of 2×10^8 Pa. *pl*, plagioclase; *af*, alkali feldspar; *βq*, beta quartz; *αq*, alpha quartz; *L*, liquid; *V*, vapor. [With permission, simplified after Fig. 22 of Luth, 1976, in Bailey, D. K., and Macdonald, R., eds., *The Evolution of the Crystalline Rocks.* Copyright by Academic Press Inc. (London) Ltd.]

sume a closed system starting at point *x* with 6.1 wt % H_2O at 1140°C and a total pressure of 2×10^8 Pa. Allow it to cool. Initially, the system is in the field where liquid and vapor coexist. However, as the temperature drops, more and more vapor (assumed to be pure H_2O) dissolves in the liquid. With further cooling, the magma crosses the plagioclase liquidus (at a temperature about midway between points *y* and *z*). But with further cooling below point *z*, vapor appears again, as crystallization of anhydrous plagioclase leaves water behind in the liquid, which is being depleted in the components of plagioclase. The magma therefore eventually becomes water-saturated, crossing the heavy curve in Figure 9-6 to enter the field of plagioclase + liquid + vapor at point *z*. At still lower temperatures, alkali feldspar and then beta quartz join plagioclase as crystallizing phases, until the solidus is reached at about 710°C. Below the solidus, the only mineralogical change is the inversion of beta quartz to alpha quartz; the system is still closed, retaining the original 6.1 wt % H_2O, which is entirely held in the vapor phase (because all the crystalline phases are anhydrous).

The heavy curve in Figure 9-6, which is the boundary between the part of the diagram in which vapor is absent (to the left) and that in which vapor is present, starts on the solidus near 710°C at very low water content and sweeps sharply to the right until the water-saturation level of the silicate liquid at 2×10^8 Pa (and 6.3 wt % H_2O) is reached at point p; with higher temperature, the boundary becomes a straight line with steep negative slope. Cooling a hydrous magma as a closed system through this vapor-saturation curve has important effects that we will consider later in this chapter.

Another way of looking at phase relations in hydrous silicate systems is with a series of isothermal sections (Figure 9-7), in this example, three slices at 1000, 900, and 850°C through the system $NaAlSi_3O_8$–$KAlSi_3O_8$–H_2O at a total pressure of 2.5×10^8 Pa. Note that the scale is in mole percent; if it were in weight percent, the liquid field (labeled L) would lie so close to the base of each triangle that the relationships would be almost impossible to read. In addition to the field of liquid, each diagram contains two two-phase fields (feldspar + liquid and liquid + vapor). The two isothermal sections at the lower temperatures also show a three-phase field (feldspar + liquid + vapor). At the given temperature represented by each section, the composition of each of the three phases is fixed. The feldspar composition is indicated by the apex of the ($F + L + V$) triangle that falls on the $NaAlSi_3O_8$–$KAlSi_3O_8$ join, the liquid composition by the apex that touches the liquid field, and the vapor composition by the apex closest to H_2O. Another two-phase field, for feldspar plus vapor, appears for sodium-poor compositions at 900 and 850°C.

Assume that a system with bulk composition x in Figure 9-7 cools from 1000 to 850°C at 2.5×10^8 Pa total pressure. Initially, composition x lies in the field of liquid alone, at 1000°C. After the system has cooled to 900°C, the

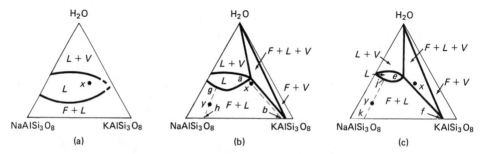

Figure 9-7. Isothermal sections through the system $NaAlSi_3O_8$–$KAlSi_3O_8$–H_2O at 2.5×10^8 Pa: (a) 1000°C; (b) 900°C; (c) 850°C. Scale is in mole percent. (After Fenn, 1977, Figs. 1 and 2, with permission of the Mineralogical Association of Canada.)

liquid field has contracted so that composition x now lies in the field of feldspar plus liquid, and consists of the phases with compositions at the ends of the dashed tie line a-b, with only a few percent of feldspar (b) and mostly liquid (a). Further cooling to 850°C moves the field boundaries so that composition x lies in the field of feldspar (with composition f) plus liquid (with composition e) plus a little vapor (of composition very close to pure H_2O). Any bulk composition falling in the field ($F + L + V$) at 850°C will be represented by these three phases of fixed compositions but varying proportions. Any bulk composition falling in a field ($F + L$), on the other hand, will be represented by phases that vary both in compositions and proportions; for example, bulk composition y in Figure 9-7 at 900°C will consist of nearly equal amounts of liquid (g) and feldspar (h), and at 850°C it will consist of roughly two thirds feldspar (k) and one third liquid (j). Whether a vapor phase forms in this system at a fixed total pressure and a particular temperature depends not only on the amount of water in the system but on the ratio of $NaAlSi_3O_8$ to $KAlSi_3O_8$, showing that variation in other components besides the volatile ones can also be important in determining the solubility of gas in silicate liquid.

For more complete discussion of liquids that contain some water, but not enough for saturation, see Wyllie (1971, p. 173–181) and Luth (1976, p. 359–368). Luth points out the lack of experimental data on hydrous silicate liquids at P_{H_2O} less than 0.5×10^8 Pa, and cautions against applying phase relations for water-saturated systems to those that are water-undersaturated.

9.4 OTHER VOLATILE COMPONENTS

In addition to H_2O, magmas hold variable but significant amounts of carbon dioxide, sulfur, chlorine, and fluorine as volatile components. Carbon dioxide is probably second in importance to water in most magmas, and dominates in others. In rock-forming minerals, the carbonate ion, $(CO_3)^{2-}$, is an essential component only in carbonates, scapolite, and some rare feldspathoids such as cancrinite. These are, except in very unusual igneous rocks such as carbonatites, only minor phases. In contrast, water is an essential component, generally as hydroxyl, $(OH)^-$, in some of the major rock-forming minerals (amphiboles and micas), as well as many accessory minerals. Carbon dioxide therefore less readily finds a home in the subsolidus mineral assemblages of most igneous rocks. In addition, the solubility of CO_2 in silicate liquids is much less than that of H_2O (Mysen and Virgo, 1980) at pressures expected in the crust and upper mantle. At higher pressures, however, the solubility of CO_2 in silicate liquids exceeds that of H_2O (Wyllie, 1979). The consequent effects on the solidus are shown in Figure 9-8.

The lower solubility of CO_2 compared to H_2O at crustal pressures is shown clearly by the observations of Moore and others (1977). Basalt magma erupting at oceanic ridges is not saturated with H_2O, and at water depths

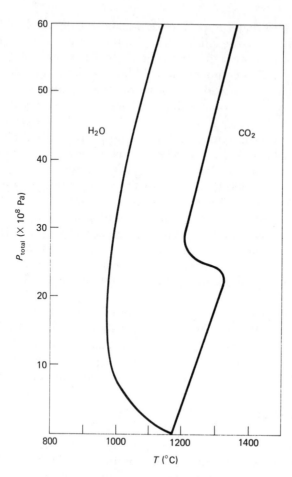

Figure 9-8. Contrasting shapes of the solidus for a peridotite coexisting with pure H_2O and with pure CO_2, in part inferred but based on experimental data. At a total pressure of approximately 2.5×10^9 Pa, equivalent to a depth of around 75 km, the peridotite solidus temperature decreases due to the increased solubility of CO_2. (After P. J. Wyllie, *American Mineralogist, 64,* 469–500, Fig. 3, 1979, copyrighted by the Mineralogical Society of America.)

exceeding approximately 500 m, confining pressure (P_{total}) is enough to keep all the H_2O dissolved in the magmatic liquid, so no water enters a vapor phase. However, CO_2 is much less soluble in the basaltic liquid and is liberated as gas bubbles as deep as 4.8 km below sea level. Consequently, oceanic basalts erupted at depths of 0.5 to 4.8 km generally have vesicles filled with a gas phase that is at least 95% CO_2. Moore (1979) further discusses vesicles in oceanic basalts, and offers an empirical equation relating depth of eruption to volume percentage of vesicles.

Under the same conditions of temperature and pressure, more CO_2 dissolves in hydrous than in anhydrous liquid, probably because of the decreased polymerization caused by the presence of water. Once it enters the liquid, CO_2 reacts with SiO_4^{4-} tetrahedra to form CO_3^{2-} according to this model:

$$2(SiO_4)^{4-} + CO_2 = (Si_2O_7)^{6-} + CO_3^{2-}$$

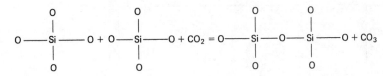

Figure 9-9. Model showing the polymerization of silicate liquid by CO_2, to be contrasted with Figure 9-1, showing the effect of H_2O.

Note that the effect of CO_2 is to increase polymerization, in contrast to the effect of H_2O (Figure 9-9). We should not be surprised to find that addition of CO_2 changes the solidus temperature of a silicate liquid very little (Figure 9-10), unless the pressure is so high that the enhanced solubility shown in Figure 9-8 becomes important.

The marked effect of small amounts of CO_2 on hydrous liquids is shown in Figure 9-11; in the presence of a fluid representing 10% of the mass of the system and containing 96% H_2O and 4% CO_2, the liquidus was raised by almost 100°C, but the solidus only by 10°C, compared to the liquidus and solidus in the system without CO_2 but otherwise the same. At low pressures carbon dioxide counteracts the effects of water, in addition to diluting H_2O in the vapor phase.

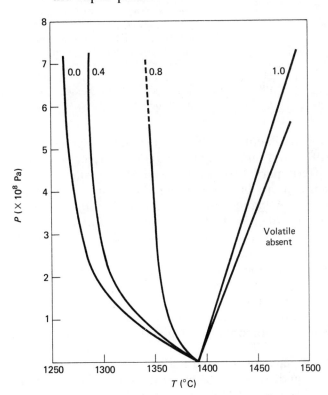

Figure 9-10. Temperature–pressure diagram showing change in the solidus temperature of diopside. Numbers at the tops of the four curves to the left of the volatile-absent solidus indicate the mole fraction of CO_2 in $CO_2 + H_2O$ mixtures. (From Rosenhauer and Eggler, 1975, Fig. 47, with permission of the Carnegie Institution of Washington.)

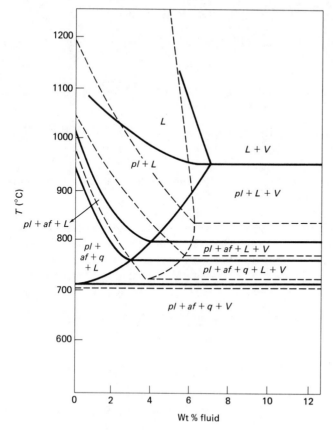

Figure 9-11. Effect of varying composition of $H_2O–CO_2$ fluid on the phase relations of a synthetic granite at a total pressure of 2×10^8 Pa. Dashed lines indicate relations without CO_2 in the fluid; they are the boundaries shown in Figure 9-6. Solid lines indicate relations when 4% CO_2 is present in the fluid. The relative positions of phase stability fields (labelled for the CO_2-bearing system) are preserved when CO_2 is absent. *pl,* plagioclase; *af,* alkali feldspar; *q,* quartz; *L,* liquid; *V,* vapor. [With permission, after Fig. 27 of Luth, 1976, in Bailey, D. K., and Macdonald, R., eds., *The Evolution of the Crystalline Rocks.* Copyright by Academic Press Inc. (London) Ltd.]

At higher pressures experimental study (for examples, Eggler, 1978; Wendlandt and Mysen, 1978) has shown that the ratio of water to carbon dioxide in a fluid phase affects partial fusion in the earth's mantle, not only controlling the solidus temperature but influencing the composition of the silicate liquid that is generated. A higher mole fraction of H_2O in the $H_2O + CO_2$ mixture produces a higher degree of silica saturation in the liquid. This result is thought to be due to the contrasting behavior of the two volatile components on polymerization in silicate liquids; carbon dioxide promotes polymerization and enhances the stability of more "silica-rich" phases (for example, orthopyroxene), leaving the liquid enriched in the components of "silica-poor" minerals such as olivine and nepheline. The composition of a fluid phase containing water and carbon dioxide, even if present in only trace amounts, may thus influence the composition of liquid generated by partial fusion in the mantle.

Eggler and others (1979) have investigated the role of carbon monoxide; the effect of CO on silicate liquids seems to be about the same as that of CO_2, probably because CO also dissolves as CO_3^{2-} in the liquid.

Other, less abundant volatile components are important in local situations and differ greatly in their solubilities in silicate liquids. Fluorine and HF are more soluble than chlorine and HCl in granitic liquids. The contrasting effects, apparently caused by their differing solubilities, of hydrochloric and hydrofluoric acids on the melting temperature of granite are shown in Figure 9-12, along with the results of adding other volatile components. Like H_2O, HF and HCl react with bridging oxygens in the liquid to promote depolymerization (Burnham, 1979a, 1979b). Chlorine, whether added as HCl, NaCl, or some other chloride, is only slightly soluble in silicate liquids but strongly concentrated in a coexisting vapor phase (Kilinc and Burnham, 1972; Barker, 1976). When the liquid crystallizes, chlorine is not accommodated within most rock-forming minerals. Among the silicates, only scapolite and sodalite contain chlorine as an essential constituent, whereas amphiboles, micas, and apatite allow entry of small proportions of chlorine for more abundant fluorine and hydroxyl.

One more volatile component deserves special mention because of its economic significance. Sulfur is only sparingly soluble in most silicate melts. It preferentially combines with metals to form insoluble sulfides as crystalline phases or, at high enough temperature, sulfide liquids that are immiscible with silicate liquids. Many metals of economic interest, including Ni, Pt, Cu, Zn, and Co, are efficiently sequestered by combining with sulfur (Chapter 16).

Basaltic liquids, and perhaps others as well, are commonly saturated with sulfur, so that an immiscible sulfide liquid separates in small amounts, subse-

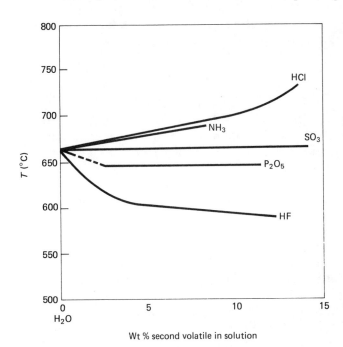

Figure 9-12. The solidus temperature variation is shown for a mixture of 50 wt% granite and 50 wt% "fluid"; the fluid varies from pure H_2O (at the left edge of the diagram) to over 85 wt% H_2O plus nearly 15 wt% of the second volatile component (the identity of which is indicated at the right-hand end of each curve). At this total pressure of 2.75×10^8 Pa, addition of HCl or NH_3 increases the solidus temperature, SO_3 has little effect, and P_2O_5 and HF decrease the solidus temperature. (From Wyllie and Tuttle, 1964, Fig. 3, with permission of the American Journal of Science.)

quently crystallizing at lower temperatures. At low total pressure, a sulfur-bearing gas (mostly SO_2) can be liberated. Anderson (1975) estimated that the magmatic gas in oceanic basalts at 1200°C and 0.5×10^8 Pa is 60% SO_2, 20% H_2O, 10% H_2S, and 10% CO_2. The saturation level of most basaltic liquids is less than 1000 ppm sulfur (Mathez, 1976). The solubility of sulfur in mafic silicate liquids increases with greater temperature and with increasing ferrous iron content. Shima and Naldrett (1975) conclude that sulfur replaces oxygen that was formerly linked to ferrous iron in the liquid, by the reaction

$$FeO \text{ (in liquid)} + 0.5S_2 = FeS \text{ (in liquid)} + 0.5O_2$$

This reaction accounts for the observed dependence of sulfur solubility on ferrous iron content.

In their effects on the structure of silicate liquids, SO_2 resembles CO_2 but H_2S acts like H_2O (Burnham, 1979a, 1979b).

The presence of sulfide inclusions in olivine phenocrysts in mafic and ultramafic lavas suggests that magmas are saturated with sulfur when they are generated in the mantle, and that the sulfur content in the mantle and in the initial liquid may be considerably higher than that in the erupting liquids (Anderson, 1974). In magmatic liquids at the site of their birth, sulfur content may reach 1 wt %. During ascent of the magma, the sulfur separates as immiscible sulfide droplets, removing most of the nickel and copper from the silicate liquid. At equilibrium, the ratios of Ni and Cu in sulfide liquid to Ni and Cu in coexisting silicate liquid exceed 300. Much of the dense sulfide liquid may coalesce into larger blebs and lag behind in the ascending silicate liquid, which therefore reaches the earth's surface with much lower nickel and copper concentrations than it started with.

9.5 OXYGEN FUGACITY

Oxygen is one of the most significant volatile components, because it exerts control on the mineral assemblage that ultimately crystallizes from a magma. Although extremely abundant as an anion in silicate liquids, oxygen contributes a surprisingly small mole fraction to a vapor phase; almost all the oxygen in a magma forms coordination polyhedra around cations, leaving very few oxygen atoms free to link up as diatomic oxygen or to participate in other volatile components, such as H_2O, CO_2, and SO_2. The dissociation of water at high temperatures,

$$H_2O = H_2 + 0.5O_2$$

does produce small amounts, relative to water, of diatomic hydrogen and oxygen that can enter a vapor phase.

Iron is the most abundant element occurring in more than one valence state in igneous rocks. Rock-forming silicates that contain iron as an important element tend to incorporate ferrous iron in preference to ferric; the oxide minerals accommodate both valences, but especially ferric. The sodic clinopyroxene acmite ($NaFe^{3+}Si_2O_6$) and some amphiboles and micas contain high proportions of ferric to ferrous iron. The mafic mineral assemblage in an igneous rock therefore reflects the oxygen fugacity in the magma (or the subsolidus rock) at the time the minerals grew, with higher oxygen fugacity tending to promote the oxidation of ferrous to ferric iron because oxygen is the most abundant donor or recipient of electrons in valence-change reactions.

The oxidation state of iron in igneous rocks is not merely due to weathering or to the interaction of hot lava with atmospheric oxygen, but is internally controlled within the rocks themselves. This is indicated by two lines of evidence. First, basaltic lavas erupted on the deep seafloor show the same range of oxygen fugacity as those erupted subaerially. Second, in basaltic flows that are more than 3 m thick, iron is more highly oxidized in the interiors of the flows, not in the flow tops and bottoms that were brecciated and in intimate contact with the atmosphere while still hot (Haggerty, 1978).

Oxidation of iron has important effects on the CIPW norm (Chapter 4), because greater oxidation means less ferrous iron for the normative constituents *fs*, *hd*, and *fa*; consequently, less silica is encumbered in these femic constituents, and the norm takes on a more silica-oversaturated aspect. The norms of mafic rocks may be changed from *ne*- to *q*-normative simply by increasing the ratio of ferric to ferrous iron. Hughes and Hussey (1979) propose normalizing an analysis to a standard Fe_2O_3/FeO ratio before calculation of the norm.

The fugacity of oxygen has been investigated experimentally for several silicate–oxide systems of geologic interest, and the oxygen fugacity of magmas and rocks, at the time their mafic mineral assemblage was last in equilibrium, can be estimated by combining thermodynamic data and analyses of the coexisting mafic minerals. Of the many possible reactions involving oxygen and mafic minerals, only two examples are given here.

$$3Fe_2SiO_4 + O_2 = 2Fe_3O_4 + 3SiO_2$$

At a given temperature, the oxygen fugacity can be calculated for an assemblage of pure fayalite, magnetite, and quartz that is in equilibrium. For real rocks, in which neither the olivine nor the magnetite is a pure endmember, the compositional variation can be taken into account, and f_{O_2} can be calculated for an assumed temperature. The oxygen fugacity as a function of temperature for the fayalite–magnetite–quartz assemblage is shown as the curve FMQ in Figures 9-13 and 9-14.

$$4Fe_2TiO_4 + O_2 = 4FeTiO_3 + 2Fe_2O_3$$

Ulvöspinel component in a magnetite solid solution oxidizes to yield an ilmenite–hematite solid solution. This reaction is the basis for the magnet-

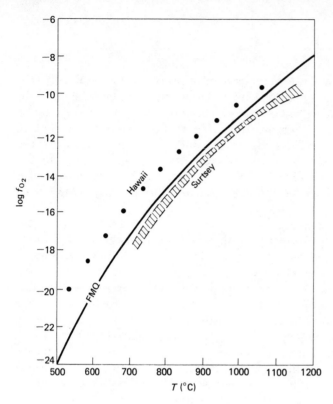

Figure 9-13. Variation in oxygen fugacity with temperature as measured in volcanic gases from lavas in Hawaii and Surtsey (Iceland). FMQ traces the equilibrium oxygen fugacity for fayalite + magnetite + quartz. (After Haggerty, 1976, Fig. Hg-47, with permission of S. E. Haggerty.)

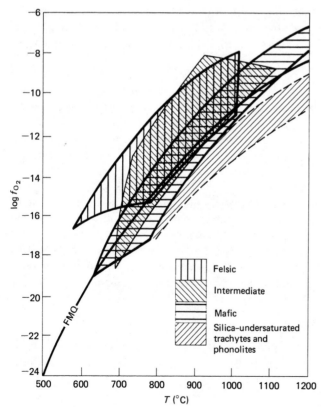

Felsic

Intermediate

Mafic

Silica–undersaturated trachytes and phonolites

Figure 9-14. Variation in oxygen fugacity with temperature as estimated from compositions of coexisting minerals in extrusive rocks. The FMQ curve from Figure 9-13 is shown for comparison. The extrusive rocks are grouped in four broad compositional types. (After Haggerty, 1976, Fig. Hg-46, with permission of S. E. Haggerty.)

ite–ilmenite geothermometer. Compositions of coexisting magnetite–ulvöspinel and ilmenite–hematite solid solutions can be measured to estimate the temperature and oxygen fugacity at which these phases were last in equilibrium (Buddington and Lindsley, 1964; Haggerty, 1976; Spencer and Lindsley, 1981).

Direct observation of volcanic gases has led to estimates of f_{O_2} as a function of temperature, as shown in Figure 9-13. The Hawaiian lava lakes emit gases that are more oxidized than the fayalite + magnetite + quartz assemblage, whereas those at Surtsey are parallel to and below the FMQ curve. These results for volcanic gases are comparable to the indirect estimates calculated from the compositions of coexisting minerals in reactions involving olivine, magnetite, ilmenite–hematite, and iron-rich biotite solid solutions (Haggerty, 1976, 1978); see Figure 9-14. Although there is considerable overlap, felsic magmas tend to have higher oxygen fugacities than mafic magmas, and silica-undersaturated magmas have the lowest oxygen fugacities of all terrestrial magmas. Lunar basalts have f_{O_2} values that are 4 or 5 log units lower than basalts on earth (Haggerty, 1978).

9.6 RESURGENT BOILING OF MAGMAS

As Figure 9-6 implies, crystallization of anhydrous phases from a hydrous liquid causes the remaining liquid to become enriched in H_2O. Ultimately, as at point z in that figure, the residual liquid becomes saturated with respect to water, and a separate vapor phase forms upon further cooling or additional crystallization, even though the total pressure remains constant. In effect, the liquid can boil while the temperature remains constant or decreases. This phenomenon of *resurgent boiling* is important in explosive volcanism.

So far, the most complete data for a hydrous silicate liquid are those for the system $NaAlSi_3O_8$–H_2O (Burnham and Davis, 1971, 1974). One mole (18 g) of water dissolved in albite liquid at 700°C and 1×10^8 Pa has a calculated volume of 20.1 cm³. If, at that same temperature and total pressure, the mole of water leaves the liquid and enters a vapor phase, its volume becomes 63.9 cm³ (Burnham and others, 1969), a volume increase of 3.2 times. The volume increase of the H_2O is partly compensated by decrease in volume of the other part of the system, as the liquid loses water and crystallizes to albite, which is more dense than the liquid (Figure 9-15). Note that a greater volume increase is achieved at 1×10^8 Pa than at 2×10^8 Pa, even though the amount of water involved is smaller. The mechanical energy released through expansion when the liquid changes to vapor plus crystals is at a maximum in this system when the total pressure is approximately 0.7×10^8 Pa (Burnham, 1979b), equivalent to a depth of 1 or 2 km. The release and expansion of the vapor can cause fracturing of the wall rock. If the fractures permit the total pressure to drop, say from 1×10^8 to 0.5×10^8 Pa, the volume of 1 mole of water

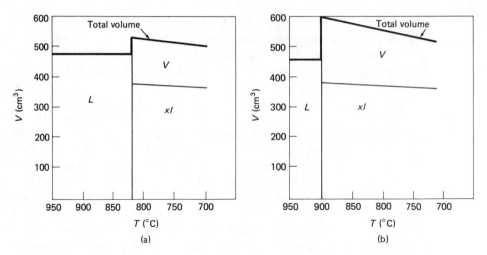

Figure 9-15. Volume–temperature relations for the system $NaAlSi_3O_8$–H_2O. (a) At a total pressure of 2×10^8 Pa, albite liquid saturated with H_2O (6.4 wt%) cools at constant total pressure in a closed system. The solidus is at 820°C. (b) At a total pressure of 1×10^8 Pa, albite liquid saturated with H_2O (4.4 wt%) cools at constant total pressure in a closed system. The solidus is at 900°C. Volume (vertical axis) is for 1 kg of total mass in the system. L, liquid; xl, crystalline albite; V, vapor. (Calculated from data in Burnham and Davis, 1971, 1974, and Burnham and others, 1969.)

at 700°C will more than double, from 63.9 to 139.3 cm³. The escape and expansion of a vapor phase from a hydrous magma is rapid and, if the pressure drops by failure of the wall rock, the expansion is self-propagating. Using the system $NaAlSi_3O_8$–H_2O as a model, Burnham and Davis (1969) and Burnham (1979b) calculate that 1 km³ of hydrous (not necessarily water-saturated) magma could release nearly 3×10^{23} ergs of mechanical energy and 10^{24} to 10^{25} ergs of thermal energy. These quantities are approximately those released during major volcanic eruptions.

For comparison, the annual expenditure of "controlled" energy for human activities is on the order of 10^{28} ergs, merely 1000 times as great as the total energy released by 1 km³ of hydrous magma upon an abrupt decrease in pressure. For further comparison, an average thunderstorm releases 4×10^{20} ergs, but there are an estimated 4×10^4 thunderstorms worldwide each day. On a global basis, therefore, the energy released each day in thunderstorms exceeds or equals that liberated far less often from 1 km³ of hydrous magma.

9.7 VOLATILES AND VOLCANISM

Gases freed from lava and by pyroclastic eruptions mix rapidly with the atmosphere, but these volatile components from magma have greatly differing residence times in the air. Some, including water and chlorine- and sulfur-bearing components, are efficiently concentrated in the hydrosphere after rapid passage through the atmosphere. Others return to the solid earth rapidly (CO_2 in carbonates, for example). Chlorides and sulfates are more leisurely recycled to the lithosphere after evaporation of seawater. On the other hand, oxygen and argon from volcanic gases probably remain, for the most part, in the atmosphere, and their proportions have increased through geologic time.

Aside from the long-term additions to, and modifications of, the atmosphere and hydrosphere, magmatism can cause more abrupt but transitory changes through large volcanic eruptions. Modification of weather close to, and downwind from, eruptive vents by water, carbon dioxide, and pyroclastic dust is abundantly documented. Widespread and longer-lasting effects on climate are another matter, subject to much speculation. Large eruptions such as that of Krakatoa in 1883 inject dust and aerosol particles high into the atmosphere, and these may take several years to settle after worldwide dispersal. Such particles provide airborne nuclei around which water droplets and ice crystals can form. Although the suspended material will tend to reduce the amount of direct solar radiation received at the surface of the earth, the increase in scattered solar radiation will nearly compensate for the blockage (Bryson and Goodman, 1980).

There is very little evidence that meteorological effects of volcanism last longer than a few years, even after the greatest eruptions. Large volcanic explosions are not sufficiently frequent to cause the cumulative cooling that might lead to ice ages or other long-term and global changes. To emphasize this point, Rampino and others (1979) argue that the available data show that rapid climatic changes actually precede large eruptions and may trigger them in some way. More study is clearly needed on the interactions between volcanoes and atmosphere, particularly in discriminating between human-made pollutants (especially fluorine-, chlorine-, and sulfur-bearing gases) and those same chemical species being added by volcanoes. Stoiber and others (1978) report that individual volcanoes emit a minimum of 500 metric tons of SO_2 daily when at low levels of activity; this output increases to at least 4000 tons per day during "mild" eruptions.

In a 1976 eruption, St. Augustine volcano in Alaska released an estimated 5×10^{11} g of chlorine into the atmosphere. Some 0.8 to 1.8×10^{11} g of this (equivalent to between 17 and 36% of the 1975 world production of chlorine in fluorocarbons) were injected into the stratosphere (Johnston, 1980) with possibly harmful effect upon the ozone layer.

This discussion of volatile components in magmas may have left the impression that most of the energy released in volcanic eruptions is derived from the expansion of gases. This is probably not true, judging from the few data available. The only energy audit of a volcano that approaches completeness is that of Stromboli (McGetchin and Chouet, 1979), and Stromboli is an atypical volcano because of its persistent but fairly low level of activity. Not counting heat transferred by convecting groundwater (and this may exceed all other energy losses combined), McGetchin and Chouet evaluated nine mechanisms of energy loss at Stromboli. By far the largest contribution (84%) was heat conducted through the ground on the slopes of the volcano. A weak second (10%) was heat carried away by escaping gas. Heat radiated directly from the vent, escaping from the surfaces of lava flows, and carried by pyroclastic material, plus seismic energy, together account for nearly all the rest. Eruption noise and the kinetic energy of escaping gas plus particles release only about 0.01% of the total energy flux. McGetchin and Chouet emphasize that eruptive styles differing from that of Stromboli in its normal activity will lead to very different proportions of energy loss among the various mechanisms. The problem needs much more study in field and laboratory.

9.8 PEGMATITES

A water-rich vapor phase has several properties that differ from those of the silicate liquid from which it separated. In addition to having lower density and viscosity, the vapor preferentially dissolves some components that had already become concentrated in the silicate liquid because they did not fit into ionic sites available in the phases that had crystallized. Such components can therefore become highly concentrated by this "double distillation" process, first in liquid, then in vapor.

Crystallization from the vapor, and from silicate liquid coexisting with vapor (Jahns and Burnham, 1969) typically produces large and euhedral crystals. *Pegmatites* are rocks of very coarse and variable grain size; the average grain diameter exceeds 1 cm. Like "porphyry," "pegmatite" is strictly a textural term conveying no information concerning chemical or modal composition. Although most pegmatites have the composition of granite (with roughly equal amounts of quartz, K-feldspar, and albite), others have modal compositions appropriate to gabbro, diorite, anorthosite, syenite, nepheline syenite, or carbonatite.

Because the number of components in significantly high concentrations can be large, a wide variety of phases can crystallize in a pegmatite. Mineralogical diversity is, however, not an essential characteristic of pegmatites. Some granitic pegmatites contain no more phases than can be seen in a typical thin section of granite, and are called "simple" pegmatites. Other, "complex" pegmatites contain high concentrations of Li, Be, Cs, B, F, Sn, Nb, Ta, U, Th, and the rare earth elements. These do not readily enter the structures of common minerals because of their unusual combinations of large ionic radii with high valence charges, for the most part. These components therefore tend to promote the crystallization of rare minerals in which they can be major ingredients (Table 9-1). As a result, some 300 mineral species may be

TABLE 9-1 SOME MINERALS REQUIRING THE LESS ABUNDANT
ELEMENTS AS ESSENTIAL CONSTITUENTS

Mineral	Formula
Autunite	$Ca(UO_2)(PO_4)_2 \cdot 10\text{--}12\ H_2O$
Beryl	$Be_3Al_2Si_6O_{18}$
Cassiterite	SnO_2
Cookeite	$LiAl_4(AlSi_3)O_{10}(OH)_8$
Lepidolite	$K(Li,Al)_3(Si,Al)_4O_{10}(F,OH)_2$
Pollucite	$(Cs,Na)_2(Al_2Si_4)O_{12} \cdot H_2O$
Samarskite	$(Y,Ce,U,Ca,Pb)(Nb,Ta,Ti,Sn)_2O_6$
Spodumene	$LiAlSi_2O_6$
Tantalite	$(Fe^{2+},Mn))(Ta,Nb)_2O_6$
Thorianite	ThO_2
Topaz	$Al_2SiO_4(F,OH)_2$
Tourmaline	$(Na,Ca)(Li,Mg,Fe^{2+},Fe^{3+})_3(Al,Fe^{3+})_6(BO_3)_3Si_6O_{18}(OH)_4$
Uraninite	UO_2

Note: The minerals listed here are typically concentrated in mineralogically
complex granitic pegmatites.

identified in a single pegmatite body. One granitic pegmatite at Ytterby, Sweden, does not contain a particularly long list of unusual minerals (Mason, 1971), but it has given its name to four elements in the periodic table (ytterbium, yttrium, terbium, and erbium).

Pegmatites may occur as dikes, sills, or irregular bodies, ranging from 1 m to more than 1 km long. Crudely concentric but discontinuous zoning of texture and mineralogy is displayed in many pegmatites, including those that are mineralogically simple as well as those that are complex. Figure 9-16 shows two examples of zoned granitic pegmatites. Generally, the average grain size increases from each zone inward to the next, and there is chemical as well as structural evidence that the zones solidified from the outside toward the center.

In many (but not all) zoned granitic pegmatites a central core of nearly pure quartz is present (Figure 9-16). In nepheline syenite pegmatites, in contrast, the core commonly consists of sodic clinopyroxene or amphibole (Figure 9-17). Although the quartz cores of many granitic pegmatites seem to indicate an unusually high percentage of quartz for the entire pegmatite, the appearance is deceptive. In the zoned pegmatite of Figure 9-16a, the outer, less quartz-rich zones wrap around the core and make up a much larger share of the volume of the pegmatite. The bulk compositions of many granitic pegmatites correspond nicely with compositions in the low-temperature region (Figure 7-3) of the system $NaAlSi_3O_8$–$KAlSi_3O_8$–SiO_2–H_2O (Norton, 1970).

The fact that the bulk composition corresponds to a low-temperature region of the pertinent phase diagram does not tell us whether the pegmatite formed as the residue from fractional crystallization or as the first liquid from

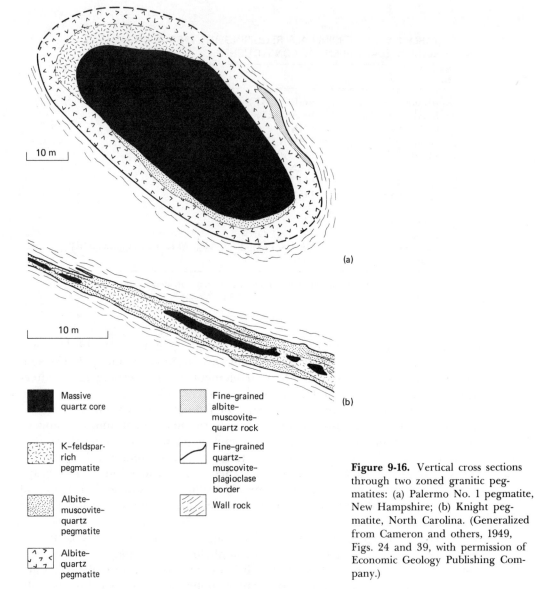

Figure 9-16. Vertical cross sections through two zoned granitic pegmatites: (a) Palermo No. 1 pegmatite, New Hampshire; (b) Knight pegmatite, North Carolina. (Generalized from Cameron and others, 1949, Figs. 24 and 39, with permission of Economic Geology Publishing Company.)

Legend:

- Massive quartz core
- K-feldspar-rich pegmatite
- Albite-muscovite-quartz pegmatite
- Albite-quartz pegmatite
- Fine-grained albite-muscovite-quartz rock
- Fine-grained quartz-muscovite-plagioclase border
- Wall rock

partial fusion, because both processes could yield essentially the same product. For some distinctive and economically significant pegmatites a decision is possible. Granitic pegmatites unusually rich in lithium (containing 1.5 wt % Li_2O) occur in many regions, accompanied by lithium-poor pegmatites and granites. The lithium-rich pegmatites make up less than 1% of the total mass of pegmatite in a region, do not occur within the granite plutons, and are located farthest of all the pegmatites from the larger granite masses, which contain less than 100 ppm lithium (Stewart, 1978). Fractional crystallization would have to be extremely efficient, deriving pegmatite with 1.5 wt % Li_2O from at

Figure 9-17. Nepheline syenite pegmatite with a core of sodic amphibole and clinopyroxene, 0.5 km northwest of Bjerke, Larvik region, Norway.

least 70 times as much initial granitic liquid with less than 100 ppm Li, and cannot account for the lithium-rich pegmatites being farthest from the plutons. Stewart (1978) concludes, on this and other evidence, that the lithium-rich pegmatites represent the first liquid formed during partial fusion of metamorphosed sedimentary rocks containing lithium. "Continued heating would yield larger amounts of granitic magma and would dilute the lithium content" (p. 979). The lithium-rich pegmatites are located on the outer, cooler, fringe of an area in which partial fusion of crust occurred, while lithium-poor pegmatites and granites toward the center of the area represent products of more advanced fusion.

Determining the mode of a pegmatite can be a challenge, particularly if one encounters "giant crystals" (Jahns, 1953); single alkali feldspar crystals reach dimensions of 2 by 4 by 10 m, and mica crystals have achieved diameters of 4 m and lengths of 11 m. Such large grain sizes indicate rapid growth from few nuclei, in a medium of very low viscosity and rapid diffusion. This coarseness is generally attributed to the presence of vapor, although Fenn (1977) and Swanson (1977) found that the appearance of a vapor phase with a silicate liquid does not produce an increase in crystal growth rate from the liquid. The progressive inward growth of zones in some pegmatites is perhaps analogous to the formation of layers in cumulates (Chapter 8), in that the bulk composition of a zone does not equal the composition of the liquid or vapor from which it precipitated.

Associated with many pegmatites, but also occurring alone as small dikes, is *aplite*. Aplite is a textural and compositional term referring to fine-grained equigranular rocks of very low color index. Many aplites apparently form by rapid crystallization from a hydrous magmatic liquid when the pressure on a coexisting aqueous vapor phase is suddenly released (Jahns and Tuttle, 1963). The rapid reduction in P_{H_2O} raises the solidus and liquidus temperatures abruptly, causing severe undercooling of the liquid and a high rate of crystal nucleation.

9.9 DEUTERIC ALTERATION

As stated before in this book, the history of an igneous rock does not end when the last drop of silicate liquid disappears. A residual vapor phase, usu-

TABLE 9-2 SOME DEUTERIC PRODUCTS AND REACTIONS

Deuteric Products

Albite, $NaAlSi_3O_8$
Calcite, $CaCO_3$
Epidote, $Ca_2(Al,Fe^{3+})_3Si_3O_{12}(OH)$
Kaolinite, $Al_4Si_4O_{10}(OH)_8$
Sericite (fine-grained white mica), $(K,Na)Al_3Si_3O_{10}(OH)_2$
Chlorite, $(Mg,Fe^{2+})_5Al(Si,Al)_4O_{10}(OH)_8$
Orthoamphibole, $(Mg,Fe^{2+})_7Si_8O_{22}(OH)_2$
Clinoamphibole, $Ca_2(Mg,Fe^{2+})_5Si_8O_{22}(OH)_2$
Serpentine, $(Mg,Fe^{2+})_3Si_2O_5(OH)_4$
Talc, $Mg_3Si_4O_{10}(OH)_2$
Leucoxene, a fine-grained mixture of titanium and iron oxides
Iddingsite, a fine-grained mixture of clays, chlorite, hydrated iron oxides, and quartz, formed from olivine

Deuteric Reactions

Orthopyroxene   Orthoamphibole
$$7(Mg,Fe)SiO_3 + SiO_2 + H_2O = (Mg,Fe)_7Si_8O_{22}(OH)_2$$

Orthopyroxene   Serpentine   Talc
$$6MgSiO_3 + 3H_2O = Mg_3Si_2O_5(OH)_4 + Mg_3Si_4O_{10}(OH)_2$$

Olivine   Serpentine
$$3Mg_2SiO_4 + SiO_2 + H_2O = 2Mg_3Si_2O_5(OH)_4$$

Anorthite   Calcite   Epidote
$$3CaAl_2Si_2O_8 + H_2O + CaCO_3 = 2Ca_2Al_3Si_3O_{12}(OH) + CO_2$$

Anorthite   Kaolinite   Calcite
$$2CaAl_2Si_2O_8 + 4H_2O + 2CO_2 = Al_4Si_4O_{10}(OH)_8 + 2CaCO_3$$

Anorthite   Quartz   Albite
$$CaAl_2Si_2O_8 + 4SiO_2 + 2Na^+ = 2NaAlSi_3O_8 + Ca^{2+}$$

Alkali feldspar   Sericite   Quartz
$$3(Na,K)AlSi_3O_8 + 2H^+ = (K,Na)Al_3Si_3O_{10}(OH)_2 + 6SiO_2 + 2(K,Na)^+$$

ally richest in H_2O but containing variable proportions of CO_2 and halogens, is still present. As the hot rock cools, the original magmatic minerals react with the vapor phase, and with one another, to form new minerals. This episode in which an igneous rock "simmers in its own juice" is called *deuteric alteration*. Table 9-2 lists some common deuteric minerals and a few of the many possible chemical reactions by which they form. Because most deuteric minerals are white or pale green, deuterically altered rock typically has a bleached aspect.

Deuteric alteration occurs in most intrusive rocks to varying degrees, in some lavas, and in thick ash-flow tuffs (devitrification, if it occurs very early in the cooling history, can be considered a kind of deuteric alteration). Although deuteric reactions tend to destroy magmatic minerals and textures, obscuring an igneous rock's original character, there are compensations; some mineral deposits owe their existence to deuteric alteration (Chapter 16).

9.10 SUMMARY

Volatile components vary widely in their abundance and in their solubility in silicate liquids. The partial pressure of a component in a gas phase, equal to the mole fraction of that component times the total pressure on the gas phase, is an important way to express the concentration of a volatile component. Of the more abundant volatile components, water is the most soluble in silicate liquids in the crust and uppermost mantle; it breaks bonds between polymerized tetrahedra, thereby lowering the viscosity and the solidus temperature. Even though the solubility of water in silicate liquids is limited, few magmas can be water-saturated because not enough H_2O is available, except at small degrees of partial fusion or extremes of crystallization. Carbon dioxide is less soluble than water except at very high pressures in the mantle. Both H_2O and CO_2 exert profound influence (discussed in later chapters) on the temperatures and compositions of magmas generated by partial fusion. Sulfur is only slightly soluble in most magmas, and tends to form either crystalline sulfides or an immiscible sulfide liquid. Although oxygen participates very slightly in a vapor phase equilibrated with magma, variations in its fugacity (or partial pressure) are important in influencing mafic mineral assemblages. As magma crystallizes, the residual liquid may become saturated with respect to water, liberating a vapor phase that is important in volcanism and in the formation of pegmatites. Finally, an igneous rock may react with volatile constituents while cooling, to undergo deuteric alteration.

For more detailed coverage of these topics, Burnham (1979a, 1979b) and Wyllie (1979) are recommended.

10

Ultramafic Rocks

Petrology is necessarily a descriptive science. The preceding chapters have emphasized processes and principles. In this and the following three chapters, chemical and physical aspects of all the more important types of igneous rocks are described in a more systematic fashion.

10.1 COMPOSITIONS

Ultramafic igneous rocks (those with color index greater than 90) consisting mostly of olivine, orthopyroxene, clinopyroxene, or amphibole are classified according to the IUGS (1973) rules shown in Figure 10-1. More unusual types, richer in garnet, mica, carbonate, or opaque oxide minerals, will be treated in better context in Chapters 11 and 13. For many purposes, rocks in Figure 10-1 are adequately described by the root names *dunite* (modal olivine at least 90% of modal olivine + pyroxenes + amphibole), *peridotite* (modal olivine at least 40%), *pyroxenite*, and *hornblendite*. More detailed nomenclature, indicated in the figure, is in wide use and will be introduced later.

Dunites can be modeled by the system Mg_2SiO_4–Fe_2SiO_4 (Figure 10-2). Like most other ultramafic rocks, dunites have high ratios of magnesium to iron, and the olivine in dunites characteristically ranges from 88 to 92% forsterite endmember. Figure 10-2 raises an important problem concerning ultramafic rocks in general; for the observed range in compositions, the solidus for dunite at 1 atm total pressure is above 1700°C. At even the highest geothermal gradient that can be reasonably postulated, such a high temperature is un-

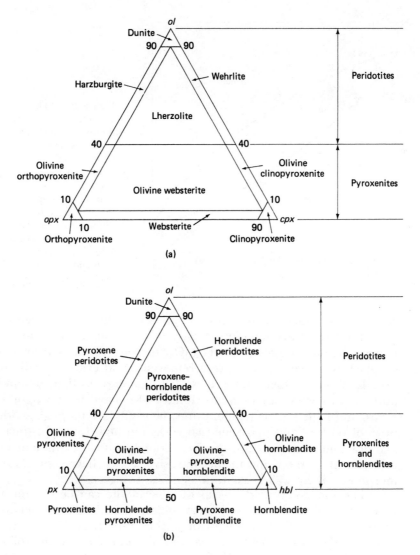

Figure 10-1. Modal classification of ultramafic rocks, excepting those rich in opaque oxides, garnet, micas, or carbonates: (a) rocks composed mainly of olivine and pyroxenes; (b) rocks composed mainly of olivine, pyroxenes, and amphibole ("hornblende"). (After IUGS, 1973, Fig. 2, with permission from the American Geological Institute.)

likely in the outer 200 km of the earth. The effect of pressure on the system Mg_2SiO_4–Fe_2SiO_4, even with sufficient water to saturate the liquid, would not lower the solidus temperature into the range expected for the uppermost mantle and the crust. The evidence of Figure 10-2 strongly indicates that dunite does not crystallize from a liquid of its own composition, and that

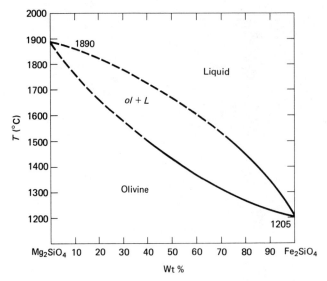

Figure 10-2. The system Mg_2SiO_4–Fe_2SiO_4 at 1 atm pressure, ignoring the incongruent melting of fayalite to iron plus liquid. The solidus and liquidus are dashed over 1500°C because experimental data are sparse and the curves are interpolated from that temperature up to the melting point of pure forsterite at 1890°C. (Simplified from Fig. 2-2 of H. S. Yoder, Jr., 1976, *Generation of Basaltic Magma*, by permission of the National Academy of Sciences, Washington, D.C.)

crystal–liquid fractionation must be involved. Dunites must either be cumulates (formed by precipitation of olivine from magmas containing other components besides Mg_2SiO_4 and Fe_2SiO_4) or solid residues from partial fusion, by which other components were removed. Field evidence indicates that both cumulate and residual dunites exist. Ultramafic rocks that contain higher proportions of pyroxenes than do dunites also have high magnesium to iron ratios, and high solidus temperatures. Not only are the high solidus temperatures of olivine- and pyroxene-rich rocks too high for the estimated geothermal gradients, but they are in direct conflict with textural and mineralogical evidence that most ultramafic intrusive rocks have not thermally metamorphosed their wall rocks to a significant degree.

The water-saturated solidus of a peridotite can be several hundred degrees cooler than the dry solidus at the same total pressure (Figure 10-3), but the effect of additional water diminishes rapidly as the total pressure increases. Other components can have profound effects on the solidus temperature. Seifert and Schreyer (1968) found that the water-saturated solidus for a mixture of 76 mole % Mg_2SiO_4, 20 mole % $MgSiO_3$, and 4 mole % K_2O between 1×10^8 and 5×10^8 Pa is more than 500°C lower than the water-saturated solidus for the same ratio of forsterite to enstatite without potassium. Most ultramafic rocks, however, are extremely low in potassium, generally containing no more than a few tenths of a weight percent K_2O. The real value of their experiment, probably, lies not in emphasizing the effect of potassium specifically but the importance of trace constituents in general. We should constantly remind ourselves that our experiments usually involve a much smaller number of components than real rocks or magmas.

Water is an essential component in those ultramafic rocks that contain

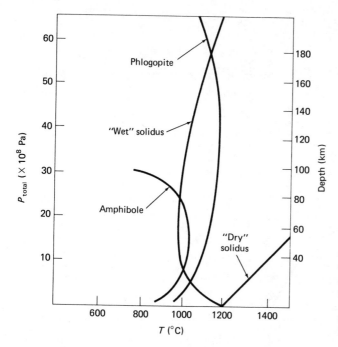

Figure 10-3. Schematic temperature-pressure diagram for a peridotite with excess water and no carbon dioxide, showing the solidus and the stability limits for phlogopite and an amphibole (each is stable to the left of the labeled curve). (After Wyllie, 1977, Fig. 2, with permission of the Geological Society of London and P. J. Wyllie.)

amphibole (Figure 10-1b). The stability of amphibole is more strongly limited by pressure (Figure 10-3) than is the stability of phlogopite, the other common hydrous silicate in ultramafic rocks. In the earth's mantle, phlogopite (K–Mg mica) is the more stable phase, but its abundance is restricted by the low potassium content in the mantle.

Plagioclase is unstable at pressures equivalent to those deeper than the base of continental crust, and reacts to form clinopyroxene and an aluminous phase. At moderately high pressure the aluminous phase is spinel with the idealized formula $MgAl_2O_4$, but at higher pressure still, garnet, with an idealized endmember formula $Mg_3Al_2Si_3O_{12}$, becomes stable. These changes can be expressed in simplified fashion by the following reactions:

Plagioclase plus olivine yields aluminous pyroxenes and spinel:

$$NaAlSi_3O_8 + Mg_2SiO_4 = NaAlSi_2O_6 + 2MgSiO_3$$

$$CaAl_2Si_2O_8 + Mg_2SiO_4 = CaAl_2SiO_6 + 2MgSiO_3$$

$$CaAl_2Si_2O_8 + 2Mg_2SiO_4 = CaMgSi_2O_6 + 2MgSiO_3 + MgAl_2O_4$$

Spinel plus orthopyroxene yields olivine plus garnet:

$$MgAl_2O_4 + 4MgSiO_3 = Mg_2SiO_4 + Mg_3Al_2Si_3O_{12}$$

These reactions represent increasing rock density in response to increased pressure; the density increase is in part accomplished as aluminum changes

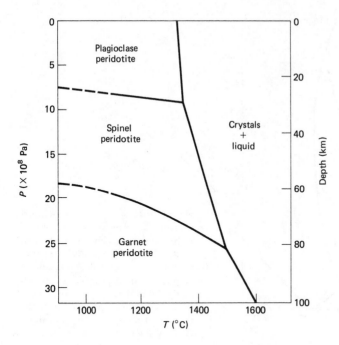

Figure 10-4. Pressure–temperature stability fields (highly approximate) for peridotites containing plagioclase, spinel, and garnet. The anhydrous solidus is also shown. (Modified from Fig. 2-10 of H. S. Yoder, Jr., 1976, *Generation of Basaltic Magma*, by permission of the National Academy of Sciences, Washington, D.C., and from Fig. 5 of O'Hara and others, 1971, with permission of Springer-Verlag.)

from fourfold coordination in plagioclase to combinations of fourfold and sixfold in pyroxenes and sixfold in garnet. Figure 10-4 is a schematic estimate of the approximate depth limits, as functions of temperature, for plagioclase-, spinel-, and garnet-bearing peridotites. Production of garnet and clinopyroxene rich in sodium and aluminum can also occur at high pressures in mafic rocks, as well as ultramafic, so that a basaltic rock can convert to *eclogite*, a rock containing MgAl-rich garnet and NaAl-rich clinopyroxene as the most abundant phases. Although eclogites are ultramafic in terms of their modes, their chemical compositions are those of basalts, and therefore their treatment is deferred until Chapter 11.

By hydration during cooling, or long afterward, many ultramafic rocks change into *serpentinites* or else into talc-bearing rocks. Among the many possible reactions that produce serpentine and talc are

$$\text{olivine} + \text{orthopyroxene} = \text{serpentine}$$

$$Mg_2SiO_4 + MgSiO_3 + 2H_2O = Mg_3Si_2O_5(OH)_4$$

and

$$\text{orthopyroxene} = \quad \text{serpentine} \quad + \text{talc}$$

$$6MgSiO_3 + 3H_2O = Mg_3Si_2O_5(OH)_4 + Mg_3Si_4O_{10}(OH)_2$$

The mineralogy and textures of altered ultramafic rocks are thoroughly discussed by Wicks and Whittaker (1977).

Table 10-1 lists some average compositions of ultramafic rocks, as weight percentages of oxides and as CIPW norms. The proportions of normative constituents listed in the table do not always agree well with the proportions of modal phases defined in Figure 10-1. There are at least two reasons for this. One is that clinopyroxene and, to a lesser extent, orthopyroxene can accommodate aluminum in tetrahedral sites; this aluminum, combined with sodium and calcium in the clinopyroxene, yields albite and anorthite in the norm. Therefore, normative salic constituents need not reflect modal felsic constituents. A second, more important reason for the discrepancy is that each average was calculated for samples that were called "dunite," "peridotite," and so on, by their original describers, regardless of whether the sample fit the definitions in Figure 10-1. The low totals of oxide weight per-

TABLE 10-1 AVERAGE COMPOSITIONS (WT %)
AND CIPW NORMS OF SOME ULTRAMAFIC ROCKS

	Dunite[a]	Peridotite[b]	Pyroxenite[c]
SiO_2	38.29	42.26	46.27
TiO_2	0.09	0.63	1.47
Al_2O_3	1.82	4.23	7.16
Fe_2O_3	3.59	3.61	4.27
FeO	9.38	6.58	7.18
MnO	0.71	0.41	0.16
MgO	37.94	31.24	16.04
CaO	1.01	5.05	14.08
Na_2O	0.20	0.49	0.92
K_2O	0.08	0.34	0.64
H_2O^+	4.59	3.91	0.99
H_2O^-	0.25	0.31	0.14
P_2O_5	0.20	0.10	0.38
CO_2	0.43	0.30	0.13
Total	98.58	99.46	99.83
c	0.80	—	—
or	0.47	2.02	3.75
ab	1.69	4.15	7.76
an	1.17	8.32	13.54
di	—	11.22	42.09
hy	14.48	15.79	6.26
ol	67.38	46.39	15.12
mt	5.20	5.23	6.20
il	0.18	1.19	2.79
ap	0.47	0.23	0.90
cc	1.00	0.67	0.28

Source: LeMaitre, 1976, with permission of Oxford University Press.

[a]Average of 78 samples.

[b]Average of 103 samples.

[c]Average of 106 samples.

centages are due to the presence of appreciable NiO and Cr_2O_3 which were not included in LeMaitre's (1976) compilation. Later in this and the following three chapters you will be invited to compare the compositions of these "average" ultramafic rocks with those of other types.

10.2 FORM AND STYLE OF OCCURRENCE

The question of the existence of hot silicate liquids with ultramafic compositions in nature was settled when Viljoen and Viljoen (1969) recognized lava flows with the composition of peridotite in the Precambrian rocks of South Africa (other geologists had earlier claimed to have found ultramafic lavas, but their field evidence was not as clear). The South African flows, although metamorphosed to varying degrees, retain quenched tops and pillow structures. Their textures (to be described later) strongly resemble those obtained in the laboratory by slowly cooling olivine-rich liquids. Such ultramafic lavas, called *komatiites*, have subsequently been recognized in Precambrian terrains in Zimbabwe, India, Australia, and Canada (see, for example, Arndt and others, 1977).

Komatiite flows are generally thin, yet they may show cumulate zones rich in olivine phenocrysts. Both features indicate low viscosity, as implied by the composition and by experimentally determined high solidus and liquidus temperatures. Green and others (1975) estimate that komatiites were extruded at $1650 \pm 20°C$. The significance of this high temperature and the steep geothermal gradient it demands are topics of continuing research. For several years it was thought that ultramafic lavas occur only in Precambrian rocks. However, Paleozoic examples have been recognized, and Echeverria (1980) reports Mesozoic or younger komatiites on Gorgona Island, Colombia, as flows interlayered with basaltic pillows.

Komatiite lavas are distinguished from basalts by having higher Ca/Al, higher Mg, Cr, Ni, and lower Ti. A precise chemical definition is difficult, especially since Arndt and others (1977) extended the term "komatiite" to cover the entire suite of rocks, including basaltic and andesitic lavas and mafic and ultramafic cumulates, that are found in association with ultramafic lavas. For the present, the term should be used, and interpreted, with caution.

Lava flows make up an exceedingly minor fraction of all ultramafic rocks. A much larger portion occurs as cumulates in intrusive bodies (Wager and Brown, 1967; Irvine, 1974). As discussed in Chapter 8, thick sequences of ultramafic rock may form from an overlying mass of less mafic magma during slow crystallization, either by gravitational segregation or by crystallization from a stagnant boundary layer of magma between convecting liquid and already crystallized material. Specific examples of ultramafic cumulates are described in Chapters 11 and 16.

A still larger fraction of ultramafic rocks is exposed as sheared, fault-bounded masses lacking cumulus texture and generally showing higher Mg/Fe ratios than cumulates. Pervasive granulation and foliation, and alteration to talc or serpentine, are characteristic. These *alpine-type ultramafic* bodies were emplaced as cold solids, and commonly lack any accompanying vestige of less mafic rocks. It is these bodies that generated most of the argument against the existence of ultramafic liquids. Current theory ascribes many alpine-type ultramafic bodies to splinters of suboceanic mantle that were rammed into continental crust at destructive plate boundaries (Chapters 11 and 15).

By far the greatest portion of ultramafic rock, indeed of any terrestrial rock, occurs in the mantle. Much of our knowledge concerning the petrology of the upper mantle comes from ultramafic nodules (xenoliths) that are carried by ascending magmas. Specific ultramafic nodule varieties are remarkably uniform the world over. *Lherzolite*, with the assemblage olivine + orthopyroxene + clinopyroxene + garnet or spinel, in decreasing order of abundance; *harzburgite* (olivine + orthopyroxene); and *dunite* (nearly all olivine) are the most common. *Wehrlite* (olivine + clinopyroxene) and *websterite* (orthopyroxene + clinopyroxene) are more rarely found. Lherzolite appears chemically to be the most likely parent from which most magmas are generated by partial fusion in the mantle. Figure 10-5 shows, schematically, some phase relations compared with ultramafic rock compositions. A lherzolite with composition X would melt to produce a liquid of composition Y; as Y escapes from the system, the bulk composition of the remaining solid migrates toward Z. When all the clinopyroxene has dissolved in the liquid and been removed, melting at the eutectic Y must stop, and the solid residue at Z is harzburgite. If the temperature continues to rise, melting might later begin at an invariant point a between Mg_2SiO_4 and $MgSiO_3$, driving the composition of the final solid residue toward that of dunite.

Of more significance than the modal proportions of minerals in ultramafic xenoliths are the compositions of the various phases. Wilshire and Shervais (1975) and Frey and Prinz (1978) emphasize that spinel-bearing ultramafic xenoliths can be divided into two groups according to clinopyroxene composition, among other criteria. Most commonly, the clinopyroxene is chromium-rich diopside, bright emerald-green in hand sample. More rarely, it is black augite containing the component $CaTiAl_2O_6$. The more widespread chromian diopside-bearing assemblage generally has higher ratios of $Mg/(Mg + Fe)$ of 0.88 to 0.92 (although lower ratios have been found; Wilkinson and Binns, 1977), whereas the Al–Ti augite-bearing assemblage has ratios of 0.62 to 0.85, with the compositions of all minerals showing wider variation than in the chromian diopside assemblage. The two suites also differ in the relative abundances of rock types found as xenoliths. The chromian diopside suite shows, roughly in order from most abundant to least, lherzolite, harzburgite, dunite, orthopyroxenite, clinopyroxenite, websterite, and wehrlite. The Al–Ti augite

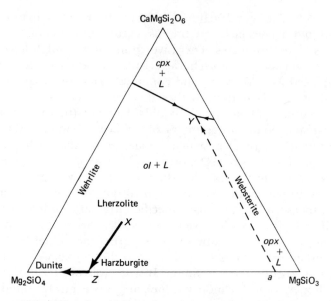

Figure 10-5. Schematic phase diagram of the anhydrous system Mg_2SiO_4–$CaMgSi_2O_6$–SiO_2 at 2×10^9 Pa. Probable configurations of boundaries between liquidus fields are shown, with a ternary eutectic at Y. Lherzolite at X melts to produce liquid at Y and solid residue tending toward harzburgite at Z. After all clinopyroxene has been removed in the liquid Y, renewed melting at invariant point a could yield liquid a with dunite as the ultimate solid residue. Mineral phases in natural peridotites are solid solutions rather than the simple compounds shown here, and the actual phase relations must be more complicated. (Inferred from experimental work by Kushiro, *Mineralogical Society of America Special Paper 2*, 179–191, Fig. 6, 1969, copyrighted by the Mineralogical Society of America, and from Fig. 2-14 of H. S. Yoder, Jr., 1976, *Generation of Basaltic Magma*, by permission of the National Academy of Sciences, Washington, D.C.)

suite is, in decreasing order of abundance, clinopyroxenite, websterite, wehrlite, dunite, and lherzolite. These relative abundances, and the chemical and mineralogical data, suggest that the chromian diopside group are solid residues from partial fusion, and the Al–Ti augite group are cumulates from the liquids (Frey and Prinz, 1978). In agreement with this inference, Irving (1980) describes composite xenoliths in which dikes of Al–Ti augite-bearing rocks cut or enclose chromian diopside-bearing spinel peridotites; he interprets these dikes as conduits through which basaltic liquid flowed and deposited clinopyroxene, spinel, and olivine, and more rarely orthopyroxene, on the walls.

Garnet-bearing peridotite xenoliths are less commonly found than spinel-bearing, but this may be because they are almost exclusively carried up by the rare rock kimberlite (Chapter 13). Evidence from experiments and from mineral compositions (Section 10.4) indicates that garnet-bearing peridotites must underlie spinel-bearing rocks in the upper mantle.

10.3 TEXTURES

Many samples of ultramafic lavas show a distinctive texture called *spinifex* (because it resembles an Australian grass of that name). In spinifex texture "large plates, needles or complex skeletal grains of olivine or clinopyroxene are oriented randomly or parallel to one another" in a finer groundmass (Arndt and others, 1977, p. 324). This texture (Figure 10-6) has been produced experimentally by slowly cooling liquids rich in normative olivine that have been undercooled sufficiently to have low rates of nucleation and high rates of crystal growth. Spinifex texture implies the former existence of a liquid with low viscosity that was emplaced at high temperature. Caution is needed in identifying spinifex texture; during alteration or metamorphism of ultramafic rocks, amphibole can crystallize in large blades superficially resembling spinifex olivine or clinopyroxene.

Alpine-type ultramafic rocks and ultramafic xenoliths share the same range of textural types, which grade into each other as the effects of repeated shearing and recrystallization become superimposed (Harte, 1977). One textural extreme has equant grains with fairly straight boundaries intersecting at approximately 120° angles when viewed in thin section (Figure 10-7); olivine and orthopyroxene tend to be larger than clinopyroxene and garnet or spinel. Plastic flow makes the grains more elongated and imposes a rough parallelism of their long dimensions. Olivine crystals show the effects of deformation more markedly than pyroxenes. Mortar texture (Figure 6-16), in which large anhedral grains of olivine and orthopyroxene survive in a finer groundmass of the same minerals plus clinopyroxene and garnet or spinel, marks a more advanced stage of deformation, but recrystallization may erase the strain effects and the preferred orientation to produce a new simple mosaic texture.

Figure 10-6. Two views of spinifex texture in a sample of ultramafic lava (komatiite) from the Barberton Mountain Land, South Africa. The long depressions form by weathering of bladed olivine crystals.

1 mm

Figure 10-7. Photomicrograph (cross polarized light) of dunite, Balsam Gap, North Carolina, showing simple mosaic texture.

10.4 GEOTHERMOMETRY AND GEOBAROMETRY OF ULTRAMAFIC XENOLITHS

An exciting and controversial treatment of ultramafic xenoliths deserves a digression here. Experiments in the system $MgSiO_3$–$CaMgSi_2O_6$ indicate that the ratio $Ca/(Ca + Mg)$ in clinopyroxene in equilibrium with orthopyroxene is a function of temperature, but is relatively insensitive to pressure. Coexisting clinopyroxene and orthopyroxene in iron-poor rocks thereby provide a geothermometer (Wells, 1977). Compositions of clinopyroxene–garnet pairs furnish other estimates of the temperature at which these phases were last in equilibrium (Ellis and Green, 1979). Furthermore, the aluminum content of orthopyroxene equilibrated with garnet is a function of both temperature and pressure (Lane and Ganguly, 1980). In the assemblage orthopyroxene–clinopyroxene–garnet, the orthopyroxene–clinopyroxene pair can give an estimate of the temperature of equilibration, and the orthopyroxene–garnet pair can then be used to estimate the pressure of equilibration, assuming that the temperature is the same as that given by the pyroxene pair. If pressure is equated to depth, the mineral compositions can be used to estimate the temperature and depth at which the ultramafic assemblage was last in equilibrium. If a range of compositions is available in a suite of garnet peridotite xenoliths from the same locality, the data may define a *geotherm* or "fossil" geothermal gradient for the upper mantle at that place and time (Boyd, 1973).

Pressure and temperature estimates for several xenolith suites show excellent correlation between texture and apparent depth of equilibration; "granular" xenoliths indicate shallower depths and lower temperatures than "sheared" xenoliths. Grain size increases, but irregularly, with depth (Avé Lallemant and others, 1980). Furthermore, the nodules with sheared textures trace steeper geothermal gradients than the granular samples, thereby defining an inflected geotherm (Figure 10-8) according to one interpretation (Boyd, 1973); the inflection was postulated to mark the base of the lithosphere, with sheared textures indicating plastic deformation in the asthenosphere. The deformation of sheared peridotite, however, is so intense

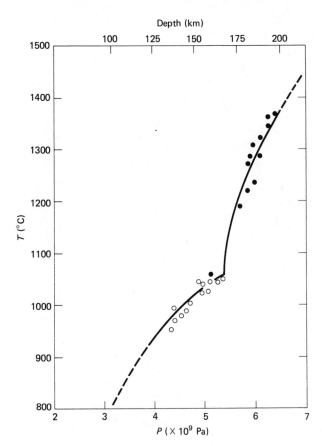

Figure 10-8. Calculated geotherm for garnet peridotite xenoliths, Lesotho. Data points for individual samples are discriminated according to texture; circles represent "granular" and solid dots "sheared" xenoliths. (After Boyd, 1973, Fig. 7A, with permission of Pergamon Press Ltd.)

that it is not likely to have formed simply in response to stress in the asthenosphere. Instead, it may represent deformation at the margin of a diapir or the wall of a magma conduit just before eruption (Mercier, 1979). Recalibration of the orthopyroxene–garnet geobarometer by Lane and Ganguly (1980) erases the inflection in a geotherm drawn through two clusters of data points for sheared and granular xenoliths (Figure 10-9).

Geotherms from garnet-bearing ultramafic xenoliths have now been estimated for several localities of different ages and tectonic settings; Meyer and Tsai (1979) provide a compilation. Garnet-bearing ultramafic xenoliths are extremely rare in oceanic regions. Figure 10-10 gives the only "oceanic geotherm" so far available, for Malaita in the Solomon Islands (Delaney and others, 1979); it indicates higher temperature than "continental geotherms" at the same depth.

In spite of the fairly consistent and reasonable results, these geotherms are of uncertain significance. Although the temperatures calculated may be accurate to within 100°C, uncertainties in the estimated pressures exceed

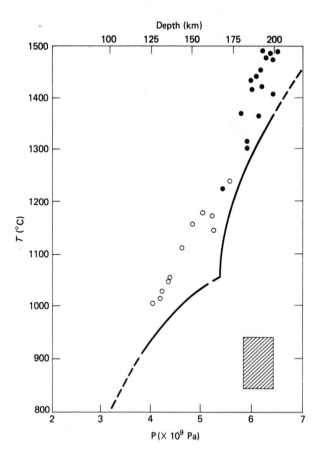

Figure 10-9. Temperature and pressure estimates for the same Lesotho samples as in Figure 10-8 using a revised geobarometer based on orthopyroxene coexisting with garnet. For comparison, the geotherm from Figure 10-8 is shown. The shaded rectangle indicates the estimated uncertainty of each data point. Symbols as in Figure 10-8. (After Lane and Ganguly, *Journal of Geophysical Research, 85,* 6963–6972, Fig. 5, 1980, copyrighted by the American Geophysical Union.)

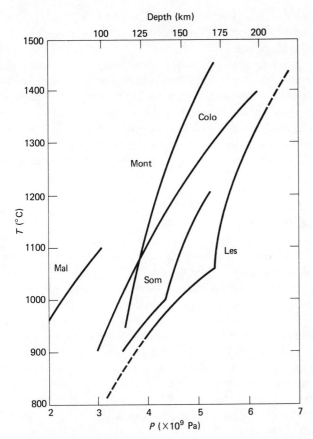

Figure 10-10. Geotherms estimated from garnet peridotite xenoliths from several localities. Les, Lesotho (Figure 10-8); Mal, Malaita, Solomon Islands; Mont, Montana; Colo, Colorado; Som, Somerset Island, Canadian Arctic. (In part from Fig. 7 of Meyer and Tsai, 1979, *Physics and Chemistry of the Earth, 11,* 631–644, copyright 1979, Pergamon Press Ltd. Malaita geotherm from data in Delaney and others, 1979.)

5×10^8 Pa. Each data point in Figures 10-8 and 10-9 therefore has a minimum uncertainty indicated by the rectangle in Figure 10-9.

Even if the temperatures and pressures are accurate, what do they represent? There seems to be little support for the ideas that ultramafic xenoliths are solid residue from the partial fusion generating the magma that carried them upward, or that they are cumulates from that magma, although many probably are residues and cumulates from earlier magmatic events. Instead, most if not all are accidental inclusions detached from the conduit wall by magma ascending from greater depth. Mitchell and others (1980) calculate that a xenolith with a radius of 10 cm or less will reach thermal equilibrium with the surrounding magma in a little over an hour, so that the center of the xenolith will attain the temperature of the magma. This does not mean, however, that the phases in the xenolith will reach chemical equilibrium by changing compositions in response to the new temperature; such reequilibration by exchange of components would require far longer times than the hours required for ascent of the xenolith-bearing magma from its source to the upper crust. Xenoliths probably do record conditions in the mantle before their incorporation in the magma, but perhaps just before, and there is dissent (for example, Wilshire and Jackson, 1975).

To accept the estimated temperatures and pressures at face value raises more questions than it answers. In several individual intrusions and flows, a wide variety of rock types as xenoliths seems to have been derived from a narrow depth range, implying great heterogeneity in the mantle. More study on a broader geographic and compositional range of samples, applying several geothermometers and geobarometers as cross-checks on the same samples (Carswell and Gibb, 1980) will modify mantle geotherms and their interpretations.

10.5 STERILE VERSUS FERTILE MANTLE

Major element and trace element analyses of ultramafic xenoliths support some of the inferences in Section 10.4. The "granular" xenoliths, from apparent depths of less than 150 km, contain chromian diopside and generally are more depleted in the components that are enriched in basalt liquid than are the "sheared" xenoliths from apparently greater depths. These components include Ti, Al, Fe, Ca, and Na. Table 10-2 compares the compositions of a "fertile" ultramafic rock (one capable of generating basalt liquid by partial fusion) and a "sterile" one (depleted in the components of basalt liquid, presumably because it has already passed through an episode of partial fusion). Both are xenoliths from the same locality. These two samples (Boyd and McCallister, 1976) are selected to illustrate the compositional differences, because data are more complete from them; there is actually variation within both the sterile and fertile types.

The table also lists one estimate for a "model" mantle composition (Ringwood, 1975). This model, termed *pyrolite*, is widely used in experiment, calculation, and meditation concerning magma genesis. Pyrolite composition is calculated by combining a Hawaiian basalt composition (assumed to represent the liquid produced by partial fusion) with the composition of a sterile ultramafic rock assumed to represent the solid residue from that partial fusion. The proportions of basalt and ultramafic residue are adjusted so that ratios of nonvolatile major components in the pyrolite mixture agree with those ratios in chondritic meteorites, which are assumed to represent the primordial material of which the earth was made. Agreement between the calculated pyrolite composition and that of ultramafic xenoliths that are sheared and contain garnet, and therefore are likely representatives of fertile parent mantle, are quite good (Table 10-2).

Calculated densities for the fertile and sterile mantle samples in the table indicate that the sterile is 2.7% less dense than the fertile. This agrees with predictions that the residue from partial fusion, with a higher $Mg/(Mg + Fe)$

TABLE 10-2 COMPOSITIONS OF "STERILE" AND "FERTILE" PERIDOTITES AND A PYROLITE MODEL

	"Sterile"[a]	"Fertile"[b]	Pyrolite[c]
SiO_2	47.52	43.70	46.1
TiO_2	< 0.02	0.25	0.2
Al_2O_3	0.63	2.75	4.3
Cr_2O_3	0.27	0.28	—
Fe_2O_3	2.37	1.38	—
FeO	3.50	8.81	8.2
MnO	0.10	0.13	—
MgO	43.21	37.22	37.6
CaO	0.49	3.26	3.1
Na_2O	0.06	0.33	0.4
K_2O	0.02	0.14	0.03
H_2O^+	1.67	1.94	—
H_2O^-	0.07	0.05	—
P_2O_5	trace	trace	—
Total	99.93	100.24	99.9
Modes (wt %)			
Olivine	52.5	58.8	—
Orthopyroxene	45.3	11.1	—
Clinopyroxene	1.4	19.7	—
Garnet	0.8	10.4	—
Density (g/cm³)	3.30 ± 0.02	3.39 ± 0.02	

[a]Granular garnet harzburgite PHN 1569, Thaba Putsoa kimberlite, Lesotho (Boyd and McCallister, *Geophysical Research Letters, 3,* 509–512, Tables 1 and 3, 1976, copyrighted by the American Geophysical Union).

[b]Sheared garnet lherzolite PHN 1611, Thaba Putsoa kimberlite, Lesotho (Boyd and McCallister, 1976).

[c]Model pyrolite calculated by combining 83 wt % residual harzburgite and 17 wt % basalt (Ringwood, 1975, p. 198, with permission of McGraw-Hill Book Company).

ratio and less garnet, should be less dense than the initial fertile rock. Boyd and McCallister (1976, p. 511) concluded that "fertile peridotite will not float in sterile peridotite even in the presence of a large temperature differential, provided both are crystalline. However, partial fusion of the fertile peridotite may overcome the density difference. . . . A diapir-like intrusion of fertile peridotite might be expected to float in depleted mantle if the degree of partial fusion were sufficient to eliminate most of the garnet." However, partial fusion would probably have to achieve 25% liquid before the density difference would be reversed, and the liquid would quite likely escape before that high percentage would be reached.

"Because its density is lower and because its resistance to deformation is increased by elevation of its solidus and by the loss of volatiles" (Jordan, 1979, p. 9), sterile (depleted) mantle rock probably rises to attach itself to the base of the lithosphere above the site of partial fusion. Through geologic time, partial fusion of fertile upper mantle in the asthenosphere should not only lead to production of crust (a significant portion of which is recycled by subduction) but to thickening of the lithospheric upper mantle as a more rigid, less dense, and gravitationally stable layer.

Unanswered questions of fundamental importance remain. Does only the liquid rise after partial fusion, or does the liquid remain temporarily trapped in residual peridotite, so that both move upward as a diapir? How fast, and on what scale, can the solid residue (now sterile) also move upward if no liquid is trapped? Through how many episodes of partial fusion can mantle rock pass before it becomes sterile? Answers to these questions will ultimately come from study of the mantle-derived liquids that yield mafic rocks, to which we turn in the next chapter.

10.6 SUMMARY

Ultramafic rocks make up the earth's upper mantle and form minor portions of the crust. Mantle rock is brought to the surface as xenoliths by some magmas, and in larger masses by diapirism and faulting.

Lava flows of ultramafic composition are widespread in very old Precambrian (Archaean) terrain. Special conditions of high heat flow and steep geothermal gradient were apparently necessary for high degrees of partial fusion and for eruption of ultramafic liquids; the high liquidus temperatures have largely restricted post-Archaean ultramafic rocks to cumulates, derived from less mafic magmas by crystal fractionation, and to solid intrusions.

Presumed temperatures and pressures at which orthopyroxene, clinopyroxene, and garnet were equilibrated in ultramafic xenoliths permit tentative reconstruction of geothermal gradients within the upper mantle. Although many ambiguities remain to be resolved, these inferred geotherms are in good agreement with experimental and theoretical models for generating basaltic liquids by partial fusion in the upper mantle. "Sterile" and "fertile" mantle rocks have been distinguished according to their ability to yield basaltic liquid upon partial fusion; "fertile" rocks contain more Ca, Fe, Ti, Al, and Na.

11

Mafic Rocks

Although ultramafic rocks are volumetrically by far the most abundant in the earth, mafic rocks are the most widespread on the surface. Basaltic magma has been voluminous in space and time, and subtle variations in basalt compositions have even been correlated by some geologists with differences in tectonic setting (Chapter 15). Because so much significance is accorded to varieties of mafic rocks, nomenclature is important. Unfortunately, the classification and nomenclature of these rocks are complex, redundant, misleading, and haphazard, and the situation further deteriorates each year. The task of considering different sets of names for the same rocks cannot be avoided for these as easily as it can for other, less important, types.

11.1 COMPOSITIONS

Chapter 4 made the point that the composition of any rock may be expressed in terms of the mode, the CIPW norm, or the proportions of oxide components. All three methods are used in classifying mafic rocks.

The one essential phase in basalt (or gabbro, its coarser equivalent) is plagioclase containing at least 50 mole % anorthite. Other prominent phases, none of which is demanded by the definitions of basalt or gabbro, are clinopyroxene, with or without orthopyroxene; quartz or olivine, the latter perhaps accompanied by a feldspathoid; small quantities of alkali feldspar (less than 10% of the total feldspar); iron–titanium oxides; and apatite. Amphiboles and biotite generally are confined to the intrusive rocks, although some amphiboles do occur as large crystals in basaltic lavas, and biotite can occur in the groundmass.

The modal and normative incompatibilities and associations among the major phases and components are efficiently summarized in the *basalt tetrahedron* of Yoder and Tilley (1962); see Figure 11-1. Nepheline cannot coexist with quartz or with orthopyroxene, and magnesian olivine cannot coexist with quartz; normative albite and diopside can coexist with all the other constituents. Rock compositions falling to the quartz side of the plane diopside–albite–enstatite (the plane of silica saturation) are *silica-oversaturated*. Those falling to the nepheline side of the plane diopside–albite–forsterite (the critical plane of silica undersaturation) are *critically silica-undersaturated*. Compositions falling in the intermediate volume diopside–albite–forsterite–enstatite are called *silica-undersaturated*. Hence the diagnostic normative criteria of silica saturation are the presence of *q*, *ne*, or *fo* + *en*.

Yoder and Tilley (1962, p. 352) divided basaltic rocks into five groups, depending on position relative to the two planes diopside–albite–forsterite and diopside–albite–enstatite:

1. *Tholeiite*; normative *q*
2. *Hypersthene basalt*; normative *en*, without *q* or *fo* (falling exactly in the plane *di–ab–en*)

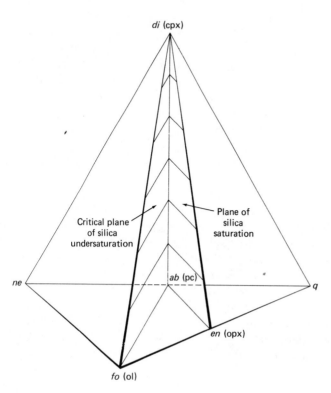

Figure 11-1. The basalt tetrahedron. Basalt compositions are classified in terms of normative quartz, nepheline, albite, and the magnesium endmembers of olivine, orthopyroxene, and clinopyroxene. Abbreviations for modal phases are in parentheses. (Figure 1 of Yoder and Tilley, 1962, used with permission from Oxford University Press.)

3. *Olivine tholeiite*; normative *en* plus *fo*, without *q* or *ne*

4. *Olivine basalt*; normative *fo* (exactly in the plane *di–ab–fo*)

5. *Alkali basalt*; normative *ne*

Unfortunately, these names are hard to keep distinct, and they were earlier used wth different meanings. The history of some of these terms is traced by Chayes (1966) and by Carmichael and others (1974, p. 41–44). Fortunately, the two types that fall exactly on the planes *di–ab–en* and *di–ab–fo* are rare, but there is complete gradation from one type to the next throughout the diagram.

The basalt tetrahedron has some shortcomings, not the least of which is that it requires three dimensions. Another objection is that some Ca- and all Fe-bearing components are ignored. Without distorting the geometric relations of the planes in the tetrahedron, we can accommodate Ca and Fe merely by renaming the normative constituents with which we label the tetrahedron, generalizing *di* to clinopyroxene, *en* to orthopyroxene, *fo* to olivine, and *ab* to plagioclase. Then we have the renamed tetrahedron with parenthetical modal labels in Figure 11-1. Now we can attack the other problem, and flatten the tetrahedron into a two-dimensional diagram. This is simply accomplished by observing that plagioclase is always present, by definition, in any basalt, so we do not need a point to represent it; then we can move the clinopyroxene apex (clinopyroxene is almost invariably present in basalts) down onto the plagioclase point on the base of the tetrahedron. We also (partly for aesthetic reasons) pull the orthopyroxene point toward us, so that the collapsed and distorted base of the original tetrahedron becomes the largest diagram in Figure 11-2, consisting of three equilateral triangles with the apices nepheline–olivine–clinopyroxene, clinopyroxene–olivine–orthopyroxene, and

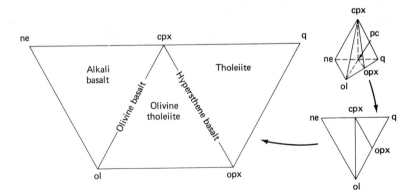

Figure 11-2. Collapse of the basalt tetrahedron into the two-dimensional diagram ne-ol-cpx-opx-q. In most published versions, the normative constituents *di* and *hy* are indicated, rather than the modal cpx and opx.

clinopyroxene–orthopyroxene–quartz. Note that the original incompatibilities and associations of the CIPW norm remain unchanged.

To return to the distinction between alkali basalts and tholeiites; the diagnostic modal minerals (quartz or its polymorphs, or feldspathoids) may be extremely fine grained groundmass constituents. Wilkinson (1967) has summarized other petrographic criteria for distinguishing basalt types. Olivine forms phenocrysts in almost all alkali basalts, but may only be a groundmass constituent in olivine tholeiites. Silica-oversaturated tholeiites can also contain olivine, but only as phenocrysts, and these are usually surrounded by rims of calcium-poor pyroxene formed by the reaction of olivine with q-normative liquid. Alkali basalts do not contain, either as phenocrysts or groundmass phases, orthopyroxene or *pigeonite* (a monoclinic pyroxene containing more calcium than orthopyroxene but much less than clinopyroxenes rich in diopside). Tholeiites and olivine tholeiites can contain orthopyroxene, pigeonite, or both, in addition to clinopyroxene. In fine-grained tholeiitic rocks, this complex assemblage of two or even three pyroxene phases can be difficult to resolve with the petrographic microscope, and electron probe microanalysis is commonly necessary. Fortunately, alkali basalts generally contain a distinctive groundmass clinopyroxene, called *titanaugite*, with more than 2 wt % TiO_2 and showing a distinctive purple-brown or mauve color in transmitted plane-polarized light.

Clinopyroxene, pigeonite, and orthopyroxene compositions in igneous rocks vary as shown in Figure 11-3. The complexity is due not only to the large number of components involved in clinopyroxenes (including Ti-, Al-, and Na-bearing components not indicated in Figure 11-3) but to instability at

Figure 11-3. Variations in pyroxene compositions in igneous rocks. Trend a-b represents calcic augite, common in slowly cooled mafic intrusive rocks. Trend c-d is for most common clinopyroxenes in both intrusive and extrusive rocks. Trend e-f is for clinopyroxenes in some rapidly cooled lavas. Pigeonites follow trend g-h, and orthopyroxenes follow i-j. Arrows indicate the direction of compositional change with falling temperature (and, in general, with decreasing color index of the rock). The more iron-rich compositions (dashed) are less stable than assemblages of quartz plus fayalitic olivine. Thin dashed tie lines connect coexisting orthopyroxene or pigeonite with clinopyroxene. (After Brown, 1967, Fig. 6, with permission of John Wiley & Sons, Inc.)

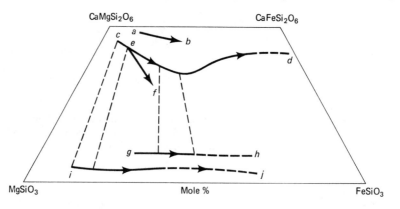

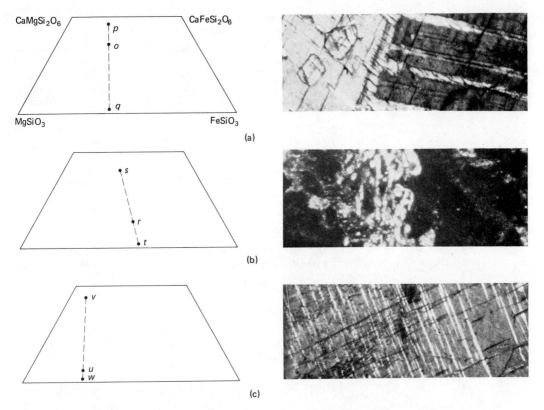

Figure 11-4. Exsolution in pyroxenes, shown by compositions in the $CaMgSi_2O_6$–$CaFeSi_2O_6$–$MgSiO_3$–$FeSiO_3$ quadrilateral and by photomicrographs (in cross-polarized light) of grains approximately 1 mm long. (a) Subcalcic augite (o) unmixes to orthopyroxene lamellae (q) in an augite host (p). Lamellae are parallel to (100) in both halves of the twinned augite. (b) Pigeonite (r) unmixes to augite (s) in orthopyroxene (t). In the photomicrograph, the now exsolved and inverted pigeonite forms an overgrowth on a larger grain of orthopyroxene (at extinction). (c) High-temperature calcic orthopyroxene (u) unmixes to augite lamellae (v) parallel to (100) in an orthopyroxene host (w).

low pressure of orthopyroxene rich in $FeSiO_3$ (Figure 4-5a), to polymorphism and exsolution, and to growth rate. Pyroxenes commonly show progressive zoning from magnesian cores to more iron-rich rims. With slow cooling, pyroxene solid solutions unmix to produce a rich variety of exsolution textures (Figure 11-4) that record a significant part of the rock's cooling history. Deer and others (1978, p. 330–342) thoroughly review the crystallization of pyroxenes in igneous rocks.

Olivine compositions in mafic rocks usually range from 90 to 60% forsterite, with the relatively more iron-rich olivine confined to the groundmass. Olivine phenocrysts in alkali basalts quite commonly show zoning (Wilkinson, 1967).

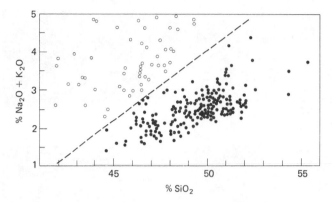

Figure 11-5. Weight percent plot of total alkali oxides versus silica for Hawaiian basalts. Solid dots represent tholeiitic rocks, defined as lacking titanaugite and groundmass olivine; circles represent alkalic rocks, defined as containing titanaugite and groundmass olivine. The dashed line separates the two groups. (From Macdonald and Katsura, 1964, Fig. 1, with permission of Oxford University Press.)

The plagioclase in basalts ranges from An_{95} (rarely this calcic, as phenocrysts) to An_{50} (the lower limit, by definition, for the average plagioclase composition in basalt; groundmass plagioclase may be considerably more sodic).

Macdonald and Katsura (1964) divided Hawaiian basalts into two groups. Their "alkalic basalts" contain titanaugite and groundmass olivine, and are virtually the same as Yoder and Tilley's alkali basalts. However, Macdonald and Katsura's other group, lacking titanaugite and groundmass olivine, embraces all four of Yoder and Tilley's (1962) other kinds of basalt under the single name "tholeiite." Macdonald and Katsura found that a graph of weight percent ($Na_2O + K_2O$) versus SiO_2 allowed the two mineralogically defined groups of Hawaiian rocks to be separated by a straight line (Figure 11-5). Suites that are mineralogically separated according to Macdonald and Katsura's modal criteria at localities outside Hawaii generally are nicely separated by lines that are parallel to, but not precisely coincident with, the Hawaiian line on Macdonald and Katsura's plot. There has been considerable speculation as to the significance of this line. It is probable, but not proved, that the line approximates the trace of the boundary surface between the clinopyroxene plus liquid volume and the plagioclase plus liquid volume on the plane of critical silica saturation (clinopyroxene–plagioclase–olivine) in the generalized basalt tetrahedron (Figure 11-1); see DeLong and Hoffman (1975).

Chayes (1966) suggested another twofold classification, based on the CIPW norm, to replace the Yoder and Tilley (1962) classification. Pointing out that the term "tholeiite" is undesirable for several reasons, Chayes substituted *subalkaline* for *q*-normative basalts and *alkaline* for those that are *ne*-normative. There remain all those basalts containing neither quartz nor nepheline in the norm but falling in the triangle olivine–orthopyroxene–clinopyroxene (Figure 11-2). Chayes proposed that these rocks also be classified as either alkaline or subalkaline, according to their norms (Figure 11-6). This proposal has been widely accepted, and is incorporated in the volcanic rock classification of Irvine and Baragar (1971).

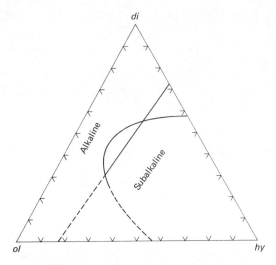

Figure 11-6. Classification of basaltic rocks that contain neither normative *q* nor *ne*. Proportions are in weight percent. Either the linear or the curved boundary may be used; the latter is preferred. (From Chayes, 1966, Fig. 2, with permission of the American Journal of Science and F. Chayes.)

Yet another basalt type, defined on the basis of oxide components rather than mode or norm, was proposed by Kuno (1960). This he named *high-alumina basalt*, but tholeiitic and alkalic basalts commonly have Al_2O_3 just as high; the most distinctive feature is the high proportion of modal plagioclase. Yoder and Tilley (1962, p. 354–355) concluded that high-alumina basalts are merely varieties of silica-oversaturated to silica-undersaturated rocks, with bulk compositions close to plagioclase in the basalt tetrahedron, in the region where all five of their basalt types converge.

In summary, such varietal terms as "tholeiite," "alkali basalt," "olivine basalt," and "high-alumina basalt" have been applied with so many shades of meaning that their significance should be suspect unless each geologist states exactly which definition is being followed.

There are also dark fine-grained igneous rocks, commonly called basalts in the field, that fall outside the definition of basalt because the average plagioclase composition (modal or normative) is more sodic than An_{50}. If silica-undersaturated, these are best called *hawaiite* (with plagioclase averaging between An_{50} and An_{30}) or *mugearite* (An_{30} to An_{15}). Silica-saturated and silica-oversaturated rocks with plagioclase averaging between An_{50} and An_{20} are *trachybasalt* if the Thornton–Tuttle (1960) differentiation index is less than 62.5, or *trachyandesite* if the index is between 62.5 and 75 (Stewart and Thornton, 1975); Wilkinson (1974) presents an alternative classification.

Finally, "basaltic-looking" rocks without modal plagioclase are voluminous in some regions; all are strongly silica-undersaturated and generally contain modal nepheline, leucite, or melilite. They are *nephelinite*, *leucitite*, and *melilitite*, grading into alkali basalts through a category called *basanite* (containing calcic plagioclase, and with more than 5% normative nepheline or leucite); see Wilkinson (1967).

TABLE 11-1 AVERAGE AND REPRESENTATIVE COMPOSITIONS FOR MAFIC ROCKS

	1	2	3	4
SiO_2	49.58	49.16	46.53	47.48
TiO_2	1.98	2.29	2.28	3.23
Al_2O_3	14.79	13.33	14.31	15.74
Fe_2O_3	3.38	1.31	3.16	4.94
FeO	8.03	9.71	9.81	7.36
MnO	0.18	0.16	0.18	0.19
MgO	7.30	10.41	9.54	5.58
CaO	10.36	10.93	10.32	7.91
Na_2O	2.37	2.15	2.85	3.97
K_2O	0.43	0.51	0.84	1.53
H_2O^+	0.91	0.04	0.08	0.79
H_2O^-	0.50	0.05	—	0.55
P_2O_5	0.24	0.16	0.28	0.74
CO_2	0.03	—	—	0.04
Total	100.08	100.21	100.18	100.05
q	2.29	—	—	—
or	2.54	3.01	4.96	9.04
ab	20.05	18.19	20.49	33.59
an	28.45	25.22	23.77	20.61
lc	—	—	—	—
ne	—	—	1.96	—
di	17.15	22.66	20.73	10.84
hy	18.90	14.44	—	0.24
ol	—	9.99	18.62	9.28
mt	4.90	1.90	4.58	7.16
il	3.76	4.35	4.33	6.13
ap	0.57	0.38	0.66	1.75
cc	0.07	—	—	0.09
Diff. index	24.89	21.21	27.42	42.63
% *an*/(*an* + *ab*)	58.65	58.09	53.71	38.02

Source: All by permission of Oxford University Press.

1. Average of 190 tholeiites (LeMaitre, 1976).

2. Sample 57364, olivine tholeiite (Yoder and Tilley, 1962), 1921 lava, Kilauea, Hawaii.

3. Sample 65992, alkali basalt (Yoder and Tilley, 1962), prehistoric flow, Hualalai, Hawaii.

4. Average of 58 hawaiites (LeMaitre, 1976).

Table 11-1 lists analyses and CIPW norms for average or representative basaltic rocks. As was also seen with the ultramafic rocks (Table 10-1), the averages in LeMaitre's (1976) compilation do not always fall within the definitions of the rock types which the averages are supposed to represent. In Table 11-1, the averages for mugearite and trachyandesite barely satisfy the

5	6	7	8	9
50.52	49.21	58.15	40.60	44.30
2.09	2.40	1.08	2.66	2.51
16.71	16.63	16.70	14.33	14.70
4.88	3.69	3.26	5.48	3.94
5.86	6.18	3.31	6.17	7.50
0.26	0.16	0.16	0.26	0.16
3.20	5.17	2.57	6.39	8.54
6.14	7.90	4.96	11.89	10.19
4.73	3.96	4.35	4.79	3.55
2.46	2.55	3.21	3.46	1.96
1.27	0.98	1.25	1.65	1.20
0.87	0.49	0.58	0.54	0.42
0.75	0.59	0.41	1.07	0.74
0.15	0.10	0.08	0.60	0.18
99.89	100.01	100.07	99.89	99.89
—	—	7.66	—	—
14.53	15.07	18.97	3.01	11.58
40.02	29.36	36.81	—	12.23
17.10	20.07	16.56	7.38	18.39
—	—	—	13.67	—
—	2.25	—	21.96	9.65
6.07	11.75	3.88	32.34	20.96
1.92	—	6.45	—	—
4.98	8.55	—	2.50	12.86
7.08	5.35	4.73	7.95	5.71
3.97	4.56	2.05	5.05	4.77
1.78	1.40	0.97	2.53	1.75
0.34	0.23	0.18	1.36	0.41
54.56	46.67	63.44	38.64	33.46
29.93	40.61	31.03	100.00	60.06

5. Average of 55 mugearites (LeMaitre, 1976).

6. Average of 155 trachybasalts (LaMaitre, 1976).

7. Average of 223 trachyandesites (LeMaitre, 1976).

8. Average of 159 nephelinites (LeMaitre, 1976).

9. Average of 138 basanites (LeMaitre, 1976).

definitions in terms of differentiation index and normative plagioclase composition, and the average for trachyandesite is silica-undersaturated. These discrepancies mean that the rock names have been loosely applied and should be interpreted with caution, especially if no analyses are given.

11.2 OCCURRENCES OF BASALTIC ROCKS

Mid-ocean ridge basalts. The greatest mass of terrestrial mafic rock is in oceanic crust. Little was known about the lavas on the ocean floor until dredging and drilling techniques were improved in the late 1950s. Analyzed samples proved to be of distinctive composition (Table 11-2). Variously called

TABLE 11-2 REPRESENTATIVE COMPOSITIONS
FOR MID-OCEAN RIDGE BASALTS

	1	2	3
SiO_2	49.20	48.53	50.81
TiO_2	2.03	0.76	1.85
Al_2O_3	16.09	22.30	14.06
Fe_2O_3	2.72	0.69	2.23
FeO	7.77	4.82	9.83
MnO	0.18	0.16	0.22
MgO	6.44	7.14	6.71
CaO	10.46	12.86	11.12
Na_2O	3.01	2.18	2.62
K_2O	0.14	0.06	0.19
H_2O^+	0.70	0.38	—
H_2O^-	0.95	0.01	0.02
P_2O_5	0.23	0.07	0.17
CO_2	—	—	—
Total	99.92	99.96	99.83
q	0.10	—	1.36
or	0.83	0.35	1.12
ab	25.47	18.45	22.17
an	29.98	50.89	26.04
di	16.58	10.03	23.02
hy	16.99	9.39	18.95
ol	—	7.86	—
mt	3.94	1.00	3.23
il	3.86	1.44	3.51
ap	0.54	0.17	0.40
Diff. index	26.39	18.80	24.66
% *an/(an + ab)*	54.07	73.39	54.02

1. Basalt dredged from 2910 m depth, Atlantic Ocean (Engel and others, 1965, Table 1, sample AD2; with permission of the Geological Society of America and A. E. J. Engel).

2. Basalt dredged from 1700 m depth, Pacific Ocean (Engel and others, 1965, Table 1, sample PVD-1).

3. Glassy rind of lava pillow dredged from Juan de Fuca Ridge, northeast Pacific Ocean (Melson and others, *Geophysical Monograph 19*, p. 352, sample 111240/52, 1976, copyrighted by the American Geophysical Union).

"oceanic tholeiite" and "abyssal tholeiite" at first, the rock is somewhat more accurately termed *mid-ocean ridge basalt* (MORB) in more recent publications. Most samples of MORB contain normative *ol + hy*, or small amounts of *q*; *ne*-normative samples are rare. The most distinctive feature of MORB, as noted by Engel and others (1965), is the strong relative depletion in K and P, and in the trace elements Ba, Zr, Rb, Sr, Pb, U, Th, and the rare earth elements, compared to other basalts. All of the elements listed are "large-ion" or "residual" elements that tend to concentrate in silicate liquid rather than in major crystalline phases (Chapter 4). Their low concentrations in MORB strongly suggest that the liquid was generated either by a large degree of partial fusion of "fertile" mantle or by a small degree of partial fusion of "sterile" mantle. The potassium contents reported for MORB in Table 11-2 are approximately equal to that in "fertile" mantle (Table 10-2), yet potassium should be substantially enriched in any liquid produced by a small degree of partial fusion of such mantle. $^{87}Sr/^{86}Sr$ ratios in MORB are consistently low (0.703), indicating that the mantle source had been depleted in Rb as well as in K. MORB therefore appear to be derived from "sterile" or depleted mantle.

There are regional variations in compositions of MORB. Lavas from the Pacific tend to be higher in Fe and Ti, and lower in Mg, than those from the Atlantic ridge, with very little overlap (Melson and others, 1976). On a smaller scale, variation in basalt composition with latitude along the Mid-Atlantic Ridge (Schilling and others, 1978) shows abrupt steps in major and trace elements, and in the strontium isotope ratio.

According to the plate tectonics model, mid-ocean ridge basalts form at constructive plate boundaries, where they cool and are carried slowly away from the ridge on the diverging lithospheric plates. Although constructive boundaries are enormously significant and stretch for tens of thousands of kilometers, they have not been well understood because of their inaccessibility (generally under hundreds or thousands of meters of water).

Cann (1974) has provided a model of magmatic processes at such a boundary. Figure 11-7 is a cross section of a magma chamber underlying a ridge crest. The chamber is elongated parallel to the ridge trend, perpendicular to the plane of the cross section, but is not necessarily continuous under the entire length of the ridge at any one time. The chamber is fed by liquid welling up from the asthenosphere, which comes closest to the earth's surface under the ridge crest. The chamber is an open system that erupts through its roof via dikes feeding lava flows. Some magma crystallizes on the walls and roof, forming unlayered gabbro, and some, especially near the floor, forms cumulates. On the assumption that mafic minerals will tend to crystallize before plagioclase, the model shows cumulate peridotite grading upward into cumulate gabbro. The walls of the chamber, belonging to diverging plates, move apart and fresh magma wells up into the chamber. Freezing of magma on the walls approximately balances the spreading, so that in this model the

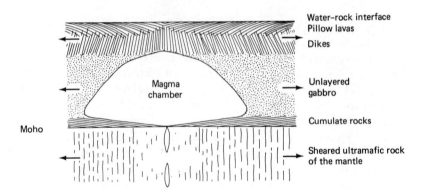

Figure 11-7. Hypothetical cross section through an elongated magma chamber under the crest of an oceanic ridge. Arrows indicate plate motion. Horizontal and vertical scales are unspecified. (From Cann, 1974, Fig. 6, with permission of Blackwell Scientific Publications Ltd.)

chamber maintains a roughly constant width. Cumulate layers dip toward the center, because it is assumed that accumulation should be most rapid near the walls. The basement under the cumulates is suboceanic mantle consisting of noncumulus, probably sheared, ultramafic rock. The magma vents to the surface through a narrow zone of dike injection. The walls of this dike zone also move apart as the plates diverge, so that younger dikes should concentrate near the centerline of symmetry. Lava that erupts on the seafloor will be chilled, forming a thin prism of flows and pillows that subsides as it is buried by younger lavas. Cann suggested that such a "steady-state" magma chamber of constant width could not be maintained if the spreading rate fell much below 1 cm/yr; crystallization would outpace the widening.

Drill core samples of lava and thin intrusive sheets recovered by the Deep Sea Drilling Project reveal more complexity. Distinctly different chemical types of lava were found to be interlayered, as packages of many thin flows with each package totalling 50 to 200 m in thickness, and volcanic successions could not be matched in drill holes only 50 m apart. Furthermore, closely spaced faults and diapirs brought unrelated lavas and coarse, brecciated ultramafic rocks into juxtaposition.

The first thorough investigation of processes taking place at a ridge was Project FAMOUS (French–American Mid-Ocean Undersea Study), for which field investigations culminated in the summer of 1974. A segment of the median rift valley at the crest of the Mid-Atlantic Ridge between the latitudes of 36°30′ and 37° north was surveyed using many methods. Rock samples were obtained by dredging and drilling and, in an aspect of Project FAMOUS that was unique at the time, samples were detached from submarine outcrops using the mechanical arms of manned submersible craft.

Bryan and Moore (1977) describe 50 basalt samples collected by the U.S.

submersible *Alvin*; sample sites were located with a precision of 5 to 10 m (better than most geological studies on dry land) at water depths exceeding 2.5 km. These samples of fresh young basalt showed almost as great a span of compositional variation as all previously studied MORB samples from the entire Atlantic. Bryan and Moore found that the basalts displayed progressive changes in composition; outward from the axis of the rift, SiO_2, K_2O, TiO_2, FeO/MgO, and vesicle content increased, age generally increased, and the abundance of olivine phenocrysts diminished. Chromite (very similar to the chromium-rich spinel in ultramafic rocks) occurred only in the olivine-rich lavas near the middle of the rift valley. Using rock and mineral compositions, Bryan and Moore calculated that the more highly differentiated lava erupted on the flanks could have been derived from the most "primitive" lava erupted at the axis by removing 16.5% plagioclase, 7.3% olivine, and 5.2% clinopyroxene (the phases observed as phenocrysts in the glassy groundmass). This crystal–liquid fractionation could not, however, fully account for the enrichment of K_2O, TiO_2, and H_2O.

The lateral differences in the compositions of erupted lavas suggest the tapping of stagnant, more highly fractionated magma near the sides of the chamber, and of "fresh" magma that has just entered the chamber, or mixtures of primitive and more evolved magmas, closer to the rift axis. Walker and others (1979) conclude that the effects of mixing superimposed on those of low-pressure crystal–liquid fractionation account for the observed variations in MORB.

The term "primary magma" has long been used for a magma that undergoes no change in composition between its generation and its emplacement or eruption. The widespread occurrence of MORB, and the narrow range of its compositional variation, suggest that it might be a primary magma. A liquid that formed by partial fusion of peridotite must have olivine and orthopyroxene, with the same highly magnesian compositions as observed in ultramafic xenoliths, as the liquidus phases (unless one or both react with the liquid at a pressure lower than that of partial fusion but higher than that of final crystallization).

In theory, the primary magma for MORB if generated from peridotite should contain between 15 and 20 wt % MgO (the so-called *picrite* or picritic basalt; O'Hara, 1968), but the most magnesian glass yet found in ocean ridges has only 12% MgO. Furthermore, although olivine is a common phenocryst phase (and presumably a liquidus phase) in MORB, orthopyroxene phenocrysts are extremely rare. Phenocryst assemblages in the FAMOUS area (Flower and others, 1977) are few and simple; olivine (with or without spinel), olivine plus plagioclase, and more rarely olivine plus plagioclase plus clinopyroxene. These are listed in order of decreasing proportion of liquid (glass), and increasing concentrations of residual (large-ion) elements in the glass. The sequence suggests crystallization in a multicomponent system

"olivine–clinopyroxene–plagioclase–SiO$_2$" (with orthopyroxene as an intermediate compound), first in an ($ol + L$) volume, then on an ($ol + plag + L$)
surface, and finally on an ($ol + plag + cpx + L$) cotectic line.

Stolper (1980) heated fragments of MORB glass enclosed in mixtures of
olivine (Fo$_{90}$) and orthopyroxene (En$_{90}$) to find the composition of liquid in
equilibrium with peridotite at various combinations of temperature and pressure; the initial MORB liquid changed composition until it became saturated
with the components of olivine and orthopyroxene. After quenching, the glass
was analyzed, yielding the position of the ($ol + opx + L$) surface (Figure 11-8).
The projected compositions of most MORB samples lie on or close to the
($ol + plag + cpx + L$) cotectic. Stolper concludes that MORB could reach its
observed compositions by fractionation of olivine (and perhaps of other
phases) from primary picrite magma, but asks (p. 24): "Why, if picritic magmas are the primary liquids of MORB suites, are they so rare (none have yet
been found among MORB samples) and why do they always fractionate approximately the same amount prior to eruption?" In a companion paper,
Stolper and Walker (1980, p. 7) offer an answer: "During fractionation of a
picritic liquid, or indeed of any liquid crystallizing only magnesian olivine at
its liquidus, the densities of residual liquids decrease continuously until pyroxene and plagioclase join the crystallization sequence. After this, residual liquid
density increases with fractionation. Thus, there is a minimum in the liquid
density vs fractionation trend of picritic compositions."

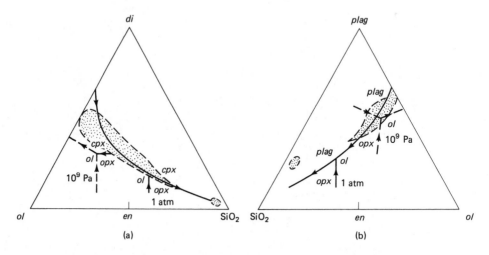

(a)   (b)

Figure 11-8. (a) The system olivine–diopside–silica with liquidus phase boundaries at 1 atm and 10^9 Pa pressure, projected from the plagioclase composition. (b) The system olivine–plagioclase–silica with liquidus phase boundaries
at one atmosphere and 10^9 Pa, projected from diopside. The two triangles are
faces of the tetrahedron olivine–diopside–plagioclase–silica. Stippled areas indicate highest concentrations of projected compositions of MORB glasses.
(Simplified from Stolper, 1980, Figs. 8 and 9, with permission of Springer-
Verlag.)

Huppert and Sparks (1980) had, independently from Stolper and Walker, recognized the correlation between minimum density and maximum ease of eruption. According to their model, the magma chamber under the ridge is periodically recharged with primary picrite magma (15 to 20% MgO), which is more dense than the already fractionated liquid occupying the chamber, and displaces some of this less dense liquid upward as flows and dikes. The composition of the primary magma causes its density to be high, even though its temperature may be some 200°C higher than that of the fractionated liquid. Because of the density contrast, the primary and fractionated liquids do not mix; the primary melt pools at the bottom of the chamber, and heat transfer causes vigorous and turbulent convection in both liquid layers. Olivine crystallizes in abundance from the lower liquid, but is kept in suspension by the turbulence until the temperature difference between the two liquids becomes very small (within a few years after the primary liquid was injected). The bulk density of the lower layer has increased while olivine crystallized, as long as the olivine has remained in suspension, but the density of the residual liquid, from which the components of the olivine have been subtracted, decreases. The liquid reaches a minimum density and, when the temperature contrast between the lower and upper layers becomes small, turbulent convection ceases, the suspended olivine crystals now settle, and the residual liquid rises buoyantly to mix with the slightly more dense liquid of the upper layer. MORB in general, Huppert and Sparks conclude, are mixtures of the residual liquid from the lower layer with more fractionated and iron-enriched liquids of the upper layer. Only rarely can the more dense primary magma breach the surface to form picrites, and then only in places where it does not encounter less dense liquid that prevents its ascent. This model must be tested, using trace element concentrations and phenocryst compositions to estimate temperatures of crystallization and degrees of magma mixing.

Ophiolites.   In addition to the direct but understandably myopic glimpses into oceanic crust provided by drilling, dredging, and diving, there is another view, provided by *ophiolites*. The term refers to an assemblage of rocks, not to a single rock type. The complete suite (rarely seen in its entirety at a single locality, but reconstructed from many incomplete but internally consistent assemblages) consists, from bottom to top, of sheared ultramafic rock (alpine-type peridotites, characteristically of depleted composition) overlain by ultramafic cumulates, then by gabbro cumulates, then by unlayered gabbro, which is topped by a "sheeted dike complex" in which dikes intrude each other with few or no intervening screens of wall rock. The dikes merge upward into pillow lavas, which are interbedded with sedimentary rocks deposited in deep marine water. Comparison of this ophiolite sequence with that in Figure 11-7 shows an obvious relationship between ophiolites and the postulated structure of oceanic lithosphere.

Coleman (1977) has provided a detailed review of ophiolites. These stacks of rock are slabs, several kilometers thick, bounded above and below by faults, that were jammed into continental crust at destructive plate boundaries. Apparently, under certain rather unusual conditions that are still topics of conjecture, the oceanic lithosphere is not entirely subducted; splinters are peeled off and thrust into the opposing plate. Perhaps the very fact that ophiolites have not been subducted should make us reluctant to equate them with "normal" oceanic lithosphere. Not all petrologists believe that the ophiolite sequence forms only in oceanic ridge crests; there are persuasive arguments that some, at least, formed in the subsidiary spreading centers of back-arc (marginal) basins, between an island arc and a continental margin (Upadhyay and Neale, 1979). However, Griffin and Varne (1980) describe a mid-Tertiary ophiolite sequence exposed on Macquarie Island (about midway between New Zealand and Antarctica) that is still on the oceanic ridge crest. Coleman (1981), and companion papers, provide the most comprehensive account of an ophiolite, that exposed on the northeast coast of Oman (Figure 11-9). In Middle Cretaceous time, seafloor spreading widened the Tethys ocean between the African and Eurasian plates. In Late Cretaceous time, Tethys closed, and a detached slab of oceanic lithosphere, 15 to 17 km thick, was thrust southward onto Arabia, forming the Samail ophiolite. The lowest 9 to 12 km is sheared harzburgite averaging 74% olivine (Fo_{90}), 24% orthopyroxene (En_{90}), 2% spinel, with traces of chromian diopside (Boudier and Coleman, 1981), cut by dikes of gabbro, pyroxenite, and dunite. The sheared harzburgite is refractory residue from a partial fusion episode, but probably not from the one that supplied the overlying gabbros and basalts. Seismic properties of the Samail ophiolite (Christensen and Smewing, 1981) are in excellent agreement with those of oceanic crust and uppermost mantle. Although the "seismic Moho," defined by abrupt change in wave velocities, occurs at the contact between cumulate gabbro and the underlying cumulate ultramafic rocks, the "petrologic Moho" between crust and mantle is lower, at the contact between the cumulate ultramafic rocks and the sheared harzburgite (Hopson and others, 1981). Gabbro crystallized in the magma chamber, which was perhaps 36 km wide with inward-sloping floor; more evolved, plagioclase-rich rocks crystallized "from stagnant residual magma at the outer edges of the chamber, beyond reach of replenishment from the center" (Pallister and Hopson, 1981, p. 2593). Above the rocks that crystallized in the chamber are 1.5 km of sheeted dike complex, overlain by about 0.5 km of pillow lava. "Dike splitting was not localized along the center of each successive dike" (Pallister, 1981, p. 2668–2669), but about 95% were injected in a zone less than 20 m wide at any one time.

The documentation in this series of papers leaves little doubt that the Samail ophiolite, at least, had formed at a constructive plate boundary. Because of their tectonic and economic significance, ophiolites are further discussed in Chapters 15 and 16.

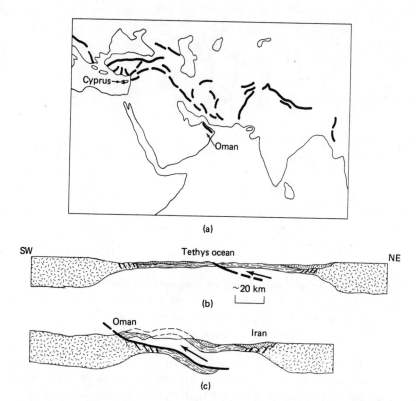

Figure 11-9. (a) Location of the Oman and other ophiolites (black) marking remnants of the Tethys seafloor. (b) Schematic cross section showing detachment of oceanic lithosphere at the constructive plate boundary 90 million years ago, as the Tethys ocean began to close. (c) Continued closure has thrust part of the oceanic lithosphere onto Arabian continental crust, where much of the thrust plate is removed by erosion. Random dash pattern, continental crust; lined pattern, oceanic crust; unornamented, mantle; open stipple, sedimentary cover. (After Coleman, 1981, Figs. 1 and 6.)

Oceanic islands.   Until the advent of deep-sea dredging and drilling, nearly all our knowledge of igneous rocks in ocean basins came from studies on islands. This information turned out to be strongly biased, because most oceanic islands expose few rocks with the composition of MORB; islands do not form merely by rapid, localized, or prolonged accumulation of the same lava that extrudes in the ridges.

Iceland, the largest subaerial exposure on the Mid-Atlantic Ridge, provides one example. Nearly all of the island consists of volcanic, intrusive, and pyroclastic rock, with subordinate sedimentary material weathered from the igneous rock and transported by water and glacial ice. Figure 11-10 shows the distribution of the youngest volcanic rocks and the silica saturation of lavas at eruptive centers. The Mid-Atlantic Ridge enters the island at the southwest

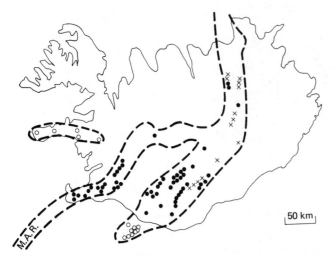

Figure 11-10. Dashes outline volcanic areas in Iceland with eruptions in the last 12,000 years. Basalt compositions at eruptive centers are shown; solid dots, *ol* + *hy*-normative; circles, *ne*-normative; ×'s, *q*-normative. M.A.R., Mid-Atlantic Ridge. (After Jakobsson, 1972, Fig. 4, *Lithos*, *5*, 365–386, with permission of Universitetsforlaget.)

and retains its *ol* + *hy*-normative character onshore, but is deflected to the east and joins a zone of much more diverse composition that is *ne*-normative offshore, as in the recently active volcanoes of Surtsey (1963–1967) and Helgafell (1973), but shows *ol* + *hy*-normative lavas onshore, interspersed with *q*-normative eruptions farther to the northeast and north. Another *ne*-normative belt stretches to the west. Over the entire island, including small islands just off the southern coast, eruptions occur on the average every five years (Jakobsson, 1972). Compared to MORB, Icelandic basalts are low in Mg and Al, but richer in Fe, Ti, K, and P. Seafloor spreading has imposed a rough symmetry on the age distribution of lavas and intrusive rocks on Iceland, with older rocks (dating back to Miocene for the oldest exposed) on the surface at the extreme east and northwest, farthest from the currently active areas. In the older terrain, erosion permits the study of internal features of large and complex central volcanoes. As an example of these, Thingmuli (Carmichael, 1964) is a pile of *q*-normative and subordinate *ol* + *hy*-normative basalts and more felsic differentiates (grading all the way to rhyolites), cut by radial dikes and cone sheets largely of the felsic rocks. Rhyolites and other felsic rocks in Iceland are known only in the older, large central volcanoes, not from the younger fissure eruptions, and appear to have formed by unusually efficient and protracted fractional crystallization.

Basaltic magmatism in Hawaii, in contrast, is unrelated to any plate boundary (Chapter 15), and is more typical of oceanic islands than is the magmatism at Iceland. In the Hawaiian chain and several other island groups in both the Pacific and Atlantic, there is a progression through time toward greater silica undersaturation. The bulk of each volcano in these examples is built up from the seafloor by *q*- or *ol* + *hy*-normative basalts. Late in the eruptive career of a "mature" volcano, the lava may change to slightly *ne*-normative. In a last gasp before extinction, some volcanoes have erupted even more silica-undersaturated lavas (basanites and nephelinites, and their felsic differ-entiates).

The active Hawaiian volcanoes of Kilauea and Mauna Loa have been intensively studied. Wright and Tilling (1980) summarize the recent work on Kilauea. A "magma batch," generated by partial fusion at a depth of approx-imately 60 km, moves in less than 10 years into a shallow storage volume within the volcano. Distinct fusion events, each producing its own magma batch, occur at intervals of months to years. Compositional variations between batches probably reflect variations in the degree of fusion, primarily, although source heterogeneity cannot be ruled out. As it ascends from source to shal-low reservoir, the magma can fractionally crystallize olivine. Within the reser-voir (certainly a labyrinth of inefficiently connected sills, dikes, and irregular bodies rather than a simple tank), a batch may be isolated, fractionally crystallizing olivine, plagioclase, and clinopyroxene, before it is erupted, or it may mix with another batch in a different stage of fractionation. Develop-ment of rapid and precise analytical techniques, coupled with computer re-duction of the data, permits the mass and residence time of each magma batch to be estimated, as well as its contribution to any given lava sample.

On the basis of trace element and isotopic differences, Leeman and oth-ers (1980) conclude that individual Hawaiian volcanoes (even neighboring Mauna Loa and Kilauea) must have separate magma sources of differing compositions. Furthermore, isotopic and major and trace element evidence, especially relatively high $^{87}Sr/^{86}Sr$, K, Rb, Fe, and P, indicates that basalts of oceanic islands come from more fertile mantle than do MORB. This disparity between fertile and sterile mantle sources for islands and mid-ocean ridges suggests convective stirring of the suboceanic mantle; otherwise, mantle di-rectly under islands farther from constructive plate boundaries would be more depleted, having passed through a cycle of voluminous partial fusion as it underlay the ridge before spreading to the sides. In the Pacific, basalts from islands within 2500 km of the East Pacific Rise have $^{87}Sr/^{86}Sr$ of 0.7034 ± 0.0004; farther from the ridge, the ratio increases, and may reflect mantle heterogeneity that has persisted for at least 10^9 years (Hedge, 1978).

Continental flood basalts.   Although hardly rivaling in scale the extent of basalt on the seafloor, *flood basalts* cover large areas of South America, Africa, Antarctica, India, Siberia, and the northwestern United States with thick accumulations of flows that are individually thin (a few meters or tens of meters thick, usually) but widespread. The flood basalts of the Columbia River Basalt Group in Washington, Oregon, and Idaho (Figure 11-11) cover 2×10^5 km^2 with an average thickness of 1 km. Baksi and Watkins (1973) calculated that 80% of the Columbia basalts were extruded in less than 3 million years: "This accumulation rate is therefore a minimum of 100×10^6 m^3/year, which is two or three times the rate of production of lavas on oceanic islands and at least four to six times that rate in a typical spreading ridge

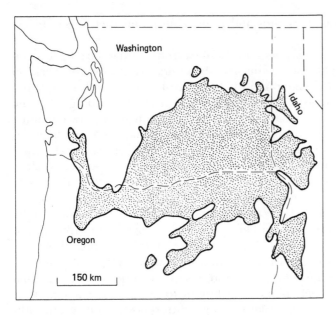

Figure 11-11. Extent of the Columbia River Basalt Group (stippled). (From Swanson and Wright, 1981, Fig. 1.)

system 600 km in length" (p. 495). Individual flows (mostly a few tens of cubic kilometers) were erupted very rapidly, with long uneventful spans, 10^3 to 10^4 years, between them in any one locality. Swanson and others (1975) showed that one basalt unit was fed by linear vents over a dike system that was at least 200 by 450 km in extent. The eruption rate was approximately 1 km³ of lava per day per kilometer of vent length, and eruptions may have lasted only a week, while feeding sheetlike flows that traveled more than 300 km from the vents, probably at 5 to 15 km/hr. The thermal energy released during such an eruption exceeded the average annual heat flow conducted to the earth's total surface.

Flood basalts are generally *ol + hy-* or *q*-normative, but *ne*-normative basalts are also represented in some regions.

For the Columbia River Basalt Group, Swanson and Wright (1981) summarize the abundant new data. Magma compositions, they conclude, were mainly controlled by partial fusion of heterogeneous source rocks in the mantle, and then were little modified during storage at depth followed by rapid transport to the surface, probably as superheated liquids (the lavas are rarely porphyritic). Shallow storage in the crust is not likely; calderas are absent, and liquid compositions do not vary in ways that would suggest low-pressure crystal–liquid fractionation. "The presence of large, deep storage reservoirs, possibly in the upper mantle, may be a principal and distinguishing characteristic of flood-basalt provinces in general" (Swanson and Wright, 1981, p. 11).

Layered gabbroic intrusions. Basaltic magma has in many places pooled at shallow depths in continental crust; rapid crystallization of finer-grained, relatively impermeable rock along the intrusive contacts provided an insulating sheath within which the remaining magma could cool slowly. Some of the most completely documented case histories of crystal–liquid fractionation have been provided by such gabbro bodies. Wager and Brown (1967) described many examples. Of these, the most thoroughly studied remains controversial.

The Skaergaard intrusion on the east coast of Greenland is a Tertiary body containing at least 500 km^3 of mostly layered rocks. Conclusions from the classic study by Wager and Deer (1939), amplified by Wager and Brown (1967), are being drastically modified by field and laboratory studies conducted since 1971 (for example, McBirney and Noyes, 1979; Taylor and Forester, 1979; Norton and Taylor, 1979). Figure 11-12 summarizes the structure and compositional trends.

The Skaergaard body is divided into three principal units. The Marginal Border Group is a finer-grained zone that crystallized from the wall inward and shows indistinct vertical layering, probably caused by variations in nucleation and crystal growth rates rather than gravitational segregation of crystals from liquid. The Layered Series is a thick sequence that crystallized from the floor upward; the Upper Border Group is a sequence of more indistinctly layered rocks that crystallized from the roof downward. As discussed in Chapter 8, McBirney and Noyes (1979) minimize the role of crystal settling or flotation in producing the layers. Instead, they build a strong case for abrupt and rhythmic variations in rock and mineral compositions by nucleation and growth from stationary liquid that was trapped between solid rock and vigorously convecting, nearly crystal-free liquid. Crystallization on a solid substrate established steep local compositional gradients in the adjacent static liquid.

Both the Layered Series and the Upper Border Group are divisible into three subunits on the basis of olivine. In the Layered Series, the Lower Zone contains magnesian olivine, the Middle Zone has no olivine, and the Upper Zone contains fayalitic olivine. In the Upper Border Group, similarly, the α Zone at the top contains magnesian olivine, the β Zone lacks olivine, and the γ Zone contains Fe-rich olivine (Figure 11-12b). Between the Upper Zone of the Layered Series and the γ Zone of the Upper Border Group, but not everywhere present, lies the thin Sandwich Horizon, which is most highly enriched in iron and sodium and probably is a product of the final residual liquid during crystallization of the Skaergaard magma.

Layering is least well developed just above and below the Sandwich Horizon. It is in these poorly stratified and young levels that lateral variations appear in plagioclase compositions (Figure 11-12c). Plagioclase becomes more

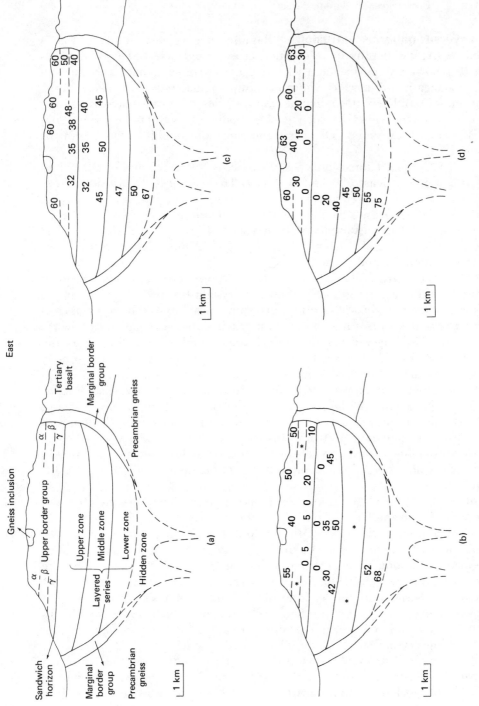

Figure 11-12. Schematic cross sections of the Skaergaard intrusion, east Greenland. (a) Internal divisions. (b) Variation in olivine composition (mole percent forsterite). Asterisks indicate units lacking olivine. (c) Variation in plagioclase composition (mole percent anorthite). (d) Variation in clinopyroxene composition [mole percent Mg/(Mg + Fe)]. (In part modified from Naslund, 1976, Fig. 89, with permission from the Carnegie Institution of Washington.)

West

East

Gneiss inclusion

α β
γ Upper border group

Sandwich horizon

Tertiary basalt

Marginal border group

Precambrian gneiss

Upper zone
Middle zone
Lower zone

Layered series

Hidden zone

Marginal border group

Precambrian gneiss

1 km

(a)

1 km

(b)

1 km

(c)

1 km

(d)

calcic toward the east, where basalt rather than granitic gneiss forms the wall rock at the level of the Sandwich Horizon. Assimilation appears to have produced lateral heterogeneity that could not be erased in the thin lens of liquid remaining after some 80% of the pluton had crystallized (Naslund, 1976).

Although mineral compositions in the Upper Border Group and the Layered Series progressively change as the Sandwich Horizon is approached, the symmetry is far from exact, and there appear to have been significant differences in the conditions under which the two units advanced into the residual liquid. Hoover (1978) has shown that the Marginal Border Group exhibits an inward progression of changing olivine and plagioclase compositions analogous to, and perhaps partly contemporaneous with, those in the Upper Border Group and the Layered Series.

Through the compositional variations in minerals and residual liquids, the Skaergaard pluton provides an often-cited example of initial iron enrichment through fractional crystallization, followed by strong alkali enrichment in the very last stages (Figure 11-13). Granite occurs as small intrusive bodies cutting the layered units, mostly above the Sandwich Horizon. Experiments (McBirney, 1975) indicate that this last, granitic, liquid (composition 5 in Figure 11-13) and the iron-rich but less silicic liquid of the Sandwich Horizon (composition 4 in Figure 11-13) coexisted as immiscible phases. According to other interpretations, the granitic rocks formed by assimilation or partial fusion of Precambrian granitic gneisses of the wall rock.

Why did olivine stop crystallizing, then resume after the residual liquid was iron-rich? The initial Skaergaard magma was either slightly q-normative or $ol + hy$-normative (Hoover, 1978), and early crystallization of olivine drove the remaining liquid toward increasing silica oversaturation. Magnesian

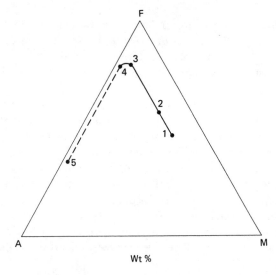

Figure 11-13. Weight percent AFM diagram (A = Na_2O + K_2O, F = FeO + $0.9Fe_2O_3$, M = MgO) showing iron and alkali enrichment in successive Skaergaard liquids. Point 2 is chilled gabbro from the Marginal Border Group (Hoover, 1978). Points 1, 3, 4, and 5 (from analyses in Wager and Brown, 1967) are respectively chilled Marginal Border Group, γ Zone of Upper Border Group, Sandwich Horizon, and granitic rock intruding Upper Border Group.

orthopyroxene melts incongruently to olivine plus siliceous liquid at pressures less than 1.3×10^8 Pa, so olivine can be a liquidus phase precipitating from magmas that are slightly silica-oversaturated. However, after silica oversaturation reaches a slightly higher level, the olivine reacts with liquid to form orthopyroxene. The residual liquid becomes more rapidly depleted in magnesium than in iron (through the precipitation of orthopyroxene as well as earlier olivine), but iron-rich orthopyroxene is not stable relative to iron-rich olivine plus silica. As a result, fayalitic olivine begins to crystallize after the interruption caused by orthopyroxene crystallization. The hiatus in olivine crystallization has been found in other layered gabbroic intrusive bodies, and was predicted by Bowen and Schairer (1935) from experimentally determined phase relations.

Lunar basalts.   An account of basaltic rocks cannot end without mention of lunar rocks. Among the 381.69 kg of samples returned from the moon by Project Apollo, basalts are conspicuous. In mineralogy and texture, lunar basalts display some gratifying similarities to basalts on earth. The major minerals in lunar basalts are olivine, clinopyroxene, and pigeonite (all distinctly more iron-rich than in common terrestrial basalts), plagioclase (commonly An_{75} to An_{96}, rarely occurring as phenocrysts and considerably more calcic than in many earth basalts), and ilmenite.

There are additional marked differences between lunar and terrestrial basalts. On the moon, oxygen fugacity in magmas was so low that iron existed almost entirely in the 0 and +2 valence states. Magnetite is absent, although aluminum- and chromium-bearing spinels are prominent. Furthermore, no hydrous mineral has been identified in lunar samples. The absence of water poses a problem, because many samples of lunar basalts are vesicular, yet the volatile components that permitted a gas phase to form remain unknown or unrecognized.

Table 11-3 lists analyses for five lunar basalt samples. Like the MORB on earth, they are very low in K, P, and other "residual" elements, and most are *ol* + *hy*-normative or slightly *q*-normative. However, the lunar basalts are higher in Fe, lower in Si, Al, Na, and K, and much more variable in Ti (Papike and Bence, 1978).

As might be expected for these anhydrous lunar lavas, there is good correspondence between the mode and the CIPW norm. The most significant departure from the norm, albeit a minor one, is the presence of modal pigeonite rather than orthopyroxene. The extremely mafic and calcic nature of the rocks is apparent in both mode and norm. In the mode, normative quartz is expressed as tridymite or cristobalite, which form clusters lining vesicles or form interstitial patches in the groundmass.

Lunar basalts cover the dark, low-lying areas (maria) that make up approximately 17% of the moon's surface area. They probably accumulated, like flood basalts on earth, as a succession of thin, widespread flows separated by

TABLE 11-3 COMPOSITIONS OF LUNAR BASALTS

	1	2	3	4	5
SiO_2	45.6	44.2	48.8	37.6	46.6
TiO_2	0.29	2.26	1.46	12.1	3.31
Al_2O_3	7.64	8.48	9.30	8.74	12.5
FeO	19.7	22.5	18.6	21.5	18.0
MnO	0.21	0.29	0.27	0.22	0.27
MgO	16.6	11.2	9.46	8.21	6.71
CaO	8.72	9.45	10.8	10.3	11.82
Na_2O	0.12	0.24	0.26	0.39	0.66
K_2O	0.02	0.03	0.03	0.08	0.07
P_2O_5	—	0.06	0.03	0.05	0.14
Total	98.9	98.7	99.0	99.2	100.1
q	—	—	1.64	—	0.55
or	0.12	0.18	0.18	0.47	0.41
ab	1.02	2.03	2.20	3.30	5.58
an	20.25	21.97	24.12	21.86	30.94
di	18.95	20.50	24.47	23.90	22.55
hy	30.21	31.89	43.56	21.07	33.43
ol	27.81	17.71	—	5.49	—
il	0.55	4.29	2.77	22.98	6.29
ap	—	0.14	0.07	0.12	0.33
Diff. index	1.13	2.21	4.02	3.77	6.55
% an/(an + ab)	95.22	91.54	91.64	86.88	84.71

Source: S. R. Taylor, 1975, *Lunar Science: A Post-Apollo View*, p. 136, with permission of Pergamon Press Ltd.

1. 15426, green glass, *Apollo 15,* representing the most "primitive" lunar basalts.

2. 15555, olivine-normative basalt, *Apollo 15.*

3. 15076, quartz-normative basalt, *Apollo 15.*

4. 71055, high-titanium basalt, *Apollo 17.*

5. 12038, relatively aluminum-rich basalt, *Apollo 12.*

long time intervals. Orbital photography at low sun angles reveals lobate flow fronts that are generally only a few tens of meters high. Lunar basalts, as predicted from their compositions and confirmed by experiment, had very low viscosities, lower than any terrestrial lava.

Generalizations concerning basalt magma generation on the moon are hampered by the suspicion, founded upon observations made in lunar orbit, that "less than half of the basalt types has been sampled on Apollo missions" (Mutch, 1979, p. 1696). Another hindrance in understanding their genesis is that "no samples of the lunar mantle have been recognized" (Bence and others, 1980, p. 251).

Lunar basalts range in age from 3.9×10^9 to 3.2×10^9 years. After freezing of the magma ocean (discussed in Chapter 12), volcanism was perhaps triggered exclusively by meteorite impact and little, if any, igneous activity has been manifest on the moon for the last 3×10^9 years.

There are even older basalt samples from a third body in the solar system; *eucrites* are meteorites composed of calcic plagioclase and pyroxene, with bulk compositions and textures strongly resembling those of basalts. Most dated samples of eucrites yield ages of 4.5×10^9 years. Although it is not certain that all eucrites came from one body, there are consistencies pointing to derivation of these meteorites from a single parent with a radius less than 200 km. The bulk compositions of eucrites indicate origin by partial fusion of plagioclase peridotite at low pressure, followed by small degrees of crystal–liquid fractionation (Walker and others, 1978).

11.3 ECLOGITES

In a discussion of ultramafic rocks in Chapter 10, eclogites were mentioned and then their description was deferred until now, because their chemical compositions are those of basalts. Eclogites contain as essential phases garnet and sodium- and aluminum-rich clinopyroxene. As pressure increases, plagioclase becomes unstable relative to more dense phases, and the following simplified reactions can be imagined:

$$\text{plagioclase} = \text{clinopyroxene} + \text{quartz}$$
$$NaAlSi_3O_8 = NaAlSi_2O_6 + SiO_2$$

$$\text{plagioclase} + \text{olivine} = \text{garnet}$$
$$CaAl_2Si_2O_8 + Mg_2SiO_4 = CaMg_2Al_2Si_3O_{12}$$

$$\text{plagioclase} + \text{olivine} + \text{orthopyroxene} = \text{garnet} + \text{clinopyroxene}$$
$$CaAl_2Si_2O_8 + Mg_2SiO_4 + MgSiO_3 = Mg_3Al_2Si_3O_{12} + CaSiO_3$$

$$\text{plagioclase} = \text{clinopyroxene} + \text{quartz}$$
$$CaAl_2Si_2O_8 = CaAl_2SiO_6 + SiO_2$$

$$\text{plagioclase} = \text{clinopyroxene} + \text{kyanite}$$
$$CaAl_2Si_2O_8 = CaSiO_3 + Al_2SiO_5$$

The garnet is usually deep purple and the clinopyroxene pale green, making an attractive rock (the name "eclogite" comes from a Latin word meaning "a choice selection"). The equations imply that quartz or kyanite should be present, but both are rare accessory minerals in eclogites, as are rutile, diamond, and others.

Experiments have very roughly outlined the conditions under which basalt converts to eclogite. Ringwood (1975) has summarized the data and provided a more rigorous and complete account of the reactions that take place during the conversion. At a given temperature, reaction from plagioclase-bearing to garnet-bearing assemblages takes place over a range of

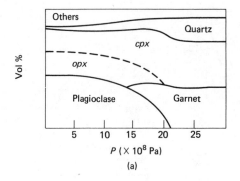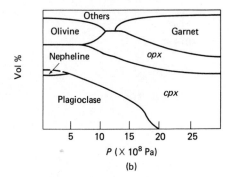

Figure 11-14. Changes in modal proportions of phases in two basalt samples at 1100°C as pressure increases: (a) q-normative; (b) ne-normative. Note that orthopyroxene appears as a high-pressure phase in (b) but not in (a). "Others" are K-feldspar, spinel, and Fe–Ti oxides. (Generalized from Fig. 6 of Green and Ringwood, *Geochimica et Cosmochimica Acta, 31,* 767–833, copyright 1967, Pergamon Press Ltd., modified with permission.)

pressure within which garnet and plagioclase coexist. Furthermore, compositional variations in the original basaltic rock impose variations on the pressure at which garnet begins to form; garnet crystallizes from nepheline-normative basalt compositions at lower pressure than from quartz-normative compositions. Figure 11-14 portrays the mineralogical changes in two basalt samples as pressure was increased at constant temperature.

Eclogite occurs as xenoliths associated with peridotite and other ultramafic inclusions, and is variously interpreted as products of basaltic magma that crystallized within the mantle at pressures too high for plagioclase to be stable, or as samples of oceanic crust that were subducted into the mantle and converted to eclogite on the way down. In addition, small masses of eclogite, rarely exceeding a few hundred meters in any dimension, have been found in highly metamorphosed continental crust; these eclogites are associated with alpine-type ultramafic rocks (dismembered ophiolites?) and are interpreted as mantle tectonically emplaced in the crust (Ernst, 1981), or as oceanic crust emplaced by buoyant rise of a serpentinized subducted slab after subduction stopped (Feininger, 1980). In addition, Figure 11-15 shows that at relatively low temperatures, basalt can be converted to eclogite within thick continental crust. Coleman and others (1965) propose criteria for distinguishing among the multiple origins of eclogites.

Eclogites are considerably more dense than basalts (3.3 versus 2.8 g/cm³). The leading edge of a subducting slab of oceanic crust, when converted to eclogite, will therefore tend to sink more rapidly, perhaps dragging the trailing portion of the slab with it. By this density increase, subduction may temporarily become self-sustaining. Eclogite in the sinking slab may be the parent of some magmas less mafic than basalt, generated by partial fusion (Chapter 12; see also Ringwood, 1975, p. 271–277).

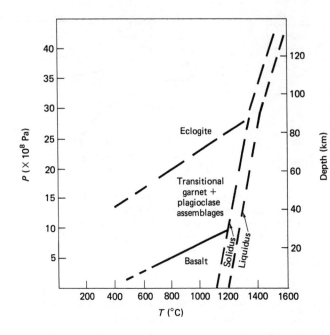

Figure 11-15. Estimated temperature–pressure conditions for the basalt–eclogite transition. The width of the transition zone parallel to the pressure axis varies with basalt composition; an example showing the maximum width was chosen for this illustration. The lower temperature limit of the transition is not known. (Modified from Yoder and Tilley, 1962, Fig. 43, with permission of the Oxford University Press, and from Fig. 1-3 of Ringwood, 1975, *Composition and Petrology of the Earth's Mantle*, with permission of McGraw-Hill Book Company.)

11.4 CAUSES OF THE DIVERSITY OF BASALTIC COMPOSITIONS

Many factors influence the composition of a liquid that is produced by partial fusion of a garnet or spinel peridotite. The first small increment of liquid to form is likely to be the richest in trace elements with large ionic radii and high valence charges, plus sodium, aluminum, calcium, iron, and titanium. As melting progresses, earlier-released components in the liquid are diluted with additional magnesium and silicon from residual olivine and pyroxenes. Therefore, the degree of partial fusion at the time of liquid escape is important. The same parent rock, at the same temperature and pressure, would yield liquid of a different composition if 5% or 1% were melted before the liquid migrated upward.

Pressure at which melting occurs has long been recognized as an important control. Other things being equal, higher total pressure tends to produce liquids of lower silica saturation. This can be illustrated using simple systems as models (Figure 11-16); Kushiro (1968) and Jaques and Green (1980), among others, showed that the same effect extends into systems with larger numbers of components. In Figure 11-16, the cotectic representing equilibrium among olivine, orthopyroxene, and liquid migrates toward the olivine composition as the total pressure increases. This means that the liquid in equilibrium with olivine and orthopyroxene (the two most abundant phases in the uppermost mantle) becomes more silica-undersaturated as it is generated at greater depth.

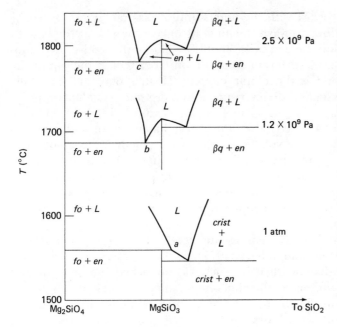

Figure 11-16. Melting relations in the magnesium-rich portion of the system Mg_2SiO_4–SiO_2 at 1 atm, 1.2×10^9 Pa, and 2.5×10^9 Pa. Small unlabeled fields contain enstatite plus liquid. Compositions *a*, *b*, and *c* represent the first liquids to form from mixtures of forsterite and enstatite at the three pressures. The forsterite plus liquid field shrinks with increasing pressure. The composition of the liquid coexisting with enstatite plus a silica polymorph, on the other hand, is not significantly affected by pressure. *fo*, forsterite; *en*, enstatite; *L*, liquid; *βq*, beta quartz; *crist*, cristobalite. (Modified from Chen and Presnall, *American Mineralogist, 60,* 398–406, Fig. 1, 1975, copyrighted by the Mineralogical Society of America.)

The effect of volatile components on partial fusion in the mantle was mentioned in Chapter 9. Silica undersaturation of liquid increases as the CO_2/H_2O ratio in the system increases. Kushiro (1975) has shown that phosphorus and titanium also lower the silica saturation, while addition of sodium or potassium raises it, all by moving the (forsterite + enstatite + liquid) boundary. Cations with charges of 4 or greater increase the polymerization of silica tetrahedra, causing expansion of the enstatite liquidus field. On the other hand, singly charged cations tend to inhibit the polymerization, thus encouraging the crystallization of forsterite over a broader composition range and pushing the cotectic toward higher silica saturation. The combined effects of these components are complicated, because water, alkalis, titanium, and perhaps phosphorus and carbon dioxide are likely to be enriched in the first liquid to form, and the monovalent cations will tend to counteract the effects of the more highly charged cations.

Within a given region, the first basaltic lavas in a succession may be most enriched in "residual" trace elements or most depleted in them, compared with later lavas. Yoder (1978) describes two simplified models for partial fusion that may explain these trends. The first is called the hot-plate model: "Heat is conducted from a hot mass depleted of its basaltic constituents into *undepleted* crystalline mantle" (Yoder, 1978, p. 307). Sterile mantle at 1800°C is instantaneously emplaced below fertile mantle at 1100°C. Fusion is assumed to occur at an invariant point, and all major phases (olivine, orthopyroxene,

clinopyroxene, garnet) coexist with the liquid until it occupies about 30 vol %
of the system. As heat flows upward, melting advances upward; at the base of
the fertile mantle, the liquid may reach 50 vol %. The hotter and more abundant liquid at the base has a different composition from that at the top of the
melting zone. The second is the diapir model: "Heat is obtained internally
from an undepleted crystalline diapir rising into a *depleted* crystalline mantle"
(p. 307). The rising diapir begins to melt at an assumed depth of 130 km,
where the solidus is encountered; after the diapir has risen an additional 35
km, melting reaches about 30 vol %. In the diapir model, the amount of
liquid, controlled by the pressure gradient, diminishes downward. "If these
sources were tapped from the top, then in the hot-plate model the magma
representing the smaller degree of melting will be intruded or extruded first,
whereas in the diapiric model the magma representing the greater degree of
melting will be first" (p. 311).

A powerful approach to estimating the depth of magma generation is
summarized by Carmichael and others (1977); if the temperature and pressure at which a lava or shallow intrusive rock was quenched can be estimated
(for example, by using the magnetite–ilmenite geothermometer mentioned in
Chapter 9), and if certain thermodynamic properties of the liquid can be estimated from mineral and whole-rock analyses, it is possible to calculate the
temperature and pressure at which that liquid would have been in equilibrium with an assumed parent rock (say, garnet peridotite). The approach is
hampered only by lack of thermodynamic data concerning some components
of the liquid at high temperatures and pressures. The depths and geothermal
gradients calculated for many samples are consistent with those estimated
from experimental and geophysical data. The high promise of the method is
emphasized here, to point out once more the dual need for more (and better)
thermodynamic data and for more petrologists with the thorough grounding
in thermodynamics that this book makes no attempt to provide.

Thermodynamic data allow predictions of the proportions and compositions of phases in systems that contain far too many components to permit
graphical representation in phase diagrams. One of the first successful applications of this approach was that of Roeder and Emslie (1970), who generalized the results of many experimental investigations of simple and complex
systems and demonstrated that the ratio of Fe to Mg in olivine is 0.3 times the
ratio of Fe to Mg in the coexisting liquid, for a wide range of liquid compositions involving many other components, and over a wide range of temperature (the effect of pressure is uncertain, but probably small). It is thus
possible, knowing the Fe/Mg ratio of a liquid, to predict the composition of
the first olivine that would precipitate from that liquid and, conversely, if the
composition of the olivine is known, the Fe/Mg ratio of the liquid from which
the olivine crystallized can be estimated. Such an approach has successfully
been applied to other crystal–liquid pairs; Hostetler and Drake (1980) review
these and propose a statistical method for predicting the distribution of major

elements among crystalline phases and liquids, analogous to the trace element distribution coefficients mentioned in Chapter 4.

To close this treatment of basalts we return to the topic of ultramafic xenoliths. All xenoliths for which a mantle origin seems likely have been found in silica-undersaturated rocks, and xenolith abundance tends to increase with increasing undersaturation of the enclosing rock; not a single garnet- or spinel-bearing chromian diopside peridotite has yet been found in a quartz-normative basalt (Sutherland, 1974). The reason for this discrimination has been widely attributed to variations in rate of magmatic ascent. According to this view, silica-undersaturated magmas originate at greater depths, and come up more rapidly, than q-normative liquids. In order for ultramafic xenoliths to be carried all the way to the surface, the magma must move upward faster than the xenoliths sink. The terminal settling velocity of a xenolith can be calculated if the magma is assumed to be a viscous fluid (Chapter 8). Kushiro and others (1976) calculated that an ultramafic xenolith with a density of 3.3 g/cm^3 and a diameter of 10 cm would sink 1.6 km/hr through a stationary liquid with a density of 3.08 g/cm^3 and a viscosity of 50 P. They concluded (p. 6355) that "magmas of such viscosity carrying peridotite inclusions greater than 10 cm in diameter may have ascended from a depth of 50 km to the surface within 31 hours." A temperature above the liquidus was explicitly assumed for the basaltic liquid, as was the lack of yield stress.

However, Sparks and others (1977) have questioned the assumption that such a magma is purely viscous or Newtonian. Field measurements on erupting lavas indicate substantial yield stress. In order to sink, a xenolith must shear the liquid surrounding it, overcoming the yield stress. For lavas so far investigated, and probably for most magmas below their liquidus temperatures, yield stress is sufficient to support xenoliths 30 cm in diameter. Yield stress is a bulk property caused by the interference of solid particles (and gas bubbles) with liquid flow. Smaller particles, such as phenocrysts, have relatively large domains of liquid surrounding them, and can sink, while large xenoliths remain suspended in the magma. Sparks and others (1977) suggested that q-normative basalts do not carry ultramafic xenoliths because these magmas ascend rapidly and remain near the liquidus temperature, and therefore behave more like viscous fluids. In contrast, according to their view, silica-undersaturated magmas ascend more slowly, lose heat to the wall rocks, and partly crystallize, thereby acquiring the greater yield stress that allows them to carry xenoliths all the way to the surface.

The completely divergent assumptions and conclusions of Kushiro and others (1976) and of Sparks and others (1977) emphasize that there is much more to be learned about the mechanical behavior, as well as the chemistry, of magmas.

An abundance of facts and ideas concerning basaltic magmatism appears in Basaltic Volcanism Study Project (1981).

11.5 SUMMARY

Basalts (and their coarser equivalents, gabbros) contain calcic plagioclase as the essential phase. In addition, most mafic rocks contain clinopyroxene, and many, olivine. Those containing normative *ol* may also be *ne*-normative; those with normative *hy* may also be *q*-normative. The three classes of silica saturation (*q*-, *ne*-, and *ol* + *hy*-normative) provide one basis for basalt classification, but others are in use.

Mid-ocean ridge basalts, usually *ol* + *hy*- or *q*-normative, result from partial fusion of relatively "sterile" mantle under constructive plate boundaries, but few seem to represent primary magmas unmodified since their generation. Ophiolites are assemblages, in ascending order, of sheared ultramafic rock, cumulate ultramafic and mafic rocks, unlayered gabbro, basaltic dikes, and pillow lavas, preserved as fault-bounded fragments at sites of former destructive plate boundaries. They are widely interpreted as slices of oceanic lithosphere generated at ocean ridges or in marginal basins and transported to the edges of continents by seafloor spreading. Oceanic island basalts generally stem from more "fertile" mantle than mid-ocean ridge basalts, and show more compositional variation. Continental flood basalts are thin but widespread flows that issued rapidly from linear vent systems supplied from very deep reservoirs. Layered gabbros are shallow intrusive bodies that record crystal–liquid fractionation, but the actual mechanisms remain unclear and deserve more study. Samples from the moon and from the body or bodies that yielded eucrite meteorites show that the same components and phases, and the same principles, combined under different conditions of total pressure and volatile partial pressures to yield rocks that still closely resemble mafic rocks in the earth. Eclogites have basaltic compositions, but crystallized at such high pressure that plagioclase was unstable; the essential minerals are garnet and Na,Al-rich clinopyroxene.

Compositional variation in mafic rocks can be caused, at the site of fusion, by variation in pressure, amount of liquid, volatile components, and parent composition, and can further be accentuated by processes during ascent, storage, and crystallization of magma.

12

Intermediate and Felsic Silica-Oversaturated Rocks

This chapter describes a broad band within the total compositional spectrum of magmatic rocks. The first group to be discussed is more closely allied to the mafic rocks of the preceding chapter than to the succeeding rock types.

12.1 ANORTHOSITES

In the IUGS (1973) classification, *anorthosite* is defined as plutonic rock in which plagioclase makes up at least 90% of the total feldspar, mafic minerals comprise less than 10% of the rock, and quartz is less than 5% of the felsic minerals. Anorthosite occurs on the earth in two styles. Nearly pure plagioclase rock has formed in some layered cumulates within large gabbroic intrusive bodies such as the Skaergaard pluton. For these layered anorthosite cumulates, the genesis was discussed in Chapter 8. More problematical are those anorthosites forming *massifs* with outcrop areas exceeding 10,000 km². These large irregular plutons generally show evidence of multiple intrusion; are restricted to old, thick, and stable continental crust; and are associated with granites and syenites. Most are Precambrian, falling in the time span of 1.5 to 1.1×10^9 years old (Herz, 1969), but some Cambrian examples, 5 to 6×10^8 years old, are known (Emslie, 1978). Gravity surveys show that some anorthosite massifs (for example, that in the Adirondacks, northeastern New York; Simmons, 1964) lack an underlying mafic or ultramafic cumulate sequence. This absence of denser rocks underneath indicates that the anorthosite had achieved its plagioclase-rich composition before ascending to the crustal level where it now sits. Furthermore, fine-grained dikes of anorthosite (Wiebe, 1979) demonstrate that liquid of anorthosite composition, rather than a suspension of crystals in liquid, was involved in the emplacement of some massifs.

No lava or pyroclastic deposit with the composition of anorthosite has yet been recognized. Several explanations have been offered for the absence of volcanic equivalents to anorthosite. The simplest states that all anorthosites are cumulates, and there is no plagioclase-rich liquid to erupt. Another, also unlikely, postulates that, because all intrusive anorthosites are relatively old, the corresponding volcanic rocks have been eroded or made unrecognizable by weathering or metamorphism. A third explanation claims that anorthositic magma is water-saturated, and therefore cools to its solidus before reaching the surface. The most satisfying to date is that offered by Wiebe (1980); anorthosite liquid is hot and relatively dry, and probably more dense than most crustal rocks. As it invades the crust, partial fusion of crustal rocks occurs (producing the voluminous granitic and syenitic rocks that accompany massif anorthosites). The anorthosite magma loses heat quite efficiently by melting the wall rocks, and is too dense to climb through the overlying more salic melt. It therefore freezes in place, commonly with crystal flotation and near-solidus flow to produce the observed cumulus and mortar textures, without ever reaching shallow crustal levels.

The anorthite content of the plagioclase in anorthosites is generally above 30%, with the cumulates generally containing the more calcic plagioclase. Plagioclase composition tends to remain fairly constant within a single anorthosite body; individual crystals may or may not show zoning. Orthopyroxene, the most common mafic mineral in massif anorthosites, generally has higher Fe/Mg ratios than in more mafic rocks. In some plutons, orthopyroxene contains oriented lamellae of calcic plagioclase. These are interpreted (Emslie, 1975) as products of exsolution from a pyroxene with a significant content of $CaAl_2SiO_6$ component, crystallized from liquid at high pressure, probably within the mantle. Clinopyroxene and olivine, which are also common in anorthosites, also are enriched in iron over magnesium. Ilmenite, magnetite, and apatite are widespread accessory minerals. In some localities, each of these three attains high concentrations within massif anorthosites or associated rocks.

Textures of anorthosites vary from fine granular to coarse or pegmatitic. Diameters of plagioclase and poikilitic pyroxene can exceed 1 m. Mortar texture is common. Plagioclase can be euhedral, and is commonly enclosed or partly surrounded by the mafic minerals (Figure 12-1). The texture suggests that plagioclase plus liquid formed a mush that was disrupted during or shortly after final emplacement.

The system $CaAl_2Si_2O_8$–$NaAlSi_3O_8$ (Figure 9-2) is not a sufficiently close approximation to anorthosites, and systems with additional components corresponding to orthopyroxene, clinopyroxene, and olivine are better representa-

Figure 12-1. Glaciated surface of a large anorthosite xenolith, Kvanefjeld, Ilimaussaq intrusion, southwest Greenland. The hammer handle for scale is approximately 30 cm long.

tions (Figures 12-2 and 12-3). With increasing pressure, with or without water, compositions of the lowest-temperature portions of the liquidus migrate toward more plagioclase-rich compositions. This change in composition is required if anorthosite magma is generated by either fractional crystallization or partial fusion at high pressures, near the base of continental crust. Although liquid rich in normative plagioclase may be generated in the mantle (and probably is, judging from the aluminous orthopyroxenes previously mentioned), anorthosites obviously crystallize within the crust; at depths exceeding 35 km, approximately, the assemblage spinel + orthopyroxene + clinopyrox-

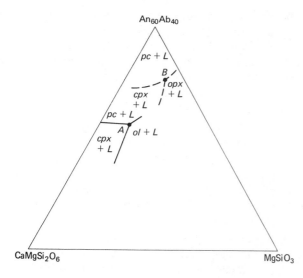

Figure 12-2. Liquidus relations in the system $CaMgSi_2O_6$–$MgSiO_3$–$An_{60}Ab_{40}$ at 1 atm (solid lines) and at 1.5×10^9 Pa (dashed lines). Invariant point A (1 atm) is at $1235 \pm 5°C$, and invariant point B (1.5×10^9 Pa) is at $1390 \pm 20°C$. L, liquid; pc, plagioclase; opx, orthopyroxene; cpx, clinopyroxene; ol, olivine (appearing as a liquidus phase in this system because enstatite melts incongruently at low pressure). (After Emslie, 1971, Fig. 4, with permission of the Carnegie Institution of Washington.)

263

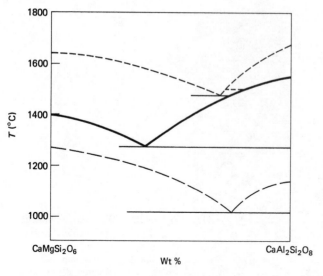

Figure 12-3. Changes in composition at the minimum liquidus temperatures in the system $CaMgSi_2O_6$–$CaAl_2Si_2O_8$ with varying pressure, in the presence and absence of water. Solid lines; anhydrous system at 1 atm. Short dashed lines; dry at 2×10^9 Pa. Long dashed lines; water-saturated at 1×10^9 Pa. (After Yoder, 1969, Fig. 2.)

ene becomes stable relative to the subsolidus assemblages plagioclase + olivine and plagioclase + orthopyroxene.

Some massif anorthosites may have formed by plagioclase accumulation near the tops of large reservoirs (Kushiro, 1980). To provide a large volume of plagioclase, the magma must already have evolved to a high aluminum content by fractionating mafic phases, either during crystallization or fusion. Thick cold crust would promote fractional crystallization by preventing mafic magma from rising rapidly (Emslie, 1978). Alternatively, a highly aluminous parental magma could be generated by partial fusion of q-normative mafic rock within the crust (Simmons and Hanson, 1978) or by partial fusion of eclogite within the mantle.

Anorthosite is much more abundant on the surface of the moon than on the earth; it occurs in the light-colored lunar highlands. In contrast, the darker, low-lying mare basalts are much more restricted in their extent, and are thin, so they make up less than 1% of the mass of the lunar crust (Head, 1976). Older than the mare basalts, and considerably more ravaged by meteorite impacts, the highland rocks are now *polymict breccias* (fragmental rocks composed of pieces of more than one kind of rock, broken and mixed by impact). Most of the fragments in lunar polymict breccias are coarse-grained and rich in extremely calcic plagioclase (An_{95-97}, unzoned) with interstitial olivine and orthopyroxene (each averaging 60 mole % of the magnesium end-member), clinopyroxene, pigeonite, and Mg–Al spinel. Impact shattering (Figure 12-4) and glassy veinlets are common. Highland anorthosites grade into more mafic rocks by the increase in olivine and orthopyroxene at the expense of plagioclase.

The plagioclase-rich lunar crust, exposed in the highlands and underlying a thin veneer of mare basalts elsewhere, extends to a depth of approximately 75 km. Beneath is dense ultramafic rock, postulated to be

Figure 12-4. Photomicrograph, in cross-polarized light, of lunar anorthosite (sample 60025,119). Polysynthetic twinning in plagioclase has been displaced along tiny faults caused by meteorite impact. Long dimension of the photograph is 2.3 mm.

olivine–orthopyroxene cumulate, to a depth of approximately 300 km. Currently under consideration is the hypothesis of a primeval magma ocean, perhaps 300 km deep, that covered the entire moon immediately after accretion and persisted for up to 2×10^8 years (Minear and Fletcher, 1978). This turbulent liquid would convect vigorously with velocities on the order of centimeters per second, allowing complete overturn in approximately 1 year (Herbert and others, 1978). The composition of the magma ocean is at least as uncertain as its existence, but olivine, spinel, and plagioclase are likely to have been the first phases to crystallize, in that order (Longhi, 1978), with orthopyroxene and pigeonite being the last major phases to precipitate. Plagioclase, as single crystals and polycrystalline clumps, is assumed to have floated to the top of the ocean to form "rockbergs" that were swept toward sites under which convection currents descended (Figure 12-5). Olivine, orthopyroxene, and traces of spinel would sink to the bottom of the ocean, forming the ultramafic cumulates. The plagioclase-rich rockbergs rapidly consolidated into an insulat-

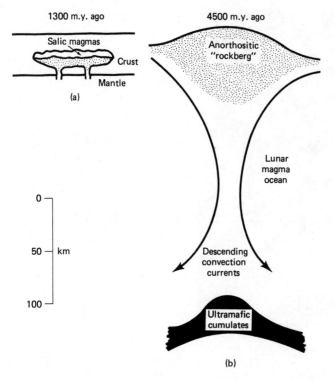

Figure 12-5. Highly schematic cross sections, at the same scale with no vertical exaggeration, showing proposed emplacement mechanisms of (a) terrestrial massif and (b) lunar highlands anorthosites (stippled).

ing crust (Longhi, 1978). Gravitational separation of mafic minerals from plagioclase may have been efficient. The final residual liquid would be trapped at the base of the feldspathic crust, above the ultramafic cumulates. A peculiar but widespread kind of lunar basalt, enriched in potassium, rare earth elements, and phosphorus (and therefore emblazoned with the acronym KREEP) may have originated by partial fusion of this residual layer that was enriched in these large-ion elements. Other mare basalts are likely to have been generated by partial fusion of isolated pods crystallized from magma that was fractionated and trapped within the cumulates, or in the less differentiated lunar material beneath the cumulate sequence.

It may be that the earth also started with a global ocean of magma; if so, the evidence has been completely obliterated or, more likely, has not been recognized because we do not know what to look for. Certainly, the much more rapid increase in pressure with depth in the earth would have led to a different crystallization history than on the moon. Also, the earth is much richer in volatile components; if a terrestrial magma ocean were richer in volatiles, it would also have had a different density profile and solidification behavior.

12.2 THE CALCALKALINE SUITE

The term "calcalkalic" was introduced in Chapter 5 (Figure 5-1) for igneous rocks in which the weight percentages of CaO and (Na_2O + K_2O) became equal somewhere in the range of silica weight percentage from 56 to 61. The term has evolved from that original definition and has more frequently been written "calcalkaline." In the IUGS classification (Figure 5-3), rocks in this broad but generally cohesive suite fall in fields 4, 5, 8*, 9*, 10*, 8, 9, and 10; many granites and rhyolites (field 3) commonly are assigned to the calc-alkaline suite, but many others are not. The course of crystallization of the calcalkaline rocks differs from that of basaltic rocks (as in the Skaergaard pluton, Figure 11-13) in lacking a strong iron-enrichment trend on an AFM diagram; in Figure 4-9, trend *a* is often called tholeiitic, trend *b*, calcalkaline.

Within the limits of the calcalkaline suite flourishes a jungle of nomenclature, replete with synonyms and near-synonyms. The following sections will concentrate less on classification than on general features and the considerable problems of genesis.

12.3 ANDESITES AND DIORITES

Fields 9, 10, 9*, and 10* in the IUGS classification (Figure 5-3) are given the plutonic root names *monzodiorite, diorite, quartz monzodiorite,* and *quartz diorite,* but all are blanketed by one volcanic root name, *andesite.* But this wide scope has not been enough for many geologists, who have tended to call any volcanic rock an andesite if it was not clearly a basalt or rhyolite. Chayes (1969) documents the carelessness with which the term "andesite" has been used, by calculating averages and standard deviations of oxide components and normative constituents for 1775 analyses of Cenozoic rocks called "andesite" (Table 12-1, column 3). Normative quartz averages more than 13 wt % and ranges from zero (*ol*-normative) to 44. Fully half of the samples of "andesite" in Chayes's compilation have been misnamed, and should have been called dacites, latites, or quartz latites, and a few were even rhyolites and basalts.

Any geologist who has attempted to classify a volcanic rock of the calcalkaline suite will be sympathetic toward those who misname a rock on the basis of hand-sample or thin-section examination (there is, however, no excuse for erroneous classification if the sample has been chemically analyzed). All of the textural and mineralogical features of andesites and their kin seem specifically contrived to thwart classification; plagioclase phenocrysts commonly show strong zoning from calcic cores to sodic rims, the groundmass is customarily either glassy or so fine-grained that quartz and alkali feldspar are not easily identifiable, and plagioclase and mafic phenocrysts commonly are altered.

TABLE 12-1 AVERAGE COMPOSITIONS (WT %) OF INTERMEDIATE AND FELSIC SILICA-OVERSATURATED ROCKS

	1	2	3	4	5	6	7	8	9	10	11
SiO_2	57.48	57.94	58.17	62.60	61.25	66.09	65.01	71.30	72.82	58.58	61.21
TiO_2	0.95	0.87	0.80	0.78	0.81	0.54	0.58	0.31	0.28	0.84	0.70
Al_2O_3	16.67	17.02	17.26	15.65	16.01	15.73	15.91	14.32	13.27	16.64	16.96
Fe_2O_3	2.50	3.27	3.07	1.92	3.28	1.38	2.43	1.21	1.48	3.04	2.99
FeO	4.92	4.04	4.17	3.08	2.07	2.73	2.30	1.64	1.11	3.13	2.29
MnO	0.12	0.14	—	0.10	0.09	0.08	0.09	0.05	0.06	0.13	0.15
MgO	3.71	3.33	3.23	2.02	2.22	1.74	1.78	0.71	0.39	1.87	0.93
CaO	6.58	6.79	6.93	4.17	4.34	3.83	4.32	1.84	1.14	3.53	2.34
Na_2O	3.54	3.48	3.21	3.73	3.71	3.75	3.79	3.68	3.55	5.24	5.47
K_2O	1.76	1.62	1.61	4.06	3.87	2.73	2.17	4.07	4.30	4.95	4.98
H_2O^+	1.15	0.83	1.24	0.90	1.09	0.85	0.91	0.64	1.10	0.99	1.15
H_2O^-	0.21	0.34	—	0.19	0.57	0.19	0.28	0.13	0.31	0.23	0.47
P_2O_5	0.29	0.21	0.20	0.25	0.33	0.18	0.15	0.12	0.07	0.29	0.21
CO_2	0.10	0.05	—	0.08	0.19	0.08	0.06	0.05	0.08	0.28	0.09
Total	99.98	99.93	99.89	99.53	99.83	99.90	99.78	100.07	99.96	99.74	99.94
q	10.29	12.39	13.61	13.98	13.06	22.33	22.70	29.07	32.90	0.86	4.95
c	—	—	—	—	—	0.26	—	0.92	1.06	—	—
or	10.40	9.57	9.51	23.99	22.87	16.13	12.82	24.05	25.41	29.25	29.42
ab	29.96	29.45	27.16	31.56	31.39	31.73	32.07	31.14	30.04	44.34	46.29
an	24.40	26.04	27.93	13.97	15.61	17.32	19.99	8.03	4.69	7.27	7.02
di	4.66	4.81	4.15	3.72	2.08	—	0.06	—	—	5.31	2.13
hy	12.60	9.52	9.84	6.19	7.16	7.46	5.83	3.36	1.44	4.18	2.15
mt	3.62	4.74	4.45	2.78	4.62	2.00	3.52	1.75	2.15	4.41	4.34
hm	—	—	—	—	0.10	—	—	—	—	—	—
il	1.80	1.65	1.52	1.48	1.54	1.03	1.10	0.59	0.53	1.60	1.33
ap	0.69	0.50	0.47	0.59	0.78	0.43	0.36	0.28	0.17	0.69	0.50
cc	0.23	0.11	—	0.18	0.43	0.18	0.14	0.11	0.18	0.64	0.20
Diff. index	50.64	51.41	50.29	69.53	67.32	70.19	67.60	84.25	88.34	74.44	80.66
% $an/(an + ab)$	44.89	46.93	50.70	30.68	33.20	35.31	38.40	20.50	13.51	14.08	13.17

Source: By permission of Oxford University Press.

1. Average of 755 diorites (LeMaitre, 1976).
2. Average of 2203 andesites (LeMaitre, 1976).
3. Average of 1775 Cenozoic andesites (Chayes, 1969).
4. Average of 252 monzonites (LeMaitre, 1976).
5. Average of 146 latites (LeMaitre, 1976).

6. Average of 723 granodiorites (LeMaitre, 1976).
7. Average of 578 dacites (LeMaitre, 1976).
8. Average of 2236 granites (LeMaitre, 1976).
9. Average of 554 rhyolites (LeMaitre, 1976).
10. Average of 436 syenites (LeMaitre, 1976).
11. Average of 483 trachytes (LeMaitre, 1976).

1 mm

Figure 12-6. Photomicrograph, in cross-polarized light, of a fritted plagioclase phenocryst in andesite, Cleopatra Wash, Clark County, Nevada.

Andesites are commonly porphyritic; plagioclase is by far the most abundant phenocryst mineral, in many samples approaching 30 vol % of the rock. Zoning of plagioclase is common, and so is "fritting" (plagioclase is riddled with inclusions of glass or groundmass; see Figure 12-6). Among mafic phenocrysts, orthopyroxene, amphibole, and biotite are much more common than in basalts, but olivine and clinopyroxene are also widespread. Magnetite, but not ilmenite, is common as phenocrysts (Ewart, 1976). In most samples, olivine shows reaction rims of calcium-poor pyroxene, as predicted for silica-oversaturated rocks. Hornblende and biotite are commonly oxidized to ferric solid solutions (amphibole to the variety called *oxyhornblende*, which is strongly pleochroic in red-brown). In many samples mafic phenocrysts are surrounded by halos of granular titanomagnetite and partly or wholly replaced by chlorite or by intimately mixed aggregates of clays, carbonates, oxides, sphene, and minor alkali feldspar (Figure 12-7).

The coarser equivalents of andesite (diorite, monzodiorite, quartz diorite, and quartz monzodiorite) may be equigranular rather than porphyritic, but even in these the plagioclase tends to be more euhedral and larger than the other phases (Figure 12-8).

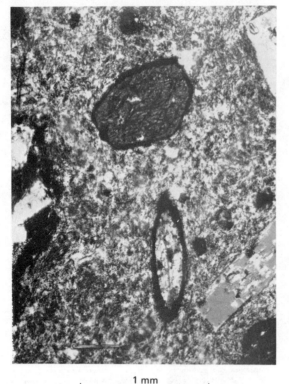

Figure 12-7. Photomicrograph, in plane-polarized light, of altered amphibole phenocrysts in andesite, Cleopatra Wash, Clark County, Nevada. Both grains are surrounded by rims of granular opaque Fe–Ti oxides, and the interior of the less equant grain has been entirely replaced by chlorite.

1 mm

Figure 12-8. Photomicrograph, in cross-polarized light, showing typical texture of a calcalkaline intrusive rock (quartz monzodiorite). Clinopyroxene, quartz, alkali feldspar, and magnetite form a groundmass surrounding large zoned plagioclase.

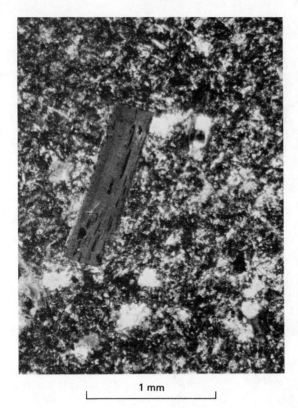

Figure 12-9. Photomicrograph, in cross-polarized light, of quartz latite, Peloncillo Mountains, New Mexico. A biotite phenocryst is set in a groundmass of alkali feldspar, plagioclase, clinopyroxene, opaque oxides, and quartz. Some quartz, probably secondary, forms the clear irregular patches. Subhedral plagioclase and clinopyroxene phenocrysts are also present, but do not appear in this view.

1 mm

12.4 LATITES, DACITES, AND THEIR COARSER EQUIVALENTS

Fields 8 and 8* of the IUGS classification are defined by subequal proportions of plagioclase and alkali feldspar; modal and normative quartz are in the same range as in andesites and their equivalents. As in andesites, plagioclase is commonly the most abundant phenocryst mineral in *latites* and *quartz latites;* alkali feldspar, usually sanidine, may form phenocrysts or may be confined to the groundmass. Quartz rarely forms phenocrysts. Mafic minerals as phenocryst phases include fairly magnesian clinopyroxene and orthopyroxene, hornblende, biotite, and Fe–Ti oxides. The groundmass in latite and quartz latite commonly shows quartz patches (Figure 12-9), giving a misleading impression of the silica content of the rock. The coarser equivalents of latite and quartz latite, respectively *monzonite* and *quartz monzonite,* usually have alkali feldspar and quartz as groundmass phases rather than phenocrysts. Among lavas of the calcalkaline suite, the total percentage of phenocrysts tends to diminish as the silica content increases toward that of rhyolite (Ewart, 1979). No such tendency has been recognized in the coarser equivalents; some quartz monzonites and granites contain 50 vol % phenocrysts.

Fields 4 and 5, with more modal quartz than fields 9* and 10*, hold *granodiorite* and *tonalite*; their fine-grained equivalents are both called *dacite.* In addition to phenocrystic quartz, which is less common than in rhyolites, dacites show phenocrysts of plagioclase (Figure 12-10), pyroxenes, amphibole, and biotite.

271

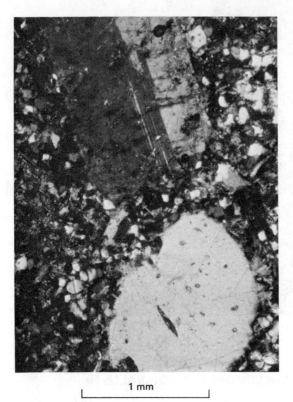

1 mm

Figure 12-10. Photomicrograph, in cross-polarized light, showing phenocrysts of quartz and plagioclase in dacite, Kis Sebes, Romania. The fine holocrystalline groundmass consists mostly of plagioclase, quartz, and amphibole.

12.5 COMPOSITIONS AND PHASE RELATIONS IN THE CALCALKALINE SUITE

Average compositions of some calcalkaline rock types are listed in Table 12-1, columns 1 through 7. These (and compositions 8 through 11, to be discussed later) are projected in Figures 12-11 and 12-12 into the systems $CaAl_2Si_2O_8$–$NaAlSi_3O_8$–$KAlSi_3O_8$ and $NaAlSi_3O_8$–$KAlSi_3O_8$–SiO_2. These two systems form the front and base of the tetrahedron shown schematically in Figure 12-13. Within the tetrahedron, the volume in which plagioclase is the liquidus phase dominates, occupying the top and most of the front space in this orientation. Second largest is the volume in which quartz is the liquidus phase, low in the back of the tetrahedron. Alkali feldspar coexists with liquid alone in a low wedge at the right front. Surfaces where liquidus phase volumes meet represent compositions of liquids coexisting with two crystalline phases (plagioclase + quartz, plagioclase + alkali feldspar, or quartz + alkali feldspar). All three crystalline phases coexist with liquid at equilibrium only if the liquid composition lies on the curve x-y at which the three volumes intersect.

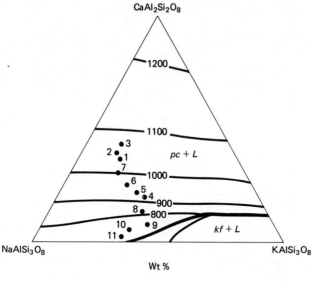

Figure 12-11. Liquidus relations in the system $CaAl_2Si_2O_8$– $NaAlSi_3O_8$– $KAlSi_3O_8$ at $P_{H_2O} = 5 \times 10^8$ Pa, and projected compositions 1 through 11 from Table 12-1. Temperature contours on the liquidus are in degrees Celsius; *pc*, *kf*, and *L* are plagioclase, alkali feldspar, and liquid. (After Yoder and others, 1957, Fig. 41, with permission of the Carnegie Institution of Washington.)

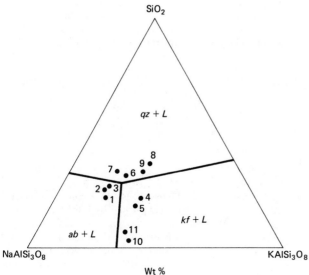

Figure 12-12. Liquidus relations in the system $NaAlSi_3O_8$–$KAlSi_3O_8$–SiO_2 at $P_{H_2O} = 5 \times 10^8$ Pa, and projected compositions 1 through 11 from Table 12-1; *qz*, *kf*, *ab*, and *L* are quartz, potassic alkali feldspar, albite, and liquid. (After Luth and others, *Journal of Geophysical Research, 69,* 759–773, Fig. 2, 1964, copyrighted by the American Geophysical Union.)

Comparison of the projected compositions in Figures 12-11 and 12-12 with the relations in Figure 12-13 is difficult, not only because of problems in perspective but because each composition lies far from the planes into which it is projected (Figure 12-14). The calcalkaline compositions 1 through 7 all contain less than 70% of (quartz + albite + alkali feldspar) and of (plagioclase + alkali feldspar), so their projections have little significance relative to the liquidus boundaries shown on the front and basal planes of the tetrahedron in Figures 12-11 and 12-12. Furthermore, those liquidus relations pertain to a

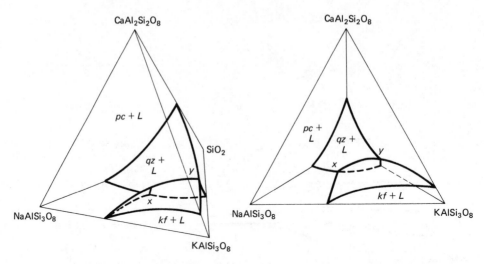

Figure 12-13. Two views of the tetrahedron representing the system $CaAl_2Si_2O_8$–$NaAlSi_3O_8$–$KAlSi_3O_8$–SiO_2, schematically showing liquidus relations at high water pressure. Volumes within the tetrahedron are labeled according to the phase that coexists with liquid under equilibrium conditions; *pc*, plagioclase, *qz*, quartz; *kf*, potassic alkali feldspar; *L*, liquid. This method of depicting phase relations among feldspars and quartz was devised by Carmichael (1963).

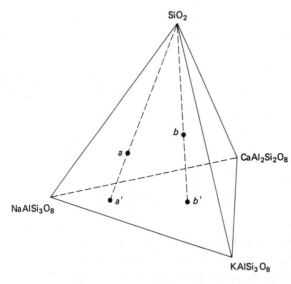

Figure 12-14. Projecting a composition expressed by four components into a three-component system. In this example, *a′* and *b′* are the compositions *a* and *b* projected from quartz onto the feldspar base. Mathematically, the projection is accomplished by ignoring the quartz and recalculating the proportions of the three feldspar components to total 100%.

water pressure of 5×10^8 Pa, chosen to avoid complications introduced by incongruent melting of $KAlSi_3O_8$, and by extensive solid solution between $KAlSi_3O_8$ and $NaAlSi_3O_8$ at higher temperatures and lower water pressures. Changes of water pressure cause the liquidus features to migrate, as discussed in the second paragraph to follow.

274

The average calcalkaline compositions 1 through 7 all fall within the plagioclase volume of the tetrahedron (Figure 12-15). With perfect fractional crystallization, the liquid will follow a curved path away from the plagioclase, which becomes progressively more sodic, moving from a' to b'; in the orientation of Figure 12-15, the liquid will move from a to b, away from the viewer, downward, and slightly to the right, gradually leveling off its trajectory to become more nearly parallel to the base of the tetrahedron as the plagioclase becomes more sodic. Eventually, the liquid will intercept a bounding surface of the plagioclase volume, either that upon which alkali feldspar or quartz begins to crystallize with plagioclase. In the figured example, the liquid reaches the (plagioclase + quartz + liquid) surface at b. Precipitation of quartz and of sodic plagioclase (b' to c') drives the liquid from b to c, where it reaches the cotectic curve x-y, representing equilibrium between quartz, plagioclase, alkali feldspar, and liquid. Alkali feldspar of composition c'' now begins to crystallize with plagioclase c' and quartz. Crystallization of all three now displaces the liquid composition down the curve x-y to the minimum liquidus temperature x, at which quartz, plagioclase x', and alkali feldspar x'' crystallize as the last liquid is consumed.

We should thus expect quartz and alkali feldspar to be more common as phenocrysts in quartz latites, dacites, and rhyolites than in andesites and la-

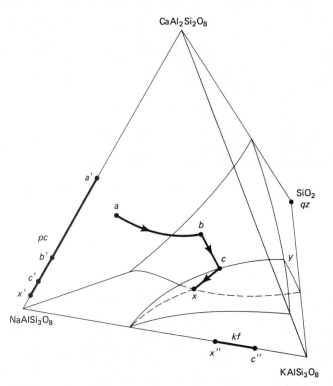

Figure 12-15. Schematic diagram showing compositions of liquids and crystalline phases during fractional crystallization in the system $CaAl_2Si_2O_8$–$NaAlSi_3O_8$–$KAlSi_3O_8$–SiO_2 at high water pressure. Labeled points are discussed in the text.

tites; in general, petrographic observations agree with our expectations. However, variation in P_{H_2O} will move the boundary surfaces, curves, and invariant points within the tetrahedron, and such variation in water pressure is likely in a hydrous magma that is crystallizing large amounts of anhydrous minerals, or that is forming by partial fusion of parent material that is unsaturated with H_2O. As a result, not only does the liquid composition migrate, but the liquidus features may move toward the liquid composition or away from it as crystallization or fusion proceeds. Thus changes in temperature and pressure, as well as the extent and varying efficiency of crystal–liquid separation, have effects that are superimposed on compositional variations produced during magma generation and ascent. All of these effects combine to produce a wide range of rock compositions and textures in the calcalkaline suite.

12.6 BATHOLITHS

Such a compositional diversity is obvious in composite batholiths; many of these trace the margins of continents adjoining the Pacific Ocean. As one example, the Coastal Batholith of Peru (Pitcher, 1978) is 1600 km long and 65 km wide, elongated parallel to the Andean Trench, which marks a destructive plate boundary (Figure 12-16). The batholith was emplaced from about 100 to

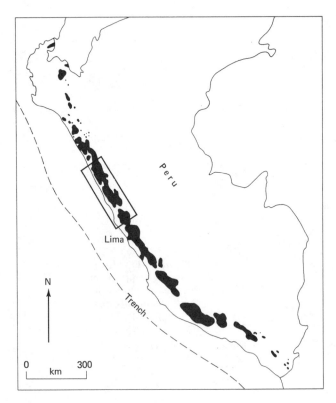

Figure 12-16. The Coastal Batholith of Peru; intrusive rocks in black. The area of Figure 12-17 is outlined by the rectangle. (Simplified from Pitcher, 1978, Fig. 5, with permission of the Geological Society of London and W. S. Pitcher.)

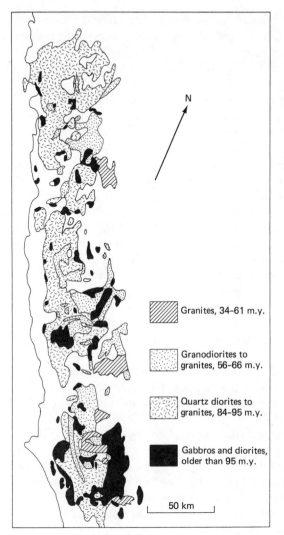

Granites, 34–61 m.y.

Granodiorites to granites, 56–66 m.y.

Quartz diorites to granites, 84–95 m.y.

Gabbros and diorites, older than 95 m.y.

50 km

Figure 12-17. Generalized geologic map of one segment of the Coastal Batholith of Peru, showing four major groups of intrusive rocks. (Simplified from Pitcher, 1978, Fig. 3, with permission of the Geological Society of London and W. S. Pitcher.)

30 million years ago, as at least 800 individual plutons (single pulses of magma). Plutons have nearly vertical sides and flat roofs (Figure 8-9 gives a cross section, and Figure 12-17 is a highly generalized map of one segment). Intrusive contacts "are flagrantly cross-cutting and the plutons so obviously cut out of the crust . . . that the emplacement can only have been accomplished by the breaking away and foundering of blocks of the crust" (Pitcher, 1978, p. 165). Magma rose to less than 8 km from the surface. Gabbro and diorite make up only 7 to 16% of the outcrop area of the Coastal Batholith, but much more probably has been obliterated by younger, more felsic, plutons. There is a general tendency for the younger intrusive rocks to be richer

in silica, but there are cycles of increasing and decreasing silica superimposed on the long-term trend. Younger plutons have successively smaller volumes. Water content increased in each cycle, reaching saturation only in the most silica-rich rocks at the end of a cycle. Atherton (1981) further describes the variations.

Complexities similar to those in the Coastal Batholith of Peru have been found in many other batholiths along the western margins of North America and South America and the eastern edges of Asia and Australia. In the Sierra Nevada Batholith of California, which is composed of a multitude of discrete plutons, there is a general progression with younger, more potassium- and silica-rich bodies toward the east (Bateman and Dodge, 1970). Emplacement style for the Sierra Nevada plutons (Bateman and Chappell, 1979) differs from the shallow cauldron subsidence of the Coastal Batholith of Peru. In the Sierra Nevada, plutons were emplaced forcibly, lifting their roofs and thrusting their walls aside; crystallization then proceeded toward the center of each pluton, but solidification was typically interrupted by a new surge of magma that disrupted the already crystalline portion. Individual plutons are postulated, on gravity and seismic evidence, to retain their identities to depths of at least 20 km (Bateman, 1979) with gradual downward increases in specific gravity and color index. Presnall and Bateman (1973), projecting Sierra Nevada rock compositions in the system $NaAlSi_3O_8$–$CaAl_2Si_2O_8$–$KAlSi_3O_8$–SiO_2 (Figure 12-15), concluded that the variations in magma compositions were caused, not by fractional crystallization, but by equilibrium partial fusion of lower crustal rocks in repeated episodes. Involvement of crustal, or at least lithospheric, material is demonstrated by eastward increase in $^{87}Sr/^{86}Sr$ (Bateman, 1979).

Another pair of parent–daughter isotopes, similar to ^{87}Rb and ^{87}Sr in some ways, is finding increased use. An isotope of the rare earth element samarium, ^{147}Sm, decays to an isotope of another rare earth, neodymium, ^{143}Nd, and has a half-life of 1.06×10^{11} years. Neodymium is present in very low concentrations in rocks, and the radiogenic isotope ^{143}Nd makes up only about 10% of the total Nd. It is therefore necessary, as with strontium, to express the daughter isotope relative to a nonradiogenic isotope, in this case by the ratio $^{143}Nd/^{144}Nd$. Relative to the parent samarium, the earth's mantle has been depleted in neodymium while the crust has been correspondingly enriched. Use of $^{143}Nd/^{144}Nd$ combined with $^{87}Sr/^{86}Sr$ on the same samples gives more precise estimates of the relative contributions from mantle and recycled crustal sources than does the strontium isotopic ratio alone, because the neodymium ratio is much less prone to later alteration by contamination than is the strontium ratio.

DePaolo (1980b) studied the neodymium ratio in the Sierra Nevada and Peninsular Ranges batholiths of California and Baja California, and concludes that the batholithic rocks were derived from mixtures of two source materials, one being the "sterile" mantle of Chapter 10 and the other being old continental crust: "A limited volume of the earth's mantle (approximately 20 to 30 percent) has been repeatedly tapped throughout earth history to produce new continental crust, and the extraction of the crustal material has caused this part of the mantle to become substantially depleted in those elements (including Nd, Sm, Rb, Sr, and K) that show marked enrichment in the crust. This suggests that the average chemical composition of new crustal additions today may differ from the composition of average Archean crust, which would have been derived from mantle which at the time was much less depleted than the present-day upper mantle" (p. 686). The neodymium data support the strontium data in indicating that rocks in the eastern portions of the Sierra Nevada and Peninsular Ranges batholiths contain higher proportions of old crustal components, and those in the western part contain higher proportions of newly added components from the mantle. Circum-Pacific batholiths thus appear to be, compositionally as well as geographically, intermediate between oceanic and ancient continental crust.

In addition to the neodymium data, much evidence is emerging from the Peninsular Ranges Batholith (Silver and others, 1979). A western belt contains a diverse assemblage (from ultramafic to quartz monzonite) emplaced between 120 and 105 million years ago in volcanic wall rocks of Late Jurassic to Early Cretaceous age. This belt is abutted sharply on the east by another, in which plutons of more restricted compositional range (mostly tonalites and granodiorites with much less gabbro, ultramafic rock, and quartz monzonite than the western belt) intruded clastic sedimentary rocks of indeterminate age at deeper crustal levels. Unlike the western belt, the eastern shows an eastward progression of ages, from 105 to 80 million years. Plutonic rocks of the eastern belt have higher Rb, Sr, Ba, Pb, U, Th, rare earth elements, and initial $^{87}Sr/^{86}Sr$ ratios. The western belt formed in a static magmatic arc fed by a source (or sources) containing mantle-inherited characteristics, and the eastern belt formed as a magmatic arc migrated eastward, fed by parent materials that involved upper crustal materials. Silver and others (1979, p. 101) conclude that "high-level assimilation and fractional crystallization cannot explain the batholithic patterns. Different original endowments and isotopic evolutions in a number of different deep source regions are indicated by various types of gradients, steps and domains."

The book edited by Atherton and Tarney (1979) treats the problems of granitic batholiths, using isotopic, trace element, and structural evidence, and is strongly recommended.

12.7 COMPOSITIONAL ASYMMETRY OF MAGMATIC BELTS

Compositional and age asymmetry across batholiths has long been recognized. Moore and others (1963) traced a "quartz diorite line" (Figure 12-18) in western North America; quartz diorite is the most abundant plutonic rock on the

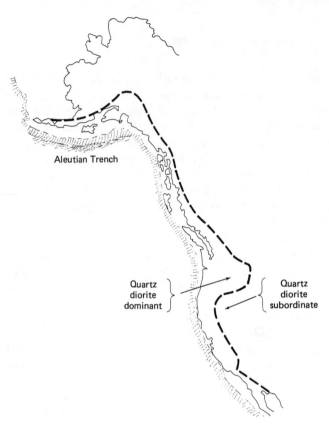

Aleutian Trench

Quartz
diorite
dominant

Quartz
diorite
subordinate

Figure 12-18. The quartz diorite line
in North America. The continental
slope (hachured) and the Aleutian
Trench are also shown. (After Moore
and others, 1963, Fig. 203.2.)

seaward side of the line, and on the landward side granodiorite or quartz
monzonite dominates. The increase in modal K-feldspar toward the continental interior is not paralleled by any significant change in sodium content,
at least in Alaska, the Sierra Nevada, and the Peninsular Ranges; calcium and
magnesium show general declines away from the ocean.

The chemical polarity across some volcanic arcs strongly resembles that
in batholiths, in that potassium abundances increase with time and with distance from the oceanic plate (Jakeš and White, 1972). The landward increase
in potassium in the lavas of some active volcanic arcs led to the hypothesis
(see Dickinson, 1975, and papers cited therein) that increasing potassium content is a function of increasing depth of magma generation. Indeed, for specific volcanic arcs over active Benioff zones, the correlation between potassium
content (at a fixed silica percentage) in lava and the depth to the underlying
seismicity is quite good. However, different volcanic arcs yield different slopes
for the potassium-versus-depth line. There are, moreover, crustal segments in
which the potassium content decreases with greater distance from the plate
boundary (McBirney, 1976; Carr and others, 1979). As a final point of uncertainty, potassium increase, where it does occur, is not uniquely related to
depth to a seismic zone; the extent of magmatic fractionation or assimilation
may be a factor, because increasing potassium also seems to correlate with
increasing crustal thickness.

12.8 SHOSHONITES

Joplin (1968) proposed that potassium-rich quartz-normative rocks (high in potassium because of unusually large proportions of $KAlSi_3O_8$ component in the groundmass, not because of potassic phenocrysts) be included in a *shoshonite* association, linked on the one extreme to calcalkaline rocks and on the other to silica-undersaturated potassic rocks (Chapter 13). Shoshonites have subsequently been accorded separate status by many petrologists, and have been interpreted as products of volcanic arcs farthest inland from plate boundaries. Whatever these rocks signify, the label "shoshonite" serves to obscure their relations to less potassic calcalkaline rocks, and should not be used. Instead, an unusually potassium-rich latite should be called an unusually potassium-rich latite.

12.9 OCEANIC PLAGIOGRANITE AND KERATOPHYRE

At the opposite extreme in terms of potassium content, and perhaps in tectonic setting, is *plagiogranite* (Coleman and Donato, 1979). Composed mostly of quartz and plagioclase (commonly zoned from An_{60} to An_{10}), plagiogranite by definition has no more than a trace of alkali feldspar, and the mafic minerals (amphibole or pyroxenes, commonly altered to chlorite) do not total more than 10 vol %. Much of the plagioclase is intergrown with quartz; the rest may be larger, euhedral to subhedral. Plagiogranite is a late-stage differentiate of q-normative basaltic magma, and occurs in the upper parts of gabbros, and in the sheeted dike complexes, of ophiolites. Generally plagiogranite makes up less than 5 vol % of the associated mafic rocks, and can be distinguished from the quartz–plagioclase rocks of continents by its low K_2O and Rb contents (less than 1 wt % and 5 ppm, respectively). The initial $^{87}Sr/^{86}Sr$ ratio of plagiogranite is usually 0.705 or 0.706, higher than that of the mafic rocks in the ophiolite; the increase is attributed to contamination with old radiogenic strontium from hot seawater that interacted with the plagiogranite. *Keratophyre* ("a fine-grained rock composed essentially of sodic plagioclase and quartz"; Coleman and Donato, 1979, p. 150) is probably the volcanic equivalent of oceanic plagiogranite, occurring as dikes, flows, and lava pillows.

12.10 TRONDHJEMITES

Slightly more potassic than plagiogranite, but still containing less than 2.5 wt % K_2O, *trondhjemite* is another quartz–plagioclase rock of low color index, essentially a leucocratic tonalite. F. Barker (1979) suggests that the definition be extended to include pure albite as the plagioclase. Trondhjemite earns its own name more through its distinctive styles of occurrence than through any intrinsic properties: "The calc–alkaline quartz diorite–tonalite–granodiorite–granite suites that form the bulk of the Mesozoic circum-Pacific batholiths are rather different from trondhjemite–tonalite suites. And, even though trondhjemite–tonalite suites in some cases are found only 5–50 km oceanward of the great batholiths, there apparently is not a compositional continuum between the two" (F. Barker, 1979, p. 7).

Most Proterozoic examples of trondhjemite were emplaced in continental crust close to destructive plate boundaries, and therefore appear to fall in a progression from plagiogranites in ophiolites through trondhjemites to increasingly potassic calcalkaline rocks farther inland. Of potentially greater significance is another style of occurrence that is almost entirely restricted to the Archean; these are bimodal assemblages of trondhjemite ("gray gneiss") and metamorphosed basaltic rocks, without any intermediate compositions (F. Barker and Arth, 1976). Because the formation of bimodal trondhjemite–basalt terrain appears to have been very important in building the nuclei for early continental crust, genesis of trondhjemite in the absence of andesitic magmatism is a significant and perplexing problem, especially because evidence of plate movement and subduction is absent, or unrecognized, in these very old rocks. Barker and Arth suggest that a thick stack of basalts accumulated, and the lower part became *amphibolite* (amphibole-rich metamorphic rock, formed essentially by hydration of basalt or gabbro). Trondhjemite liquid could then be generated by about 15 to 30% partial fusion of amphibolite, with most amphibole remaining in the solid residue. Ascent of trondhjemitic magma, while basaltic magmatism continued, would produce a bimodal suite without andesites (which would form if partial fusion of amphibolite exceeded about 30%).

12.11 GRANITES AND RHYOLITES

By definition, these rocks of fields 2 and 3 in the IUGS classification consist of quartz and alkali feldspar with subordinate sodic plagioclase and mafic minerals. Alumina saturation (Table 5-1) becomes important in granites and rhyolites, because SiO_2 is always very high (typically 70 wt %) and $MgO + FeO$ are low, so aluminum, calcium, and alkali oxides compete for the remainder of the total. The molecular proportions of Al_2O_3, CaO, and $Na_2O + K_2O$ determine which mafic minerals are present. Muscovite and biotite characterize the peraluminous intrusive rocks, in which alumina exceeds the sum of the other three oxides. In the metaluminous examples, where alumina is less than the sum of the other three but greater than the sum of the alkali oxides without calcium, amphibole, and/or biotite, without muscovite, are present. When alkali oxides exceed alumina, in the peralkaline granites and rhyolites, the mafic minerals are distinctive and include sodic clinopyroxenes and amphiboles (generally iron-rich as well as sodic), ex-

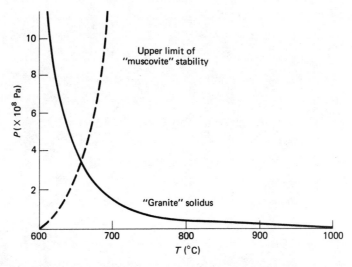

Figure 12-19. Temperature-water pressure projections of the solidus for the minimum melting composition in the system $NaAlSi_3O_8$–$KAlSi_3O_8$–SiO_2–H_2O and an approximate upper stability limit of "muscovite" in igneous rocks. The stability curve is schematic, because it is influenced by the specific reaction followed during breakdown of the mica, by the sodium and fluorine contents of the mica, and by solid solution with Mg-, Fe-, and Si-rich components in the mica (Miller and others, 1981). (After Luth and others, *Journal of Geophysical Research, 69,* 759–773, Fig. 1, 1964, copyrighted by the American Geophysical Union.)

tremely iron-rich olivine, and aenigmatite (a sodium, iron, and titanium silicate that is very dark red or brown, commonly nearly opaque). Metaluminous granites and rhyolites are by far the most common, followed by peraluminous, with peralkaline types a distant third.

Although the most common mafic phases are amphiboles and micas in these rocks, two exceptions must be noted. Muscovite does not occur in lavas, because the high P_{H_2O} required for this mica to crystallize and survive cannot be maintained at the relatively low temperature and pressure of extrusion (Figure 12-19). The second exception is that some granitic rocks, called *charnockites*, contain orthopyroxene as a dominant mafic phase.

Among accessory minerals zircon, sphene, apatite, magnetite, and ilmenite are common. These are also widespread in mafic and intermediate rocks, but are more obvious in granites and rhyolites because of the lower abundance of mafic silicates with high relief and strong colors that tend to mask them in the less felsic rocks. Garnet and cordierite, $(Mg,Fe)_2Al_4Si_5O_{18}$, appear in some peraluminous granites and rhyolites (Clarke, 1981). Other minerals, present because some essential component is enriched in these felsic rocks, include tourmaline, topaz, monazite, cassiterite, allanite, and uraninite.

In granites and in other plutonic rocks such as granodiorite and quartz diorite, quartz forms anhedral grains that commonly are interstitial to all or most other minerals. Plutonic quartz tends to show wavy extinction or to recrystallize into polycrystalline aggregates in which the individual grains have

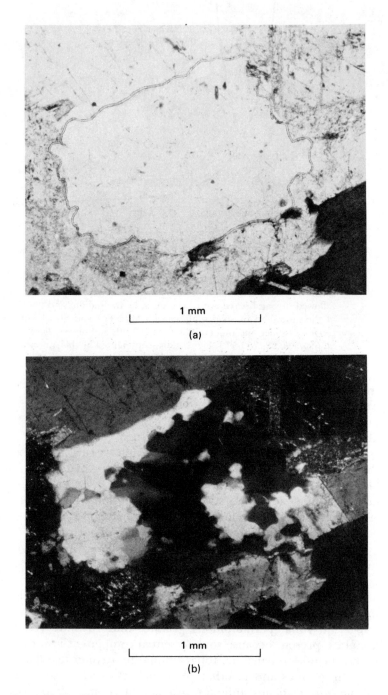

1 mm

(a)

1 mm

(b)

Figure 12-20. Photomicrographs (a) in plane-polarized light and (b) in cross-polarized light of a polycrystalline quartz aggregate in granite. The original grain boundary is outlined in (a).

differing optical orientations and highly irregular or "sutured" boundaries (Figure 12-20). These aggregates represent large single crystals of quartz that originally had the same external outlines as the aggregates but broke up into smaller domains. It is difficult to tell whether the recrystallization was a response to regionally imposed stresses, to annealing during slow cooling, or to volume change during inversion from the beta to alpha quartz polymorph (Figures 12-21 and 12-22). Probably all three causes are effective.

In rhyolites and other volcanic rocks, beta quartz morphology can be preserved in phenocrysts, distinctive in their bipyramidal shapes. Quartz in the groundmass of volcanic and shallow intrusive rocks is almost always anhedral, interstitial, and smaller than feldspars and most mafic silicates.

Alkali feldspars in plutonic rocks tend to exsolve into K-rich and Na-rich phases (Figure 3-17). Unmixing may have been complete, producing separate grains of optically homogeneous albite and K-feldspar (*microcline* or *orthoclase*), or the separation may have halted when the temperature dropped below that at which Na and K ions had sufficient thermal energy to migrate through the feldspar structure. Incomplete exsolution yields lamellae or patches of one phase within the other. *Perthite* contains lamellae of the sodic phase within the potassic host; in *antiperthite* the roles are reversed (Figure 12-23). *Microperthite* and, even worse, *microantiperthite* are used if the lamellae can only be seen with the microscope. *Cryptoperthite* refers to intergrowths that cannot be resolved except by electron microscopy.

In shallow intrusive and volcanic rocks, *sanidine* becomes the dominant K-feldspar. More complete solid solution exists between sodic and potassic

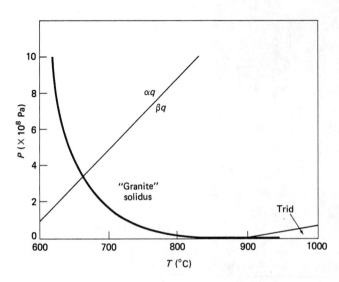

Figure 12-21. Temperature–pressure projections of the alpha quartz to beta quartz inversion, the beta quartz to tridymite inversion, and the solidus in the system $NaAlSi_3O_8$–$KAlSi_3O_8$–SiO_2–H_2O. Alpha quartz crystallizes directly from magmatic liquid only at pressures above approximately 3.5×10^8 Pa; at lower pressures the beta polymorph forms, then inverts to alpha at subsolidus temperatures. Tridymite crystallizes from rhyolitic liquids only at high temperatures and very low pressures, but commonly forms or persists metastably in the groundmass or as a vesicle lining (Figure 8-26). [With permission after Fig. 1 of Luth, 1976, in Bailey, D. K., and Macdonald, R., eds., *The Evolution of the Crystalline Rocks.* Copyright by Academic Press Inc. (London) Ltd.]

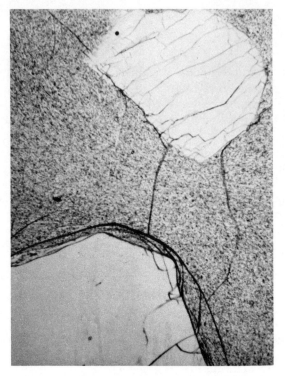

1 mm

Figure 12-22. Photomicrograph, in plane-polarized light, of vitrophyre with quartz and alkali feldspar phenocrysts. The quartz, in the lower left, is surrounded by concentric cracks but the adjacent feldspar shows none. This feature, visible throughout the thin section, is attributed to the volume decrease of approximately 0.8% occurring abruptly at the beta to alpha quartz inversion and a further thermal contraction of approximately 3.8% during cooling of alpha quartz to 20°C (four times the volume change of the alkali feldspar cooling through the same range without inversion). When quartz is firmly "locked" to the surrounding glass, its contraction causes the curving tensile cracks.

0.5 mm

Figure 12-23. Photomicrograph, in cross-polarized light, of antiperthite. Lamellae of microcline are enclosed by the albite host (at extinction).

endmembers at high temperature and low P_{H_2O}, producing *anorthoclase*. Anorthoclase and microcline are triclinic, like albite, and show polysynthetic twinning in contrast to the simple twinning of monoclinic orthoclase and sanidine. In anorthoclase, the twinning forms a grid pattern that is more finely spaced, sharply defined, and regular than that in microcline.

Plagioclase tends to be more euhedral than alkali feldspar in granites and calcalkaline rocks. Mafic silicates also tend to show their own crystal faces against quartz and feldspars, except in the peralkaline rocks, where sodic pyroxenes and amphiboles are commonly anhedral and interstitial even to quartz and alkali feldspar.

Two kinds of microscopic quartz-feldspar intergrowth are widespread and should not be confused. *Granophyre* is an intergrowth of quartz and either alkali feldspar or plagioclase, occurring in the groundmass, and uncommonly as phenocrysts, and indicates rapid and simultaneous crystallization of the two phases from undercooled liquid, vapor, or devitrifying glass, in a near-surface environment. Most granophyre contains roughly equal amounts of normative *q*, *ab*, and *or* (Barker, 1970), and therefore corresponds compositionally to liquid at the minimum temperature on the solidus in the system $NaAlSi_3O_8$–$KAlSi_3O_8$–SiO_2. In recent years, study of ophiolites has revealed a similar intergrowth between quartz and fairly calcic plagioclase in plagiogranites, and the term granophyre should cover these also. The second type of intergrowth, *myrmekite*, contains quartz "worms" in plagioclase rims or patches adjoining more calcic plagioclase cores and frequently found in contact with K-feldspar. Myrmekite rarely forms more than a few percent of the rock, in contrast to granophyre, is common in many metamorphic as well as plutonic (not shallow or extrusive) igneous rocks, and forms by subsolidus reactions (Phillips, 1980). Myrmekite and granophyre are illustrated in Figure 12-24.

The effect of varying P_{H_2O} on phase relations in the system $NaAlSi_3O_8$–$KAlSi_3O_8$–SiO_2 (Figures 12-25 and 12-26) strongly influences the texture and composition of granitic rocks. The liquid coexisting with vapor, quartz, and one or two feldspars becomes more albite-rich as the water pressure increases. Furthermore, increasing P_{H_2O} depresses the solidus until the alkali feldspar solvus is intersected, so that two feldspar phases, one rich in K and the other in Na, can crystallize from the liquid. Tuttle and Bowen (1958) distinguished hypersolvus and subsolvus granites; *hypersolvus* granites only contain plagioclase as lamellae intergrown with K-rich alkali feldspar, not as separate grains. This textural arrangement indicates that the solidus did not intersect the feldspar solvus. In *subsolvus* granites, some plagioclase forms discrete grains outside alkali feldspar, although the latter may be perthitic (Figure 12-27).

Increasing water pressure lowers the solidus but, because water is not a component in alkali feldspar, the solvus is unaffected by the amount of water in the system, but does shift to higher temperatures in response to increase in total pressure. The increase is about 15°C for every 1×10^8 Pa. As a result, in

1 mm

(a)

0.5 mm

(b)

Figure 12-24. Photomicrographs, in cross-polarized light, of quartz-feldspar intergrowths: (a) granophyre surrounding a central plagioclase grain, with alkali feldspar at extinction; (b) myrmekite rim on plagioclase (nearly at extinction).

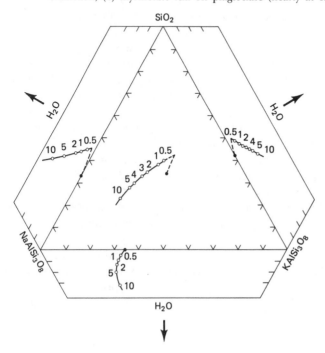

Figure 12-25. Effect of water on the composition of silicate liquid at a eutectic or minimum in the system $NaAlSi_3O_8$–$KAlSi_3O_8$–SiO_2 (in weight percent). Numbers refer to $P_{H_2O} \times 10^8$ Pa. Schematic at water pressures between 0.5×10^8 Pa and 1 atm. [With permission after Fig. 5 of Luth, 1976, in Bailey, D. K., and Macdonald, R., eds, *The Evolution of the Crystalline Rocks*. Copyright by Academic Press Inc. (London) Ltd.]

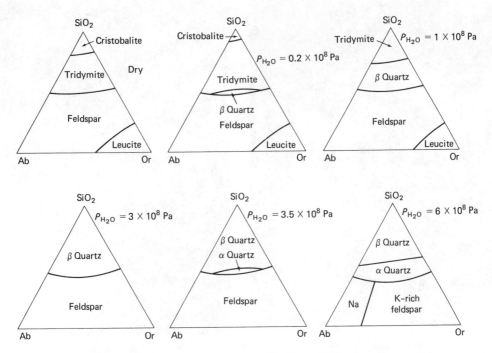

Figure 12-26. Effect of water on liquidus fields in the system $NaAlSi_3O_8$–$KAlSi_3O_8$–SiO_2. (From Tuttle and Bowen, 1958, Figs. 33 and 34.)

the system $NaAlSi_3O_8$–$KAlSi_3O_8$–SiO_2 the solution of more water in the liquid depresses the solidus to intersect the solvus so that two alkali feldspar phases crystallize directly from the liquid (their compositions being closer to end-members at higher water pressure). In dry systems, the solidus is reached at a higher temperature than the crest of the solvus, so one initially homogeneous alkali feldspar crystallizes and, with slow cooling, later unmixes below the solvus to form hypersolvus rocks. The terms "hypersolvus" and "subsolvus" are useful only for rocks with low contents of normative anorthite, because in the system $CaAl_2Si_2O_8$–$NaAlSi_3O_8$–$KAlSi_3O_8$ the solvus surface climbs to higher temperatures as small amounts of anorthite are added (Figure 12-28). Consequently, the feldspar solvus is high enough to intersect the solidus at all values of P_{H_2O} in magmas containing more than about 5 wt % normative anorthite, so hypersolvus rocks are all low in *an*.

The compositions of granites and rhyolites, in terms of *q*, *or*, *ab*, and *an*, cluster near the lowest-temperature portion of the solidus (Figures 7-3, 7-4, 12-11, 12-12, and 12-25). Such compositional constraint can result either from fractional crystallization or from partial fusion, and both origins have been demonstrated for granitic magmas. Fractional crystallization of *q*-normative basaltic magma can produce small amounts of granitic liquid compared to the mass of initial magma; if fractional crystallization is extremely efficient, a

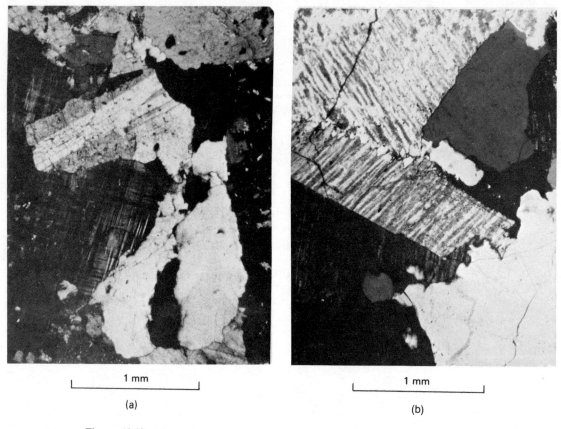

1 mm

(a)

1 mm

(b)

Figure 12-27. Photomicrographs, in cross-polarized light. (a) Subsolvus granite, Hallowell, Maine. The major phases are plagioclase (twinned, upper left quadrant), microcline microperthite (nearly at extinction), and quartz (lower right quadrant). (b) Hypersolvus granite, Agamenticus Complex, Maine. Major phases are microperthite and quartz.

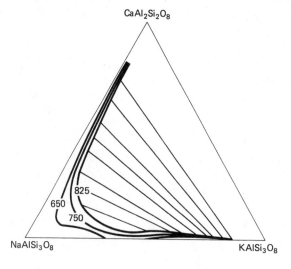

$CaAl_2Si_2O_8$

825

650

750

$NaAlSi_3O_8$

$KAlSi_3O_8$

Figure 12-28. Temperature contours at 650, 750, and 825°C on the ternary feldspar solvus at $P_{H_2O} = 1 \times 10^8$ Pa. Tie lines connect coexisting plagioclase and alkali feldspar compositions at equilibrium at 825°C at this water pressure. The slopes of the tie lines are functions of both temperature and pressure. If the pressure can be confidently estimated, the temperature of equilibration can be calculated from the mole fractions of $NaAlSi_3O_8$ component in the two feldspar phases (Stormer, 1975). (After Seck, 1971, Figs. 5 and 6c, with permission of E. Schweizerbart'sche Verlagsbuchhandlung.)

maximum of perhaps 10 wt % granite can result. The rarity of granitic rocks in oceanic crust (occurring as plagiogranites in ophiolites, and as peralkaline granites and rhyolites on some oceanic islands) conforms to expectation in an environment where their origin by partial fusion is unlikely.

The compositional variations of liquids can be modelled via computer, using techniques devised by Bryan and others (1969) and by Wright and Doherty (1970). Assuming that crystal–liquid separation is the only mechanism that operated, a generalized model for crystal fractionation of minerals X, Y, and Z from a parent liquid to yield a derivative daughter liquid is:

1 mass unit of parent liquid A yields b mass units of daughter liquid $B + y$ mass units of mineral $Y + x$ mass units of mineral $X + z$ mass units of mineral Z

Because mass must be conserved, for each component there is an equation

$$A = bB + yY + xX + zZ$$

where each term represents the number of mass units of each phase times the concentration of the component in that phase. In addition,

$$b + y + x + z = 1$$

The computer program adjusts the mass coefficients (b, y, x, z) to achieve the "best fit" for all components simultaneously. The input data are the compositions of the parent and daughter liquids and of each mineral that is removed from the liquid, and the output consists of the mass coefficients plus statistical estimates describing the closeness of fit of the solution. When this modelling is applied to felsic rocks on oceanic islands (for example, Baitis and Lindstrom, 1980), major element and trace element components agree in estimating crystal–liquid ratios during the assumed progression from basalt through dacite to rhyolite. Unfortunately, in most young volcanoes, and especially on oceanic islands, the degree of dissection by erosion has not advanced far enough to permit an estimate of the relative proportions of basalt, dacite, and rhyolite; such an estimate is needed to compare the actual proportions of liquids with those calculated. Nevertheless, in many instances crystal–liquid fractionation appears to have been capable of causing all the observed variation. This is somewhat surprising, in view of the likelihood that a magma reservoir is an open system, and that other processes beside crystal–liquid fractionation seem inevitable.

Derivation of granitic liquids by partial fusion of continental crustal rocks has long been acknowledged. Nesbitt (1980) suggested that high-grade metamorphic rocks that lack hydrous minerals are the residue after partial fusion of hydrated metamorphic rocks. Compositional differences between the presumptive parents and the residual rocks can be accounted for by extracting liquids of granitic composition, in amounts up to nearly 50% of the original parent rock. Such high degrees of fusion require repeated additions of water from outside the parent rock, probably from dehydrating lower crust that is heated by ascending mafic magma.

Reactions involved in crustal melting have been reconstructed by Lappin and Hollister (1980) in migmatite terrains. *Migmatite* ("mixed rock") contains streaks or networks of felsic material interspersed within a more mafic host of metamorphic rock (Figure 12-29), and is widely believed to represent granitic magma caught in the act of leaving its birthplace, although some migmatites probably form by intimate intrusion of magma that was generated elsewhere. Lappin and Hollister derived three general reactions for partial fusion in their samples:

(a) biotite + plagioclase (An_{28}) + quartz = hornblende + plagioclase (An_{33}) + sphene + granitic liquid

(b) hornblende + plagioclase (An_{40}) + biotite + quartz = more magnesian hornblende + plagioclase (An_{48}) + quartz + granitic liquid

(c) hornblende + quartz = more magnesian hornblende + granitic liquid

Melting stops when quartz or biotite is exhausted in reactions (a) and (b), respectively.

The mineral assemblages involved in these reactions are common in igneous and in highly metamorphosed sedimentary rocks. Chappell and White (1974) coordinate many observed features of granitic rocks in a twofold genetic classification. *S-type* granitic rocks arise from partial fusion of metasedi-

Figure 12-29. Migmatite; more mafic (biotite-rich) borders separate light granite from darker gneiss. Togus, Maine.

mentary parent material, and *I-type* from partial fusion of igneous or meta-igneous parents. S-type rocks are distinguished by their higher initial $^{87}Sr/^{86}Sr$ ratios (generally exceeding 0.708), by lower sodium content, by normative corundum exceeding 1 wt %, by the molecular ratio of alumina to calcium plus alkali oxides exceeding 1.1 and, as modal phases, muscovite, garnet, cordierite, and monazite. I-type granites contain normative diopside or less than 1 wt % normative corundum and show modal hornblende and sphene. "These chemical properties result from the removal of sodium into sea water (or evaporites) during sedimentary fractionation, and calcium into carbonates, with subsequent relative enrichment of the main sedimentary pile in aluminium. S-type granites come from a source that has been subjected to this prior chemical fractionation" (Chappell and White, 1974, p. 173).

I-type granites further differ from S-type in showing a broader range of compositions within a pluton or group of plutons, in lacking foliation, and in thermally metamorphosing the wall rocks at their generally discordant intrusive contacts. Inclusions in I-type granites commonly are hornblende-rich "igneous-looking" rocks, whereas those in S-type granites are highly metamorphosed shales or sandstones. These inclusions in many examples do not correlate with the rock surrounding the granite pluton, and are interpreted by Chappell and White as unfused residue of the parent material, carried up in the granitic magma. Much of the compositional variation in granitic plutons might be caused by varying degrees of separation of liquid from the refractory residue.

In many batholiths and plutonic chains, there is a marked tendency for S-type granites to occur farther inland than I-type (see, for example, Miller and Bradfish, 1980; see also Section 12.7), probably reflecting a greater involvement of ancient continental crust in the generation of S-type magmas. I-type granites, on the other hand, may represent the recycling of volcanic arc materials on continental margins.

Loiselle and Wones (1979) recognize the need for a third category of granitic rocks, the anorogenic or *A-type*. As will be discussed in Chapter 15, these occur in rift zones and in the interiors of stable plates, both continental and oceanic, far from destructive plate boundaries. A-type granites are higher in potassium relative to sodium and in iron relative to magnesium, and are lower in calcium than I-type and lower in aluminum than S-type. A-type rocks generally are true granites (fields 2 and 3 in Figure 5-3), whereas S-type granitic rocks tend to cluster across fields 3 and 4 (granite to granodiorite) and I-type rocks sprawl through fields 4, 5, 9*, and 10* (granodiorite, tonalite, quartz monzonite, and quartz diorite). Initial $^{87}Sr/^{86}Sr$ ratios of A-type overlap those of I- and S-types. Some A-type granites border massif anorthosites and probably formed by fusion of crust by the hot anorthositic magma. A-type granites include the uncommon peralkaline type and are characteristically hypersolvus (because the peralkaline composition precludes any anorthite component in the feldspars, thereby lowering the solvus temperatures). The peralkaline rocks form smaller intrusive and extrusive bodies than do metal-uminous and peraluminous rocks, suggesting that they originate from smaller magma batches. The viscosity of peralkaline liquid is lower than that of other rhyolites (Schmincke, 1974), probably because the superabundance of Na and K over Al produces network modifiers rather than network formers. In addition, peralkaline magmas tend to have lower water contents and oxygen fugacities, but higher fluorine.

12.12 SYENITES AND TRACHYTES

Except for smaller amounts of normative and modal quartz (Table 12-1, compositions 10 and 11), as required by their definitions in fields 6 and 7 of Figure 5-3, *syenites* and *trachytes* have textural and mineralogical features closely resembling those of granites and rhyolites, into which they grade through fields 6* and 7*. Subordinate in abundance, the alkali feldspar-rich rocks have compositions near the $NaAlSi_3O_8$–$KAlSi_3O_8$ join of Figures 7-3, 7-4, 12-11, and 12-12. Their higher solidus and liquidus temperatures, compared to granites and rhyolites, help to explain their infrequent occurrence. Some may result from partial fusion of unusual parent materials (low in quartz), but most are probably the products of extreme differentiation and, perhaps, assimilation. They occur with A-type granites and massif anorthosites in one association, and with silica-undersaturated rocks, described in Chapter 13, in a second association. Syenites and trachytes in either association may be peralkaline, metaluminous, or peraluminous. Modal corundum can occur in peraluminous syenites.

Syenites form small plutons and ring complexes. Trachytes build shield volcanoes of pyroclastic debris and lava flows, the latter showing puzzlingly large variations in apparent viscosity. Trachytes occur on some oceanic islands, associated with *ne*-normative basalts in a bimodal suite containing very little rock of compositions intermediate to the felsic and mafic extremes. For many years this compositional gap has been debated, some petrologists claiming that it merely demonstrates the difficulty of representative sampling on undissected young volcanoes, and others maintaining that the gap is not a result of faulty sampling but indicates that trachyte is far too abundant to be a differentiation product of the exposed basalt. The gap is best portrayed when the number of analyzed samples is plotted against a major element or oxide component or a sum of such components (SiO_2 or differentiation index, for example). Clague (1978) has demonstrated that in the compositional range from basalt to trachyte the major components change abruptly with crystal fractionation, or else change very little (Figure 12-30), because concentrations of some major elements in the fractionated minerals may coincide with those in the coexisting liquids during some stages of the crystal–liquid separation. Trace elements with small crystal–liquid distribution coefficients are much bet-

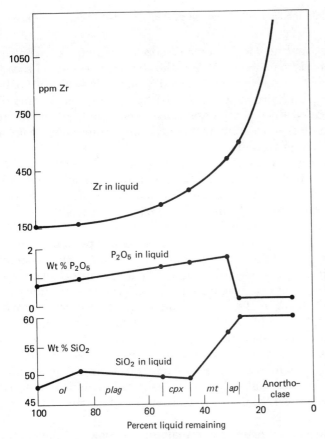

Figure 12-30. Calculated example of fractional crystallization, showing behavior of two oxide components (SiO_2 and P_2O_5) and a trace element (Zr) that does not enter any of the crystallizing phases. Starting with a basalt liquid, the following phases were successively removed: 15% olivine (Fo_{85}), 30% plagioclase (An_{55}), 10% clinopyroxene, 15% magnetite, 3.5% apatite, and 20% anorthoclase. The crystallization interval for each phase is indicated.

ter estimators of the extent of fractionation. In the example of Figure 12-30, silica contents of the rocks representing successive liquids would tend to cluster around 49 and 60 wt %, while samples representing the rapid increase in silica during magnetite and apatite crystallization might not be collected, giving an impression of compositional (and genetic) discontinuity.

The derivation of syenite from a more mafic parent can be documented with special clarity in some plutons with cumulate sequences (for a spectacular example, see Parsons, 1979).

Peralkaline syenites and trachytes commonly show pronounced iron enrichment, especially of ferric iron. The phase relations in systems pertinent to these rocks are reviewed by D. K. Bailey (1976).

12.13 SUMMARY

This chapter considered igneous rocks rich in plagioclase, alkali feldspar, and quartz in varying proportions. Anorthosites (nearly monomineralic plagioclase rocks) occur as layers in cumulates and as large plutons. Most anorthosite massifs are Precambrian, and no Mesozoic or Cenozoic examples are known; neither are lavas or pyroclastic materials of anorthositic composition. The parental magma of anorthosite massifs probably was dry, had relatively high temperature and density, and had already become enriched in the components of plagioclase before it invaded thick continental crust. Anorthosite of the lunar highlands is probably a flotation cumulate formed on a convecting magma ocean very early in the moon's history.

Calcalkaline rocks are rich in modal plagioclase, and contain variable amounts of quartz, alkali feldspars, and mafic minerals. Batholiths and volcanic arcs, parallel to destructive plate boundaries, show diversity of compositions, emplacement mechanisms, and parent materials.

Other feldspar-rich rocks contain more or less potassium than calcalkaline rocks. The more potassic tend to appear over deeper parts of subduction zones, while the less potassic (plagiogranite and trondhjemite) respectively mark fractionation in shallow magma chambers in oceanic crust and intrusion over shallow parts of subducting plates. Trondhjemite forms a bimodal assemblage, with mafic rocks, in very old parts of continental crust.

Granitic rocks contain roughly equal amounts of normative quartz, albite, and orthoclase, and represent the most highly evolved compositions among silica-oversaturated rocks. Such compositional extreme is reached by fractional crystallization of mafic magmas in oceanic and continental settings, and by partial fusion of igneous and sedimentary parents in continental crust, probably most commonly in response to invasion of the crust by mafic magma.

Syenites and trachytes comprise the remaining possible combinations, rich in normative albite and orthoclase without large proportions of quartz, anorthite, or femic constituents. These relatively uncommon rocks occur in a variety of settings, and are associated with granites, anorthosites, gabbros, or silica-undersaturated rocks.

13

Silica-Undersaturated Rocks

This chapter describes igneous rocks that were not covered in Chapter 12 because they lack normative quartz and hypersthene and that were neglected in Chapters 10 and 11 because they are compositionally more extreme, and far less abundant, than peridotites or nepheline-normative basalts. In spite of their low proportions in the earth's crust, these rocks are widespread and significant. The research effort expended by petrologists on silica-undersaturated rocks (which probably form only a few percent of the crust, at most) is reflected in the profusion of rock names that have been proposed for members of this group, nearly a majority of all names applied to igneous rocks. In the biological sciences, much is learned from study of aberrant and pathological conditions. The same is true in petrology, where many magmatic processes can be more clearly traced when they have been carried to unusual extremes than in the more abundant rock types. Silica-undersaturated rocks are the *Drosophila* of igneous petrology.

The terms "alkaline" and "alkalic" are commonly applied, but the usage should not be encouraged for two reasons. First, some of the rocks so indicated are not unusually rich in sodium or potassium. Second, the terms embrace silica-oversaturated rocks that are peralkaline or contain low proportions of calcium, as well as silica-undersaturated rocks. An "alkaline" granite commonly was so named because it contains a blue-green amphibole or because it has little or no plagioclase, but such a rock has scant relation to the ones now to be discussed.

13.1 LAMPROPHYRES

Usually forming small dikes or sills and rarely lava flows, *lamprophyres* as a group are defined by their abundant euhedral or subhedral phenocrysts of biotite and/or amphibole. Phenocrysts of olivine, clinopyroxene, apatite, or opaque oxides may also be present, but felsic minerals do not form phenocrysts in these rocks. The groundmass may be mafic or felsic, or entirely glassy. As a group, lamprophyres are heterogeneous (especially in silica content and in ratio of sodium to potassium). An average composition for one variety of lamprophyre is listed in Table 13-1. Some rocks classified as lamprophyre are silica-oversaturated, but most contain modal feldspathoids in

TABLE 13-1 AVERAGE COMPOSITIONS OF SOME NEPHELINE-NORMATIVE ROCKS

	1	2	3
SiO_2	42.96	54.99	56.19
TiO_2	2.75	0.60	0.62
Al_2O_3	14.82	20.96	19.04
Fe_2O_3	4.64	2.25	2.79
FeO	7.44	2.05	2.03
MnO	—	0.15	0.17
MgO	6.68	0.77	1.07
CaO	10.18	2.31	2.72
Na_2O	3.41	8.23	7.79
K_2O	2.02	5.58	5.24
H_2O^+	—	1.30	1.57
H_2O^-	—	0.17	0.37
P_2O_5	0.71	0.13	0.18
CO_2	2.38	0.20	0.08
Total	97.99	99.69	99.86
or	11.91	32.97	30.96
ab	23.60	29.24	35.43
an	19.20	3.77	1.51
ne	2.84	21.89	16.52
di	9.40	4.54	7.09
ol	12.54	0.65	0.00
mt	6.74	3.26	4.05
il	5.22	1.14	1.18
ap	1.68	0.31	0.43
cc	5.41	0.45	0.18
Diff. index	38.35	82.90	84.09
% *an/(an + ab)*	44.86	11.43	4.10

1. Average of 95 camptonites, a variety of lamprophyre (Rock, 1977, Table VIII, by permission of Elsevier Scientific Publishing Company and N. M. S. Rock).

2. Average of 108 nepheline syenites (LeMaitre, 1976).

3. Average of 320 phonolites (LeMaitre, 1976).

the groundmass and some contain melilite (Section 13.2). High H_2O, CO_2, P_2O_5, and S contents are reflected in the abundant amphibole and biotite, the common presence of calcite and apatite, and abundant but dispersed grains of sulfide minerals. Round aggregates of calcite and feldspathoids or zeolites, up to several centimeters in diameter, are frequently seen strongly contrasting with the dark, fine-grained groundmass, and have been attributed to liquid immiscibility.

Rock (1977) reviewed lamprophyres in detail, concluding that most are derived from nepheline-normative basalt magma unusually rich in water and carbon dioxide. Lamprophyres are typically emplaced late in a magmatic episode and many probably represent shallow crystal–liquid fractionation in "stagnant" volatile-rich mafic magmas. Judging from the phenocryst content, which commonly exceeds 50 vol %, many lamprophyres may be remobilized cumulates. Some, however, are fairly primitive, rather than fractionated, products of a small degree of fusion of mantle unusually enriched in some components, as discussed in Chapter 14. Streckeisen (1980, p. 202–204) presents a classification for these varied and complex rocks.

13.2 MELILITE-BEARING ROCKS

The melilite group consists of solid solutions among the endmembers $Ca_2MgSi_2O_7$, $Ca_2Al_2SiO_7$, and $NaCaAlSi_2O_7$. These can be viewed as high-calcium, low-silica analogues of diopside and plagioclase, just as feldspathoids are low-silica analogues of alkali feldspars. Melilites are not stable at high pressure and have been found only in meteorites and in volcanic and shallow intrusive rocks meeting stringent compositional requirements; silica undersaturation is combined with high color index, and Ca and Mg contents are abnormally high for rocks enriched in Na and K. Except in highly unusual circumstances, melilite does not occur with feldspars. Melilite can form phenocrysts, but more commonly it lurks inconspicuously in the groundmass of dark, olivine-bearing rocks without feldspars. The presence of cs in the CIPW norm of a silica-undersaturated rock is an indication that modal melilite should be found. Yoder (1979) reviews the classification and significance of melilite-bearing rocks, including some lamprophyres.

13.3 NEPHELINE SYENITES AND PHONOLITES

Rocks consisting mostly of alkali feldspars and feldspathoids are the silica-undersaturated analogues of granites and rhyolites (Figures 7-3 and 7-4), falling in fields 11 and 12 of Figure 5-3. Rocks with compositions in fields 6′ and 7′ are less common, contain smaller proportions of feldspathoids, and grade into the silica-saturated and slightly oversaturated syenites and trachytes of

Section 12.12. Table 13-1 lists average compositions of nepheline syenites and phonolites; nepheline-bearing rocks are much more common than the more potassic varieties containing leucite or $KAlSiO_4$ polymorphs, and these potassic types are considered separately in Section 13.5.

Table 13-1 shows that, on the average, normative *ab* + *or* + *ne* constitute more than 80% of the rock, and that Ti and Mg are low. Aluminum content is high, expectably so in a silica-undersaturated rock of low color index, because most other components have been depleted with the exception of alkalis. Consequently, nepheline syenites and phonolites weather to yield the aluminum ore bauxite, and the unweathered rocks have also been used as sources of aluminum.

The system $NaAlSiO_4$–$KAlSiO_4$–SiO_2 (Figure 13-1) embraces the compositions of alkali feldspars and most feldspathoids. *Nephelines* in igneous rocks are solid solutions between $NaAlSiO_4$, $KAlSiO_4$, and SiO_2 falling in the compositional range marked *ne*. *Analcime*, which varies in its Al/Si ratio from the stoichiometric formula of $NaAlSi_2O_6 \cdot H_2O$ but does not tolerate much K in place of Na, projects in the region marked *anl* in the figure. *Leucite* and the polymorphs of $KAlSiO_4$ can contain considerable Na in place of K, however. Other feldspathoids, namely sodalite, cancrinite, nosean, and hauyne, require additional components.

Figure 13-1 also shows the liquidus fields and boundaries for the system at 1 atm. Incongruent melting of $KAlSi_3O_8$, to form leucite plus liquid, dominates the phase relations for potassium-rich compositions, and the leucite field even intrudes upon more sodic compositions so that leucite can form as an

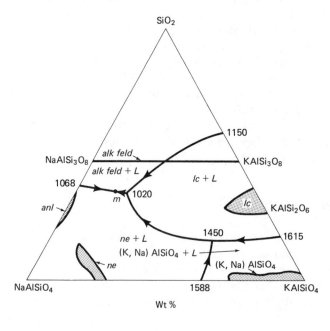

Figure 13-1. The system $NaAlSiO_4$–$KAlSiO_4$–SiO_2, showing liquidus relations at 1 atm. Numbers indicate temperatures in degrees Celsius. Point *m* is the minimum on the nepheline–alkali feldspar cotectic, and lies close to the invariant point at 1020°C where leucite is resorbed. Compositional ranges of alkali feldspars and feldspathoids in igneous rocks are shown along the alkali feldspar join and by the labeled areas. *alk feld*, alkali feldspar; *anl*, analcime; *ne*, nepheline; *lc*, leucite; *L*, liquid. (In part after Schairer, 1957, Fig. 29, with permission of the American Ceramic Society.)

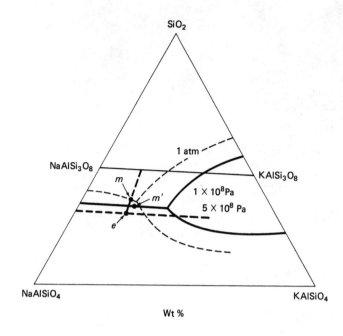

Figure 13-2. Liquidus relations in the system $NaAlSiO_4$–$KAlSiO_4$–SiO_2 at 1 atm (light dashed lines), 1×10^8 Pa P_{H_2O} (heavy continuous lines), and 5×10^8 Pa P_{H_2O} (heavy dashed lines). Point m is the minimum on the nepheline–alkali feldspar cotectic at 1 atm, and m' is the corresponding minimum at a water pressure of 1×10^8 Pa. Point e is the eutectic, at which nepheline, albite, and K-rich alkali feldspar coexist with liquid at a water pressure of 5×10^8 Pa; the two feldspars are present because the solidus has intersected the feldspar solvus. (After J. Gittins in H. S. Yoder, Jr., ed., *The Evolution of the Igneous Rocks: Fiftieth Anniversary Perspectives,* copyright 1979 by Princeton University Press, Fig. 12-2. Reprinted by permission of Princeton University Press.)

early phase in phonolites and shallow nepheline syenites, but commonly is destroyed by reaction with liquid at the ternary invariant point. Addition of water to the system shrinks the leucite field, displaces the minimum and the nepheline + feldspar + liquid cotectic, and eventually causes the solidus to intersect the alkali feldspar solvus (Figure 13-2). Gittins (1979) clearly and thoroughly describes the complexities in the hydrous system, and their effects on observable features in rocks.

As with granites and syenites, nepheline syenites can be divided into sub-solvus and hypersolvus types, depending on water pressure and normative anorthite content (Figure 13-3). Similarly, presence of plagioclase signals per-aluminous or metaluminous compositions while the absence of normative *an* and of modal plagioclase indicates peralkaline composition. Peralkaline nepheline syenites and phonolites are the less common varieties, but they have been more intensively studied because of their unusual mineralogical and economic aspects. Peralkaline silica-undersaturated rocks commonly contain aenigmatite and sodium- and iron-rich clinopyroxenes and amphiboles (as in the silica-oversaturated peralkaline rocks), but in addition can contain unusual sodium–zirconium and sodium–titanium silicates; many examples of these rare rocks are extremely enriched in such trace elements as Zr, Nb, Sn, Be, U, Th, and the rare earth elements. Other minerals prominent in nepheline syenites and phonolites are fayalite, biotite, garnet, sphene, apatite, and zircon. The last need not occur in abundance in the peralkaline varieties, in spite of high Zr contents, because zirconium remains in peralkaline liquid until extreme stages of fractionation are reached and then enters complex

1 mm

(a)

1 mm

(b)

Figure 13-3. Photomicrographs in cross-polarized light. (a) Subsolvus nepheline syenite, Lyndoch Township, Ontario. Anhedral nepheline is entirely altered to layer silicates; plagioclase and microcline make up most of the rock. (b) Hypersolvus nepheline syenite, Granite Mountain, Diablo Plateau, Texas. Anorthoclase cryptoperthite ($An_3Ab_{67}Or_{30}$) forms unoriented subhedral tablets. Interstitial nepheline has largely altered to analcime (dark, center).

silicates. Garnet and biotite tend to be iron-rich, as are olivine, clinopyroxene, and amphiboles.

The chemical and mineralogical features of these felsic feldspathoidal rocks indicate extreme crystal–liquid fractionation, especially in the peralkaline varieties, which may depart significantly from the composition plane shown in Figures 13-1 and 13-2 by enrichment in iron and sodium. To portray the phase relations in simple peralkaline liquids, Bailey and Schairer (1966) investigated the system Na_2O–Al_2O_3–Fe_2O_3–SiO_2. One part of that complex system is shown in Figure 13-4. *Acmite*, $NaFeSi_2O_6$, a major component of clinopyroxenes in peralkaline rocks, melts incongruently to hematite plus liquid at 1 atm. Two ternary eutectic points correspond to peralkaline rhyolites (746°C) and peralkaline phonolites (858°C). These eutectic temperatures are considerably lower than those of the corresponding minima (960 and 1020°C) at 1 atm in the system $NaAlSiO_4$–$KAlSiO_4$–SiO_2 (Figure 7-3), suggesting that "excess" Na (above the molecular proportion of Al) and the ferric iron act as efficient network modifiers. Peralkaline phonolite lavas, like the peralkaline rhyolites, commonly show low viscosities supporting this inference (Figure 13-5).

Although the total mass of feldspathoidal felsic rock in the crust is small, nepheline syenites and phonolites are scattered widely on oceanic and continental crust. On oceanic islands (Borley, 1974), these rocks have formed by fractionation from already *ne*-normative mafic magmas (alkali basalts, nephelinites, basanites, hawaiites, and mugearites mentioned in Section 11.1). On continents, feldspathoidal syenites and phonolites generally form small composite plutons (in which silica-undersaturated and silica-oversaturated rocks may both appear, as in the ring complex of Figure 8-7) and small volcanic piles.

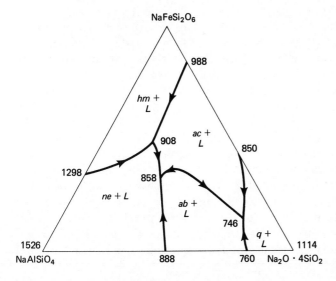

Figure 13-4. Liquidus relations in the system $NaFeSi_2O_6$–$NaAlSiO_4$–$Na_2O \cdot 4SiO_2$ at 1 atm. Numbers are liquidus temperatures in degrees Celsius; *hm*, hematite; *ac*, acmite; *ne*, nepheline; *ab*, albite; *qz*, quartz; *L*, liquid. Acmite melts incongruently to *hm* + *L*; *hm* reacts with *L* at the peritectic point (908°C). Two ternary eutectics, at 858 and 746°C, are separated by a temperature maximum on the *ac* + *ab* + *L* cotectic, where the latter crosses the projected *ac*–*ab* join (not shown, because *ab* does not lie in the plane of the diagram). (After Bailey and Schairer, 1966, Fig. 13, with permission of Oxford University Press.)

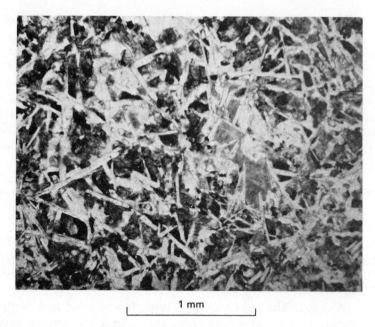

1 mm

Figure 13-5. Photomicrograph, in plane-polarized light, of groundmass, Kapiti phonolite, Sinya, Kenya. Unusually coarse for such a felsic lava, the groundmass contains tabular, unoriented anorthoclase, plus nepheline, glass, aenigmatite, clinopyroxene, and amphibole, all interstitial to the feldspar. Anorthoclase phenocrysts are absent from this view. The coarseness and lack of preferred orientation result from low viscosity of the liquid, which did not appreciably crystallize until after it stopped flowing.

An exceptionally large accumulation of phonolite lavas formed in the Kenya segment of the East African rift in Miocene time (Figure 13-6). According to Lippard (1973b, p. 217), "Kenya phonolites exceed the total volume of phonolite lava found elsewhere in the world by several orders of magnitude." The total volume of phonolite in and around the Kenya rift probably is between 4 and 5 × 10^4 km^3. Individual flows have volumes approaching 300 km^3, larger than those of other felsic lavas such as rhyolites. One Kenya phonolite flow extends for 250 km with an average thickness of only 10 or 15 m, partly filling a river valley with a very gentle gradient (Lippard, 1973a). Goles (1976) and Williams (1978) have addressed the problems of phonolite distribution and genesis. Basaltic lavas and phonolites erupted from fissures before the Kenya rift became well defined by faulting; later, central volcanoes topped by calderas emitted more localized trachytes, rhyolites, and phonolites. Gravity surveys indicate dense rock at a depth of only a few kilometers beneath the floor of the Kenya rift. The high-density rock is interpreted to be gabbroic intrusions or mantle diapirs emplaced under the axis of the rift to act as an enormous reservoir from which the phonolites (and trachytes and

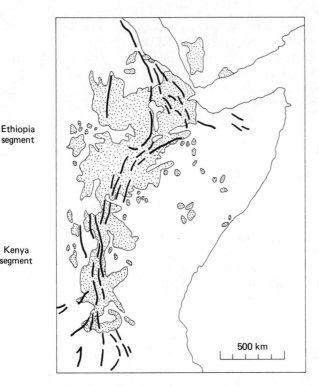

Ethiopia
segment

Kenya
segment

500 km

Figure 13-6. Distribution of Cenozoic volcanic rocks (stippled) and major faults (heavy lines) in the East African rift system. Phonolites are voluminous in the Kenya segment, but not in Ethiopia or adjacent parts of the Arabian peninsula. In these other regions, basaltic rocks abound, with subordinate trachytes and rhyolites (including peralkaline varieties). (Simplified from Baker and others, 1971, Fig. 3, with permission of the Geological Society of America and B. H. Baker.)

rhyolites) were ultimately derived by prolonged fractionation. Baker and others (1978) discuss the differentiation mechanisms and the relations between faulting and volcanism in the southern part of the Kenya rift.

The silica-undersaturated felsic rocks contain iron-bearing silicate and oxide phases that indicate low oxygen fugacities during crystallization (Figure 9-14). Apparently in agreement with low oxygen fugacity, fluid inclusions trapped in growing crystals in feldspathoidal magmas are rich in methane and other hydrocarbons (Konnerup-Madsen and others, 1979; Petersilje and Pripachkin, 1979). The hydrocarbons do not appear to have been derived from sedimentary wall rocks, but to have formed during crystallization and cooling (Gerlach, 1980b).

Aside from crystal fractionation of mafic, nepheline-normative magmas, felsic feldspathoidal rocks have also been postulated to form in other ways, including partial fusion and assimilation. Partial fusion in the crust requires unusual parent materials to generate silica-undersaturated liquids, because the "thermal barrier" separating silica-oversaturated and silica-undersaturated parts of systems (for examples, Figures 7-3 and 13-4) prevents a liquid from migrating across the barrier during either heating or cooling. Therefore, the parent material for a silica-undersaturated liquid, except under very special and unlikely circumstances, must also be silica-undersaturated. Remelting of igneous rocks derived from *ne*-normative mafic magmas is a possibility, but fusion of evaporite-rich sedimentary rocks is unlikely (Barker, 1976) because salt layers will either escape as diapirs before melting begins or will form an immiscible halide-rich liquid coexisting with silica-oversaturated liquid. Furthermore, $^{87}Sr/^{86}Sr$ ratios for silica-undersaturated felsic rocks generally are in the same range as for MORB and other oceanic basalts, strongly implicating derivation from the mantle (Powell and Bell, 1974).

13.4 NEPHELINE–CLINOPYROXENE ROCKS

Combinations of nepheline and clinopyroxene, of widely varying color index, differ from the rocks just described in lacking feldspars. Instead, they may contain garnet, melilite, wollastonite, sphene or perovskite, calcite, zeolites, apatite, and opaque oxides. Almost always they are intrusive, commonly in ring complexes with carbonatites and/or nepheline syenites. Bailey (1974) describes the chemical and classificatory complexities of this group, which boasts such names as *urtite, melteigite, ijolite,* and *jacupirangite.*

13.5 POTASSIUM-RICH ROCKS

Modal leucite is the hallmark of silica-undersaturated rocks in which potassium exceeds sodium. At low water pressures, the liquidus field of leucite stretches over the alkali feldspar join into quartz-normative space in the system $NaAlSiO_4$–$KAlSiO_4$–SiO_2 (Figures 13-1 and 13-2). In full agreement with these experimental observations, we do find rocks with modal leucite but without normative leucite and even with normative quartz (Table 13-2). The incongruent melting of potassium feldspar to leucite plus silica-rich liquid is inhibited by increasing pressure, and K-feldspar begins to melt to a liquid of its own composition at about 2.6×10^8 Pa P_{H_2O} and at about 2×10^9 Pa P_{total} (dry). Leucite itself melts at approximately 8.4×10^8 or 3×10^9 Pa (water-saturated and dry, respectively). In spite of its persistence at moderately high pressures in water-undersaturated experimental systems, natural leucite has been found in volcanics and very shallow intrusive rocks, but so far in only one plutonic body (Brooks and others, 1981). In part, this may be due to a tendency, upon slow cooling, for leucite to break up into intergrowths of nepheline plus alkali feldspar, called *pseudoleucite,* or into intergrowths of $KAlSiO_4$ plus alkali feldspar (Gittins and others, 1980).

In addition to leucite in phenocrysts or groundmass (Figure 13-7), potassium-rich rocks typically contain nepheline and $KAlSiO_4$ polymorphs along with clinopyroxene, phlogopite, or amphibole. Those most deficient in silica contain olivine and/or melilite. Plagioclase is absent in peralkaline as well as melilite-bearing varieties, and K-feldspar may not be a major constituent. Ap-

TABLE 13-2 COMPOSITIONS OF SOME LEUCITE-BEARING ROCKS

	1	2	3	4
SiO_2	40.23	50.90	54.50	55.43
TiO_2	5.02	3.42	0.27	2.64
Al_2O_3	10.94	9.86	21.70	9.73
Fe_2O_3	7.12	2.46	0.80	2.12
FeO	5.67	3.78	1.98	1.48
MnO	0.33	0.09	—	0.08
MgO	6.09	7.95	0.54	6.11
CaO	13.73	4.63	3.20	2.69
Na_2O	4.06	1.65	6.40	0.94
K_2O	4.43	11.61	9.14	12.66
H_2O^+	1.08	1.17	0.89	2.07
H_2O^-	0.15	0.05	—	0.61
P_2O_5	0.89	1.48	—	1.52
CO_2	trace	0.05	—	—
Total	99.77	99.60	99.91	99.75
q	—	—	—	5.53
or	—	51.98	54.00	53.12
ab	—	—	2.13	—
an	—	—	3.50	—
ne	17.13	—	28.18	—
lc	20.52	1.46	—	—
ac	2.41	7.12	—	6.13
ns	—	1.37	—	0.23
ks	—	4.09	—	6.01
di	31.31	10.19	7.66	0.21
wo	—	—	1.39	—
ol	2.20	11.64	—	—
hy	—	—	—	15.12
cs	7.05	—	—	—
mt	4.80	—	1.16	—
il	9.53	6.50	0.51	3.30
tn	—	—	—	2.22
hm	2.97	—	—	—
ap	2.11	3.51	—	3.60
cc	—	0.11	—	—

1. Leucite-bearing lava with phenocrysts of olivine, clinopyroxene, perovskite, and magnetite, Nabugando Crater, Katwe–Kikirongo volcanic field, Uganda (Holmes, 1952, with permission of the Geological Society of Edinburgh). Analysis total includes 0.03 Cl.

2. Average of 11 leucitites from Gaussberg Volcano, Antarctica (Sheraton and Cundari, 1980, with permission of Springer-Verlag). Analysis total includes 0.11 SO_3, 0.33 F, 0.06 Cl.

3. Pumice, A.D. 79 eruption of Vesuvius, Pompeii (Washington, 1917, p. 304–305). Analysis total includes 0.49 Cl.

4. Glassy lava with phenocrysts of leucite, clinopyroxene, and phlogopite, Steamboat Springs, Leucite Hills, Wyoming (Carmichael, 1967, p. 50, with permission of Springer-Verlag). Analysis total includes 0.46 SO_3, 0.28 ZrO_2, 0.02 Cr_2O_3, 0.27 SrO, and 0.64 BaO.

(a)

(b)

Figure 13-7. Photomicrographs, in cross-polarized light, of leucite-bearing rocks. (a) Olivine leucitite, Nabugando Crater, Katwe–Kikirongo volcanic field, Uganda (same rock as number 1 in Table 13-2). An olivine phenocryst is in the center; other, smaller, phenocrysts are clinopyroxene, perovskite and magnetite. Leucite forms equant grains in the groundmass, mingled with the same phases that make the phenocrysts and a little melilite. (b) Leucite phenocryst, showing anomalous birefringence and polysynthetic twinning, in leucite tephrite, Acquapendente, Italy.

atite and perovskite are widespread accessory phases. Generally, the olivine, clinopyroxene, amphibole, and mica have high Mg/Fe ratios, but ferric iron commonly substitutes for aluminum in alkali feldspars and feldspathoids in these rocks, leading to the anomalous situation in which most of the iron in a rock resides in the felsic minerals (for example, at the Leucite Hills, Wyoming; Carmichael, 1967).

The peculiar mineralogy merely reflects the aberrant compositions. In addition to high Mg/Fe, high K/Na, and generally low Al/Si, leucite-bearing rocks typically are enriched in Ti, P, Cr, Ni, Ba, Sr, Rb, Pb, Nb, Zr, Th, and the rare earth elements. The first six elements are preferentially incorporated in crystalline phases rather than liquid, but the other five and the rare earths tend to concentrate in the liquid. The two groups therefore give contradictory testimony for and against a small degree of crystal–liquid fractionation. Sheraton and Cundari (1980) suggest that the discrepancies are resolved if the magma is generated by a small degree of partial fusion from source mantle rich in phlogopite (to yield the K, Mg, Rb, Ba, and Ti) and in apatite (to provide the P, Sr, Pb, Nb, Zr, Th, and rare earths), and subsequently undergoes little fractionation.

Wendlandt and Eggler (1980a, 1980b) have shown with high-pressure experiments that potassium-rich, silica-poor liquids can indeed arise from small amounts of partial fusion of water-poor peridotite containing phlogopite or other K-rich silicates. From the mineral assemblage of some leucite-bearing mafic lavas, Van Kooten (1980) infers that the liquid parental to the lavas was generated by no more than 2.5% melting of clinopyroxene-rich mantle containing phlogopite and garnet at a depth of 100 to 125 km. Current speculation as to how the upper mantle may be locally enriched in phlogopite and apatite is reviewed in Chapter 14.

Leucite-bearing lavas tend to be highly vesicular, but the identities of the volatile components are uncertain. Probably H_2O is subordinate to CO_2. Eruptions are typically explosive (Vesuvius being a notorious example), so pyroclastic deposits commonly overwhelm lava flows from these volcanoes. Small dikes, sills, and diatremes of leucite-bearing rocks are rare. Leucite crystals have such low density that many leucite-rich rocks, especially tuffs, have been suspected to be flotation cumulates that were disrupted during volcanism.

Only one occurrence of leucite-bearing rocks in oceanic crust is authenticated (Gupta and Yagi, 1980). Most are found in thick, cool, and stable continental interiors, as in Wyoming, western Australia, and Uganda. For example, Van Kooten (1980) describes highly potassic mafic lavas, 3.5 million years old, in the Sierra Nevada, clearly unrelated to the much older batholith. The K_2O content in the young mafic suite increases from east to west, the opposite of the trend in the Sierra Nevada batholith rocks.

However, other leucite-bearing rocks are intimately associated with calc-alkaline rocks in volcanic arcs, as in Italy and Indonesia; Brooks and others (1981) cite the literature concerning these, and other occurrences are mentioned in Chapter 15.

13.6 KIMBERLITES

Kimberlites are defined (Meyer, 1979, p. 777; Skinner and Clement, 1979) as ultramafic rocks containing large grains of olivine and generally of at least one of phlogopite, garnet, orthopyroxene, clinopyroxene, and ilmenite, in a groundmass of serpentine, phlogopite, chlorite, calcite, monticellite, magnetite, and perovskite in widely varying proportions (Figure 13-8). Apatite is abundant in some samples. Most or all of the large crystals of orthopyroxene, clinopyroxene, and garnet, and some of the olivine, typically are regarded as

1 mm

(a)

1 mm

(b)

Figure 13-8. Photomicrographs of kimberlite. (a) In plane-polarized light. The rock, from the Sloan diatreme, Colorado, consists largely of calcite and serpentinized olivine. (b) In cross-polarized light. This kimberlite, from the Premier diamond mine, South Africa, shows large crystals of partly serpentinized olivine in a groundmass of serpentine, chlorite, calcite, olivine, and orthopyroxene.

xenocrysts, contaminating grains incorporated from the upper mantle through which the kimberlite ascended. Kimberlites form diatremes (some, at least, flaring downward into dikes) and rarely sills. The largest known kimberlite body has an outcrop area less than 2 km² (Dawson, 1980, p. 31). The largest ones probably are the shallow underpinnings of maars.

Kimberlites occur in stable continental crust, generally far from plate boundaries, as in Australia, India, Siberia, southern Africa, North America, and Brazil. These occurrences were reviewed by Meyer (1977). In a given region, kimberlites commonly have been injected at several widely separated times: in South Africa in Precambrian, Permian–Triassic, and Cretaceous; in the United States, in Silurian, Cretaceous, and Cenozoic; and in southeastern Australia, in Permian, Jurassic, and Late Cenozoic (Ferguson and Sheraton,

1979). This longevity or recurrence of a specific kind of magmatism is a topic for Chapter 15.

Elthon and Ridley (1979) interpret the porphyritic to seriate textures of kimberlites as products of two stages of crystallization. Early precipitation at high pressure yields magnesian ilmenite, olivine, and phlogopite, and the groundmass results from lower-pressure crystallization of perovskite, magnesium- and titanium-rich magnetite, more olivine and phlogopite, monticellite, calcite, apatite, and sulfides. Serpentine and chlorite probably form by hydration at low temperature, perhaps before emplacement of the kimberlite is complete; cataclasis has produced mortar texture (Figure 13-8) in many samples. Metallic iron, iron–nickel alloy, and copper are found, as are iron, nickel, and copper sulfides (Haggerty, 1975). In the South African kimberlite samples studied by Skinner and Clement (1979, p. 136), modal percentage ranges, after subtracting all garnet and orthopyroxene as xenocrysts, are: olivine, 17 to 56; phlogopite, 0 to 67; calcite, 0 to 29; serpentine, 3 to 28; clinopyroxene, 0 to 27; monticellite, 0 to 26; apatite, 0 to 7; perovskite, 0 to 6; and opaque minerals, less than 1 to 20.

In spite of the unusual compositions of kimberlites (phlogopite- and carbonate-rich peridotites), the rocks themselves have attracted less attention than the materials they carry upward, namely diamonds and xenoliths of mantle and lower crust. Ultramafic, mantle-derived fragments are the garnet and spinel peridotites and eclogites described in Chapters 10 and 11. These xenoliths do occur in other rock types (alkali basalts and their kin, plus leucite-bearing mafic lavas, phonolites, melilite-bearing rocks, and carbonatites), but usually much more sparsely. The abundance of xenoliths and xenocrysts in kimberlite, commonly exceeding half the volume, suggests rapid ascent and emplacement. After studying rates of strain and recrystallization of olivine in experiments and comparing these with textures of olivines in xenoliths, Mercier (1979) has calculated that kimberlite magma, rich in CO_2 and H_2O, climbs at an average rate of 40 to 70 km/hr, reaching the earth's surface within 4 to 6 hours after picking up the xenoliths. Mercier further concluded that channels for the ascent of kimberlite must be narrow. Expansion of exsolving gas increases as the confining pressure decreases while the kimberlite travels upward through a gradually narrowing conduit, until the mass may be accelerated to supersonic velocity near the surface and cooled to less than 100°C (McGetchin and Ullrich, 1973). Probably rapid ascent and cooling are essential to prevent diamond from inverting to graphite after the kimberlite crosses into the graphite stability field (Figure 13-9), at a depth that is likely to exceed 150 km.

Most kimberlites do not carry diamonds, and even in those from which diamonds are profitably extracted the concentration rarely exceeds 0.1 ppm (0.25 carats per ton), and even the highest grade is less than 1 ppm (Dawson, 1980, p. 66). Consequently, most diamonds are recovered, not from kimberlite diatremes, but from unconsolidated sediments in which they have locally been

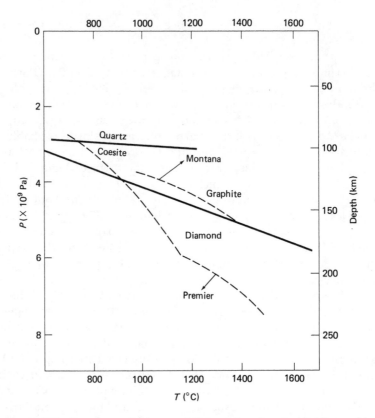

Figure 13-9. Pressure–temperature relations among the graphite-diamond inversion (Kennedy and Kennedy, 1976), the quartz–coesite inversion (Boettcher and Wyllie, 1968), and two geotherms (dashed) calculated for garnet peridotite xenoliths (Chapter 10) from a diamond-free kimberlite in Montana and a diamond-bearing kimberlite from the Premier Mine, South Africa. Coesite occurs as inclusions in diamond but quartz of undoubtedly primary origin does not. (In part from Fig. 7 of Meyer and Tsai, 1979, *Physics and Chemistry of the Earth, 11,* 631–644, copyright 1979, Pergamon Press Ltd., with permission.)

concentrated by their high specific gravity. Diamond-bearing kimberlites are restricted in occurrence, apparently only being found where the crust is so thick and the geothermal gradient so low that the diamond–graphite inversion can occur at shallower depths than those at which the kimberlite magma is generated. In the United States, kimberlites have been found in 15 states (Meyer, 1976), but diamond-bearing ones are known only in Arkansas and in a recently discovered cluster straddling the Colorado–Wyoming boundary (McCallum and others, 1979).

Mineral inclusions in diamonds have been studied in detail; they tend to be euhedral or subhedral single crystals (olivine, orthopyroxene, rare clinopyroxene, garnet, coesite, zircon, rutile, ilmenite, chromite, and iron and nickel sulfides), but some diamonds do contain multiphase assemblages, one resembling the peridotite xenoliths and the other the eclogite xenoliths (Tsai and others, 1979; Meyer and Tsai, 1979). Both assemblages of inclusions occur in diamonds from the same kimberlite, suggesting that diamonds form in both peridotite and eclogite in the mantle.

Differences of interpretation remain concerning the relation of diamonds to kimberlite. Harte and others (1980) point out that eclogite xenoliths more commonly contain diamonds than do peridotite xenoliths, but mineral inclusions in diamonds more commonly resemble the minerals in peridotites than in eclogites. In addition, more diamonds are recovered from the kimberlite matrix than from xenoliths, and the olivine, orthopyroxene, and garnet inclusions in diamonds are richer in Mg and Cr, and poorer in Ca and Al, than the corresponding minerals in the xenoliths. Harte and others suggest that diamonds for the most part are not xenocrysts, but crystallized from kimberlite magma rich in CO_2, generated by approximately 5% partial fusion of garnet peridotite. In contradiction, Meyer (1979, p. 780) concludes that diamonds are accidentally incorporated in kimberlite, which is "only the vehicle which has transported the diamond to the surface." Boyd and Finnerty (1980) have estimated temperatures of equilibration among olivine–garnet pairs included in diamonds; the temperatures are low enough to indicate subsolidus growth of diamonds before their incorporation in kimberlite.

The deep generation of kimberlite magma is a problem; there is nearly universal agreement that volatile components must be important, but little agreement on their specific role. Wyllie (1980) proposes a multistage model. An "unknown mechanism" releases vapor deep in the mantle; the vapor rises, and at about 260 km depth causes partial fusion. The lower density of the partially fused mantle rock allows it to rise as a diapir, which crystallizes at 100 to 80 km depth, releasing CO_2-rich vapor which opens channels to the surface. The reduction in pressure triggers renewed partial fusion, propagating to greater depths, until kimberlitic magma from the diamond-stable region is tapped. "It is time-honored practice for petrologists to push the ultimate origins of magmas to depths greater than the limits of their experimental knowledge" (Wyllie, 1980, p. 6905), or, to paraphrase Robert Browning, a man's reach should exceed his grasp, or what's a lower mantle for?

13.7 CARBONATITES

Carbonatites are igneous rocks containing at least 50 vol % carbonate minerals (Streckeisen, 1980, p. 204). Calcite is the carbonate usually found, but dolomite, ankerite, and siderite also occur (Figure 13-10). In order of decreasing temperature, the carbonates appear to crystallize in the sequence calcite, dolomite, ankerite, and siderite. Other common minerals of carbonatites include clinopyroxene and amphibole (both usually sodic), phlogopite, or biotite, per-

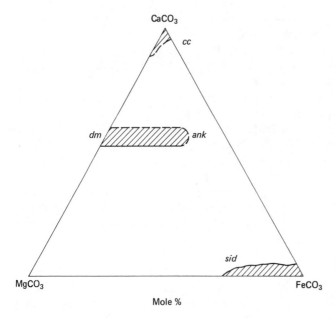

Figure 13-10. Compositions, in mole percent, of carbonate minerals in carbonatites; *cc,* calcite; *dm,* dolomite; *ank,* ankerite; *sid,* siderite. (In part from data in Nash, 1972.)

ovskite, apatite, magnetite, Mn-rich ilmenite, and sulfides (Figure 13-11). Olivine, monticellite, and nepheline appear in some carbonatites. Table 13-3 lists chemical analyses of four samples to show the variations. The CIPW normative calculation was not designed to accommodate such unusual rocks, so only the components are listed.

As a group, carbonatites are highly enriched in Ba, Sr, Nb, rare earth elements, F, Cl, and S. For many years, these high concentrations of elements that are usually present only in trace amounts in most igneous rocks were not considered particularly significant by most petrologists, who thought carbonatites to be remobilized limestones or the products of replacement of silicate rocks by carbonate-rich solutions. Few geologists were willing to imagine a carbonate-rich magma. For one thing, calcite melts at a higher temperature than basalts. Then in 1960 two events combined to cause a reversal of opinion. The first was the eruption of carbonatite lava (number 4 in Table 13-3) from the volcano Oldoinyo Lengai in northern Tanzania. The second was the publication of Wyllie and Tuttle's (1960) experimental studies of the system $CaO-CO_2-H_2O$. Addition of water at 1×10^8 Pa lowers the solidus temperature in the system $CaO-CO_2$ from over 1200°C to about 740°C, that is, into the same temperature range as granitic liquids. Demonstrating that water lowered the solidus of a carbonate melt did not solve the problem of how the melt was generated. Initial $^{87}Sr/^{86}Sr$ ratios for carbonatites (Powell and others, 1966) fall in the same range of low values as do those of MORB, and the same as the silicate rocks associated with carbonatites, averaging 0.7035 with a range from 0.7016 to 0.7057. These low ratios indicate that carbonatites, and

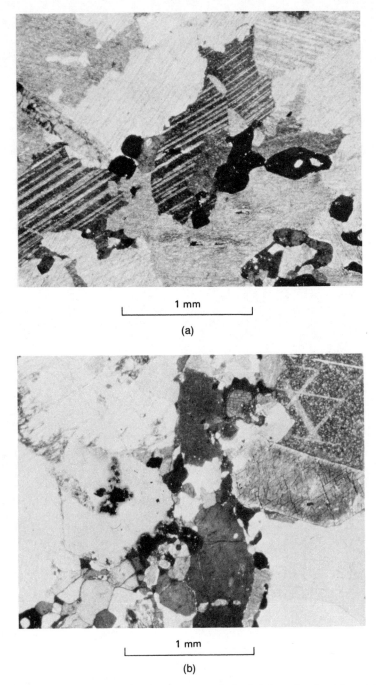

1 mm

(a)

1 mm

(b)

Figure 13-11. Photomicrographs, in cross-polarized light, of carbonatites. (a) Cappelen quarry, Fen complex, Norway; abundant calcite (with polysynthetic twinning) is accompanied by prismatic apatite and equant perovskite. (b) Oka complex, Quebec. In addition to calcite, the view shows biotite enclosing dark pyrochlore, clinopyroxene, apatite (hexagonal prisms in lower left), and monticellite (large gray areas below center).

TABLE 13-3 CHEMICAL ANALYSES OF CARBONATITES (WT %)

	1	2	3	4
SiO_2	3.36	2.22	13.53	0.05
TiO_2	0.30	0.15	1.94	0.03
Al_2O_3	1.69	2.01	2.40	0.17
Fe_2O_3	6.13	1.99	12.96	0.32
FeO	2.99	6.23		
MnO	0.31	0.90	0.45	0.24
MgO	3.10	9.40	8.45	0.26
CaO	44.35	30.24	35.33	14.13
BaO	0.10	5.47	< 0.10	0.90
SrO	—	—	0.68	1.21
Na_2O	0.04	0.26	0.87	32.11
K_2O	0.50	0.31	0.07	5.64
H_2O^+	0.16	0.15	5.83	0.12
H_2O^-	0.14	0.11	1.83	5.49
P_2O_5	3.26	1.00	3.27	1.05
Nb_2O_5	0.80	0.07	—	—
CO_2	32.80	35.96	11.64	34.08
F	0.28	0.15	—	1.49
Cl	0.02	—	—	0.90
SO_3	0.06	2.86	0.28	2.67
S	0.42	0.52	0.17	0.08
Total	100.81	100.00	99.70	100.08

1. Calcite carbonatite, Hydro's quarry, Söve, Fen complex, Norway (Barth and Ramberg, 1966, p. 238, with permission of John Wiley & Sons, Inc.).

2. Ankerite carbonatite (average), Fen complex, Norway (Barth and Ramberg, 1966, p. 238, with permission of John Wiley & Sons, Inc.).

3. Vesicular carbonatite lava, Kalyango, Fort Portal, Uganda (previously unpublished analysis by G. K. Hoops).

4. 1960 lava, Oldoinyo Lengai, Tanzania. Analysis total is corrected for oxygen equivalent to F, Cl, and S. (Gittins and McKie, 1980, Table 1, No. 5, *Lithos, 13,* 213–215, with permission of Universitetsforlaget).

the silicate rocks found with them, are derived from the mantle, and most show little contamination by crust.

Carbonatite magmas could form by partial fusion of mantle unusually rich in CO_2, Ba, Sr, and so on, or by fractional crystallization of CO_2-bearing silicate magmas, or by the splitting of a mantle-derived liquid into two immiscible liquids, one silicate-rich and the other carbonate-rich. Koster van Groos and Wyllie (1973), among others, demonstrated liquid immiscibility between carbonates and silicates. In the system $NaAlSi_3O_8$–$CaAl_2Si_2O_8$–Na_2CO_3–H_2O at 1×10^8 Pa, a peralkaline silica-undersaturated silicate liquid coexists with a calcium-rich carbonate liquid.

The only carbonatite lava seen to erupt, however (that of Oldoinyo Lengai), differed from all intrusive carbonatites in its high alkali content (Table 13-3). It did, however, resemble the carbonate liquid of Koster van

Groos and Wyllie's experiments by coexisting with silica-undersaturated silicate liquids; the cone of Oldoinyo Lengai is largely built of nephelinites and phonolites, and blocks of nepheline–clinopyroxene rocks occur in the carbonatite tuffs that followed the 1960 lava extrusion. Thus there was the question of whether carbonatite magmas in general contain alkalis which escape during crystallization (except at Oldoinyo Lengai), or have very low Na and K contents. Rankin (1977) showed that fluid inclusions, presumed to be trapped magmatic liquid, in apatite crystals from a carbonatite are alkali carbonate-rich aqueous solution, although the carbonatite itself now shows very low concentrations of alkalis and water. In contrast, Nesbitt and Kelly (1977) interpreted inclusions in monticellite from another carbonatite as representing primary liquid with negligible alkalis. Keller (1981) has found a pyroclastic deposit at the Miocene Kaiserstuhl volcano, Germany, containing lapilli that are quenched carbonatite liquid droplets rich in calcite and containing magnetite, apatite, clinopyroxene, and other phases common in the intrusive carbonatites at Kaiserstuhl; these lapilli are very low in alkalis.

Cooper and others (1975) studied the system Na_2CO_3–K_2CO_3–$CaCO_3$ and compared the phase relations with the mineralogy and texture of the Oldoinyo Lengai lavas. The 1960 lava, which was not incandescent when observed at night and therefore cooler than 650°C when extruded (Dawson and others, 1968), contained small phenocrysts of two carbonate solid solutions (one involving Na and K, and the other Na, K, and smaller proportions of Ca) in a groundmass of the same phases plus an opaque phase, probably MnS (Gittins and McKie, 1980). In other carbonatites, the alkalis are in silicates rather than carbonate minerals. Cooper and others concluded that many carbonatite magmas probably did contain higher proportions of alkalis, but these were lost with escaping water and by increasing miscibility of alkali carbonate and silicate liquids with decreasing pressure. The wall rocks surrounding many carbonatite intrusions are enriched in alkalis (Chapter 14) which probably escaped from carbonate and silicate magmas during crystallization.

Experiments by Freestone and Hamilton (1980) using Oldoinyo Lengai lavas only partly confirmed the inferences made by Cooper and others. With increasing P_{CO_2} and decreasing temperature, the immiscibility gap between silicate and carbonate liquids becomes wider and the carbonate liquids become more calcium-rich at the expense of alkalis. Alkali-poor carbonate liquids in these experiments were not immiscible with silicate liquids, even at high pressures, but alkali-rich carbonate liquids were. The carbonate liquid is enriched in Ca, Na, P, F, and Cl and the silicate liquid incorporates more Al, Ti, and Fe. In addition, sulfur is more soluble in the carbonate member of a pair of immiscible liquids in a simple system (Helz and Wyllie, 1979).

Besides very rare lavas and pyroclastic deposits, carbonatites form small intrusive bodies, usually ring complexes, intimately associated with feldspathoidal syenites and with nepheline–clinopyroxene rocks. Many such complexes, such as the ones at Magnet Cove in Arkansas and Kaiserstuhl,

Germany, are the eroded cores of volcanoes that erupted silica-undersaturated mafic to felsic lavas (nephelinites to phonolites), usually with very little if any carbonatite preserved as extrusive or pyroclastic rock. The number of known carbonatite bodies rapidly increased to several hundred during the 1960s, as they became high-priority targets of economic evaluation because of their high concentrations of niobium and rare earth elements (used as components of the phosphors in color television picture tubes, and for other electronic products), and of gold, copper, uranium, agricultural lime, and phosphate.

Although there are calcite-rich kimberlites, these and carbonatites do not seem to be genetically related (Mitchell, 1979); the two rock types are not found in the same intrusions, they share few minerals with the same compositions, and kimberlites do not occur, as do carbonatites, in ring complexes with feldspathoidal rocks. As Mitchell points out, kimberlites and carbonatites may be contemporaneous in a given region, because both magmas may use the same structural weaknesses in traversing the crust. Similarly to kimberlites, carbonatites tend to recur; in northern Ontario and western Quebec (Erdosh, 1979), some 50 carbonatites are known, and these yield ages that cluster at 1700, 1100, 570, and 120 million years.

13.8 RARE EARTH ELEMENTS IN CARBONATITES AND OTHER IGNEOUS ROCKS

One topic could have been introduced at several points in this or the preceding chapters; it was deferred until now because it provides perspective over all the rock types and processes. The rare earth elements form relatively large trivalent cations (Table 4-7), dispersed in very low concentrations, usually hundreds of ppm at most, but relatively immune to redistribution during weathering, alteration, or metamorphism. Furthermore, they can be analyzed with great precision. The distribution coefficients (Section 4.5) for rare earth elements in general are small, favoring concentration in the liquid rather than crystalline phases. However, the liquid composition and structure do strongly affect the distribution coefficients.

The generally accepted way to show rare earth concentrations is portrayed in Figures 13-12 and 13-13; the horizontal scale is the atomic number (from 57 to 71), and the vertical scale shows the ratio of the concentration of each rare earth element to its concentration in the average chondritic meteorite, on a logarithmic scale. It must be pointed out that there is not universal agreement on the rare earth element concentrations in an "average" chondritic meteorite. Such a plot (or rare earth "profile," in which the sample/chondrite ratios are connected by straight line segments) smooths out the variations in absolute abundances; in general, the rare earth elements of lower atomic number are more abundant, and those with even atomic numbers are

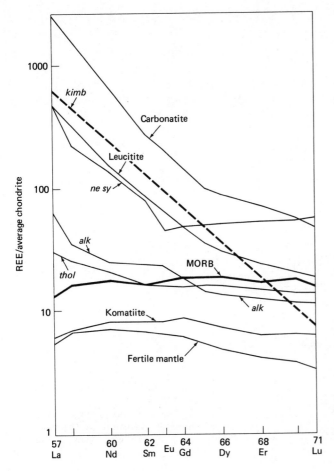

Figure 13-12. Rare earth element profiles, generalized from many sources, for fertile mantle, komatiite, mid-ocean ridge basalt, tholeiite, alkali basalt, nepheline syenite, leucitite, kimberlite and carbonatite. Horizontal scale is atomic number. Vertical scale (logarithmic) is the ratio of each rare earth element concentration to that in an average of chondritic meteorites.

more abundant than those with the next lower and higher odd numbers. Normalization to chondrites (or to any other reference material, for that matter) smooths this sloping sawtooth pattern. A rare earth profile has three significant features: the intercepts, slope, and inflections.

The intercept, or degree of enrichment of each element relative to chondritic meteorites, for these elements of small distribution coefficients reflects the degree of partial fusion or the extent of crystal–liquid fractionation. Higher concentrations of rare earth elements imply smaller degrees of fusion or higher degrees of fractionation. The slope of the profile reflects the nature of the solid phases and the components in a vapor phase. Rare earth elements are not all accepted or rejected equally by crystalline phases; zircon, amphibole, and most of all, garnet preferentially accept those of higher atomic number, and apatite incorporates more of the lighter rare earths. Wendlandt and Harrison (1979) demonstrated experimentally that a vapor phase with a

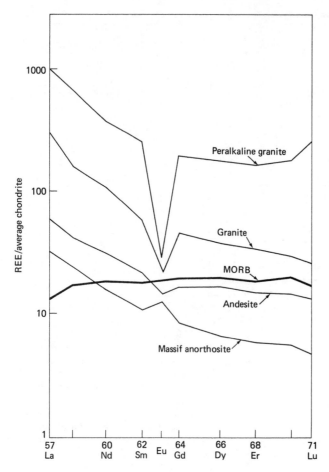

Figure 13-13. Rare earth element profiles for massif anorthosite, andesite, mid-ocean ridge basalt, granite, and peralkaline granite, generalized from many sources. Coordinates are the same as in Figure 13-12.

high $CO_2/(CO_2 + H_2O)$ ratio becomes enriched in all rare earth elements, but especially those of lower atomic number, relative to more H_2O-rich vapor. Presumably the rare earth elements form carbonate complexes. A steep negative slope for the rare earth profile therefore implies either the persistence of amphibole or garnet as a solid phase during fusion or fractional crystallization, or the intervention at some stage of a vapor with a high CO_2/H_2O ratio.

Wendlandt and Harrison also found that, in a pair of immiscible liquids, the carbonate liquid is enriched two or three times in light rare earths, five to eight times in heavy rare earths, compared to the coexisting silicate liquid.

The most prominent (and best understood) inflection in rare earth profiles occurs at europium (Eu), because that is the only rare earth element to form significant amounts of divalent cations. Eu^{2+} is preferentially incorporated in feldspars (by a factor of about 10, relative to the other rare earth elements), substituting for calcium. Dips in the profile, or negative europium anomalies, reflect fractionation of feldspar from liquid, or the persistence of feldspar as a residual solid phase during partial fusion, and a positive europium anomaly suggests a feldspar cumulate origin.

Figures 13-12 and 13-13 give schematic rare earth profiles for some igneous rock types. These were constructed from many published sources, but are not averages; they do show, with some editorial license, the major features that distinguish some rock types on such a plot. Samples of a given rock type generally vary by a factor of 3 or more in their rare earth concentrations, but show parallel slopes and the same inflections. In Figure 13-12, the profile for komatiite is closest to that for fertile mantle (implying a higher degree of partial fusion) and shows an approximately horizontal slope. The profile for MORB (shown as a reference line in both figures) is also approximately horizontal; the lack of depletion in the heavier rare earths suggests the absence of garnet in the parent material, in agreement with the presumably shallow source of magma under constructive plate boundaries. Tholeiitic and alkali basalts show progressively greater enrichment in light, and depletion in heavy, rare earths. At present it is impossible to decide whether these differences are due mostly to increasing depth of magma generation, decreasing degree of partial fusion, or greater participation of CO_2; all three are likely factors. Kimberlite and the strongly silica-undersaturated rock types (Figure 13-12) show steeper negative slopes and pronounced enrichments in light rare earths. Nepheline syenites commonly show a negative europium anomaly, implying feldspar fractionation, and share with carbonatites a great enrichment in heavy rare earths. Silica-oversaturated felsic rocks (Figure 13-13) show profiles that generally support the genetic inferences already discussed. Massif anorthosites show europium enrichment imposed on a trend of low concentrations, indicating plagioclase accumulation in a "primitive" mafic magma. Andesites show profiles that usually resemble those of quartz-normative basalts (tholeiites), with which they are commonly but not universally associated. Granites show general enrichment, with a negative europium anomaly reflecting plagioclase removal, and peralkaline granites show a more pronounced europium anomaly combined with enrichment in heavy rare earths, probably caused by fluorine or CO_2, as in nepheline syenites and carbonatites.

More complete discussion of rare earth behavior is given by Yoder (1976, p. 150–161), and in valuable reviews by Frey (1979), Haskin (1979), Hanson (1978), and Irving (1978). Numerical modelling that correlates degree of rare earth enrichment or depletion with liquid/crystal ratios depends critically on accurate data for distribution coefficients, and these are still poorly known for many combinations of minerals and liquids. In addition, we need more rare earth profiles of separate phases, in addition to whole-rock concentrations (for a good example, see Larsen, 1979). The same plotting strategy, normalizing with respect to chondrites, has successfully been used for the transition metals (Ti, V, Cr, Mn, Fe, Co, Ni, Cu, and Zn) that are preferentially incorporated by crystalline phases (Wass, 1980).

13.9 SUMMARY

Although not as abundant as the igneous rocks described in Chapters 10 through 12, silica-undersaturated rocks with more extreme compositions than alkali basalts are widespread and provide several economically important resources. Compositionally diverse, the rocks range from ultramafic to extremely felsic. Low initial $^{87}Sr/^{86}Sr$ ratios indicate mantle origins for the parental magmas. Leucite-bearing rocks, kimberlites, and carbonatites may have undergone little modification after their generation by small degrees of partial fusion in the mantle, probably at greater depths than those at which basaltic liquids are generated, and probably from mantle materials that were enriched in potassium, carbon dioxide, water, and certain trace elements before fusion. In contrast, some lamprophyres and feldspathoidal syenites and phonolites have compositions, in terms of major and trace elements, that were extensively modified from the compositions of the parent magmas. Incongruent melting of K-feldspar and of acmitic clinopyroxene strongly influenced the mineral assemblages of volcanic versus plutonic rocks. Most, if not all, strongly silica-undersaturated rocks show evidence for the intervention of CO_2 during fusion or final consolidation. The rare earth elements, preferentially held in liquid relative to most crystalline phases, are distributed in silica-undersaturated and other igneous rocks in ways that seem to confirm petrogenetic theory.

14

Metasomatism

14.1 SCALE AND DEFINITION OF METASOMATISM

Bodies of magma commonly have compositions and temperatures that strongly contrast with those of the material they intrude. The chemical and thermal imbalance causes exchange of components across the intrusive contact. *Metasomatism* (literally, change in body or substance) is change in rock composition caused by processes acting at temperatures below the solidus of the rock. As the composition changes, the solidus temperature also changes; metasomatic addition of certain components may lower the solidus sufficiently to promote partial fusion.

Metasomatism can occur far from magmatic bodies, by movement of components from one rock unit to another of different composition during metamorphism. Gains and losses of components are achieved by diffusion and by percolation of fluids (liquid and gas) through the rocks. As a consequence of compositional change, texture is also modified. Metasomatism is an important topic in the study of igneous rocks because the process can affect the wall rocks adjacent to magmatic bodies, the igneous rocks themselves, and perhaps most importantly, the parent materials of magma in crust or mantle, preceding or triggering partial fusion.

Scale is critical in evaluating metasomatic changes. Usually, we assume that components have entered or left a system that is defined as the rock under consideration, but that system may range in size from a thin section to an outcrop or a map unit. For example, growth of an albite crystal within 1 cm³ of rock may profoundly change the composition of that cubic centimeter, but the components that built the albite crystal may all have been drawn from the immediate vicinity, and the composition of the cubic meter of rock surrounding the albite crystal may remain unchanged.

Closely linked to the problem of scale is the difficulty of expressing compositional change (observed composition of a sample, compared with the inferred initial composition or the observed composition of another sample that is believed to have undergone no metasomatism). Because we usually deal in weight percentages of oxide components, and these must sum to unity, a gain in one component automatically results in apparent loss of the others simply by their dilution. To identify the components that have actually entered the system and those that have left, we need a different frame of reference.

Structural and textural studies of some metasomatized rocks indicate that compositional change can be accomplished without change in volume. For example, the thickness of a sedimentary layer may remain constant as an intrusive contact is approached, although the composition of the layer may increasingly depart from that shown by the same layer far from the magmatic body that acted as a source of added components and a sink for extracted components. In such a case, compositional changes can be expressed as gains and losses of components in terms of mass per volume (g/cm^3 or $moles/cm^3$). In other metasomatic rocks, there is textural and structural evidence that volume, and therefore density, have changed. Because most rocks can be imagined as arrays of large and abundant oxygen anions with smaller cations between them, one way to express compositional change is in numbers of cations gained or lost relative to a fixed number of oxygen anions; another is to recalculate the rock compositions to a constant density. Appleyard and Wooley (1979) evaluate the merits of these strategies, which vary with individual problems.

14.2 INFILTRATION METASOMATISM

When the medium that transports components into and out of the system is itself in motion, percolating through the rocks, the process causing compositional change is called *infiltration metasomatism*. Its efficiency is controlled by the permeabilities of the rocks. One common kind of infiltration metasomatism results from interaction between cooling intrusive rock and groundwater vigorously convecting through the intrusive and wall rocks. It is possible to estimate the ratio of surface-derived groundwater that has reacted with a given mass of intrusive rock, using isotopic compositions of oxygen and hydrogen in the metasomatized rock. Water that has evaporated and precipitated at low temperatures in the surface hydrologic cycle becomes enriched in ^{18}O relative to ^{16}O and in deuterium relative to hydrogen. In contrast, fractionation of these isotopes is very much less during high-temperature magmatic processes. Consequently, unaltered rocks that crystallized from magmas originating in the lower crust or upper mantle, well out of reach of percolating meteoric water, show much lower ratios of $^{18}O/^{16}O$ and D/H than do surface and groundwaters and sediments. By reaction with the percolating groundwater, hydrogen and oxygen in the igneous rocks can be replaced by isotopically heavier hydrogen and oxygen. Measurement of oxygen and hydrogen isotopic ratios in the altered rocks therefore permits estimates of the relative masses of water and rock involved in the reaction (for example, see Taylor, 1978). Water/rock ratios thus calculated reflect permeability and temperature, and can exceed 1 g of water per gram of rock. Igneous rocks can be sources, as well as passive sinks, for water; Ferry (1978) demonstrated that H_2O moved out of granitic magma and CO_2 moved in from carbonate wall rocks.

14.3 DIFFUSION METASOMATISM

At the other extreme is the concept of *diffusion metasomatism,* wherein the transporting medium is motionless, and components diffuse into and out of the system through this static fluid. Diffusion has a broader variety of paths to follow through a rock than does percolation; in descending scale these are joints and other macroscopic fractures, grain boundaries between crystals of the same phase or different phases, cleavages and other microscopic cracks, and lattice dislocations. Relative ease of diffusion along these paths decreases in that same order. At low temperature, diffusion along fractures and grain boundaries will dominate, but at higher temperatures diffusion within crystals becomes more efficient than diffusion around them, owing to the much larger number of possible paths offered by microscopic cracks and lattice dislocations (Figure 14-1).

For pervasive changes in the mineralogy of a rock by metasomatism, either all phases must dissolve and reprecipitate or diffusion must take place within crystals, even though the components may be carried to and from the crystal perimeter by a moving fluid. The persistence of concentric compositional zoning in plagioclase in samples that are 10^9 years old indicates that diffusion within crystals is severely limited at low temperature. Compared to diffusion rates in silicate liquids at the same temperature, diffusion rates in aqueous solutions or supercritical fluids are faster, and those in solid crystals are several orders of magnitude slower. For example, in one experimentally studied alkali feldspar at 800°C, sodium would diffuse 80 cm in a million years, whereas potassium would move only 2.5 cm (Fletcher and Hofmann, 1974).

Rates of diffusion of components through aqueous solutions and crystals increase with increasing temperature, decrease with increasing ionic radius,

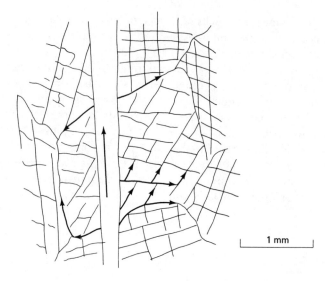

Figure 14-1. Diffusion paths along open channels and grain boundaries, and within crystals. (Suggested by Manning, 1974, Fig. 1, used with permission of the Carnegie Institution of Washington.)

1 mm

and generally increase with increasing ionic charge, although formation of coordination polyhedra (such as SiO_4^{4-} tetrahedra) can cause small, highly charged cations such as silicon to move more slowly through crystals than some larger ions with smaller charges. It is the relative immobility of Al and Si in their tetrahedral sites that ensures the survival of zoning in plagioclase, because Al and Si must migrate to balance charges if Ca and Na move. For some components, increasing pressure diminishes the diffusion rate, but it is not clear whether this is a general rule.

An example of metasomatic exchange between a crystalline phase and an aqueous fluid was studied by Orville (1963). For alkali feldspar, the exchange reaction

$$KAlSi_3O_8 + Na^+ = NaAlSi_3O_8 + K^+$$

proceeds with relative ease in either direction. K-feldspar immersed in sodium chloride solution (or in the molten salt) rapidly converts to albite without change in external crystallographic form or internal symmetry, because only the alkali cations move without disturbing the SiO_4^{4-} and AlO_4^{5-} tetrahedra. Conversely, albite surrounded by potassium-rich solution or melt is rapidly changed to K-feldspar. The rate of exchange increases at higher temperature, expectably, but the compositions of the alkali feldspar and coexisting solution or melt are also influenced by temperature. At higher temperatures, the solution in equilibrium with alkali feldspar becomes richer in potassium relative to sodium, and the feldspar becomes richer in sodium as a consequence. In a thermal gradient, sodium migrates toward the hotter portion of a static fluid coexisting with alkali feldspar and enters the crystals, whereas potassium diffuses toward the cooler part of the system and enters the feldspar there. As a result, a rock initially containing an alkali feldspar of uniform composition in all its portions can acquire different feldspar compositions in different parts of the system, in response to a temperature gradient.

The analogous reaction for plagioclase feldspars,

$$2NaAlSi_3O_8 + Ca^{2+} = CaAl_2Si_2O_8 + 2Na^+ + 4SiO_2$$

goes rapidly toward the production of anorthite-rich plagioclase with liberation of silica, but in experiments at the same temperature proceeds more slowly in the opposite direction because of the difficulty with which Al and Si diffuse through the feldspar (Orville, 1972).

14.4 METASOMATISM AT INTRUSIVE CONTACTS

Skarn is metasomatized and metamorphosed wall rock at intrusive contacts. The initial rock was commonly dolomite or limestone, but by addition of silicon and iron the rock was converted to assemblages typically containing calcium- and iron-rich garnet, clinopyroxene (hedenbergite), and other calcium-rich silicates, together with quartz, calcite, magnetite, or hematite.

Kerrick (1977) evaluated chemical criteria for infiltration versus diffusion metasomatism and concluded that some carbonate wall rocks of the Sierra Nevada batholith became skarns by percolation of aqueous fluid from the intrusive rock, coupled with diffusion of Ca and CO_2 (liberated as carbonates were converted to silicates) back into the cooling plutons. It seems logical that both diffusion and infiltration metasomatism will be effective at contacts between magma and wall rock, with the relative importance of the two mechanisms varying according to permeability, temperature, and mineralogy.

Another product of contact metasomatism is *fenite* (named for the Fen intrusive complex, Norway). Fenites differ from skarns primarily in the restriction that carbonatites and silica-undersaturated silicate magmas are their sources of metasomatizing fluids. Generally, in accordance with the relatively low silica content of the magma, the wall rocks become desilicated while alkalis, especially sodium, are added. Iron and many trace elements that are relatively abundant in carbonatite and feldspathoidal magmas also become enriched in the neighboring fenites (Figure 14-2). Sodic clinopyroxenes and amphiboles are typical of fenites; alkali feldspar increases, relative to the proportion in the unmetasomatized rock, and quartz decreases. Nepheline or other feldspathoids may appear in extreme examples of fenitization, yielding products that closely resemble the intrusive nepheline syenites.

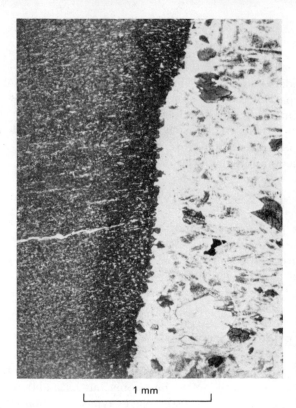

Figure 14-2. Photomicrograph, in plane-polarized light, showing small-scale fenitization of hornfels adjacent to a nepheline syenite dike, a portion of which is visible on the right. In a sharply defined band, light-colored diopside has been replaced by darker acmitic clinopyroxene (richer in sodium and ferric iron that migrated from the dike). The wall rock to the extreme left has been metamorphosed, but retains approximately its original composition except for loss of CO_2. The metasomatic zone of enrichment in sodium and ferric iron is more narrow than the zone of thermal metamorphism because heat diffuses much more rapidly than matter.

1 mm

14.5 METASOMATISM ON THE SEAFLOOR

In a sense, submarine lavas are intrusive with seawater serving as their wall rock. The compositional and thermal contrasts between lava and seawater can lead to exchange of components, producing *spilites*. These are mafic lavas and shallow intrusive rocks associated with marine sediments and containing sodic plagioclase and chlorite, epidote, or calcite. Commonly, the original texture is preserved (Figure 14-3), so that it is easy to see that phenocrysts of calcic plagioclase have been converted to albite and the glassy groundmass has become an aggregate of chlorite (with or without other hydrous silicates). Olivine phenocrysts typically have completely altered to chlorite, but clinopyroxene may still be intact. Iron–titanium oxides may survive or be replaced by fine granular sphene. Actinolitic amphibole may also be present. Spilites become depleted in calcium and ferrous iron, whereas ferric iron, sodium, H_2O, and CO_2 increase. In spite of the drastic chemical changes, textural differences between originally glassy margins of lava pillows and the coarser, more highly crystalline pillow centers are preserved.

Three hypotheses have been considered for the origin of these rocks with igneous textures but low-grade metamorphic mineralogy. The first, crystallization of low-temperature magma enriched in sodium, water, and carbon dioxide, is not supported by experiments or by the presence of olivine and clinopyroxene phenocrysts. The second, metamorphism after burial, is less

Figure 14-3. Photomicrograph, in plane-polarized light, of spilite. Plagioclase phenocrysts and microphenocrysts retain their euhedral to skeletal outlines, but only a few patches of calcic plagioclase survive, most having been replaced by albite. Calcite fills a large irregular vesicle. The groundmass is now pale green chlorite and fine granular sphene.

1 mm

easily dismissed, but the textural preservation and several lines of chemical and isotopic evidence favor the third, that these rocks form by reaction of hot but already solidified rocks with percolating brines derived from seawater.

Why, then, are not all submarine basalts converted to spilites? Many samples contain unaltered phenocrysts in a glassy groundmass without a trace of devitrification. In these, apparently the temperature of the rock dropped so rapidly that reaction was inhibited. It is in the thick piles of lavas filling the axial valleys of oceanic ridges, and in the underlying sheeted dike complex, that we should expect vigorous convection of water to be driven by heat from an underlying magma chamber. These localized convection systems permit rapid percolation of large quantities of heated water through the rocks to produce spilites (see also Chapter 16). In agreement with this model, spilitized rocks are abundant in ophiolites (Coleman, 1977, p. 108–114). Among recent papers reviewing the genesis of spilites are those by Vatin-Perignon and others (1979) and Munhá and Kerrich (1980).

14.6 THE GRANITE CONTROVERSY

From about 1935 to 1960, the outstanding controversy in geology concerned the origin of granitic rocks. Granites were held by one faction to be magmatic, by the other to be metasomatic. The topic was hotly debated and, like any worthwhile scientific disagreement, it stimulated more and better research.

During slow cooling of a magmatic granite below its solidus, mineralogical and textural changes (inversion, exsolution, granulation, and deuteric alteration) continue to occur, resulting in a low-temperature mineral assemblage (microcline, albite, micas, and recrystallized quartz) with a modified texture. Granites therefore converge with metamorphic rocks in mineralogy and texture. Those geologists favoring a metasomatic origin concluded that granites are produced in the solid state from precursors of diverse compositions, by diffusion or percolation of fluids rich in the components of granite (SiO_2, Al_2O_3, Na_2O, and K_2O).

Confusion between the mechanisms of emplacement and the origins of postemplacement features therefore led to the debate, in which some partisans on each side attributed all the features of granites to a magmatic or a metasomatic origin. Few, if any, lines of evidence for either origin were undisputed, and the same data were used by both sides with equal conviction. Papers by Read (1948) and Bowen (1948) summarized the two positions and should still be read by every petrologist. Eventually, the debate subsided, as geologists realized that subsolidus adjustments tend to obliterate mineralogical and textural evidence for magmatic crystallization, that the compositions of granites fall in the lowest-melting part of the system $NaAlSi_3O_8 - KAlSi_3O_8 - SiO_2 - H_2O$ (Tuttle and Bowen, 1958) and therefore are likely, or even inevitable, products of both partial fusion and fractional crystallization, and, finally, that metasomatism can indeed produce granite, so that the origin of each granitic body must be assessed on its individual evidence.

Although probably most petrologists today would concede that metasomatism can produce granite on a relatively small scale, and that such "granitization" might be expected at the contacts of large magmatic bodies, few occurrences of metasomatic granites have been documented since the demise of the controversy. The granitic rocks that would appear most likely to be metasomatic are those forming concordant layers interbedded with metasedimentary rocks and showing no evidence of having displaced older rocks formerly occupying the space now filled by granite. Such rocks, however, could very likely be recrystallized rhyolitic lavas or ash-flow tuffs, metamorphosed without substantial change in composition but with profound textural modification (for example, see Carl and Van Diver, 1975).

14.7 METASOMATISM IN THE MANTLE

It is ironic that metasomatism, now in disfavor as a mechanism for producing large quantities of granite in the continental crust, should be so widely invoked as a means of promoting the generation of silica-undersaturated magmas, and carbonatites and kimberlites, in the mantle. Partial fusion of normal fertile mantle (garnet or spinel peridotite) can only yield silica-undersaturated liquids enriched in Ca, Al, Fe, Na, H_2O, and CO_2 if a small degree of fusion (1 or 2%) is reached before the liquid escapes. Larger degrees of fusion, involving dissolution of larger proportions of olivine and orthopyroxene in the liquid, lead to rapid dilution with additional Si and Mg. However, other elements that are enriched in alkali basalts and other silica-undersaturated silicate magmas, carbonatites, and kimberlites include K, Ti, P, Ba, light rare earth elements, U, and Th; and the concentrations of these in the major phases of peridotites (olivine, orthopyroxene, clinopyroxene, and garnet or spinel) are so very low that a *large* degree of partial fusion is required for them to be liberated in the amounts observed in the magmas. A way out of this dilemma is to derive these elements from accessory phases in the peridotite that dissolve in the first liquid to form.

In many localities, the ultramafic xenoliths carried up by silica-undersaturated magmas are not mineralogically simple and anhydrous peridotites, but contain variable proportions of phlogopite, apatite, and amphibole. These phases are repositories of the trace elements enriched in silica-undersaturated magmas. Metasomatism of normal mantle could have introduced the components of these minerals (for example, see Boettcher and O'Neil, 1980; Menzies and Murthy, 1980). Elements that are enriched in hydrated and apatite-bearing ultramafic xenoliths include the light rare earths, Ti, Ba, Zr, P, F, Cl, and Rb. Introduction of these and of Fe, Na, H_2O, and CO_2 can lower the solidus temperature of the rock; simultaneously, addition of K, U, and Th provides radiogenic heat sources to promote melting.

There is widespread conviction that metasomatism in the mantle is an essential precursor to the generation of silica-undersaturated and carbonatite liquids, and that these liquids arise from substantial degrees of partial fusion of mantle previously enriched in amphibole, phlogopite, and apatite. Furthermore, regional heterogeneities in mantle $^{87}Sr/^{86}Sr$ ratios imply differences in rubibium–strontium ratios caused by metasomatism. The nature of the metasomatizing fluid (vapor or magma) is in doubt, as is its source. Wyllie (1980) proposes that mantle metasomatism is a byproduct of kimberlite magmatism, although it may then serve as a prerequisite for the generation of other kinds of liquid.

14.8 SUMMARY

Addition and subtraction of components in rocks under subsolidus conditions by diffusion and infiltration can be effective on a broad range of scales, and can act across intrusive contacts, between submarine rocks and convectively percolating seawater, and within the mantle. Metasomatism may be a cause, as well as effect, of magmatism by modifying the composition of parent rock in the mantle to promote melting, and perhaps by refertilizing sterile mantle.

15

Magmatism and Tectonism

15.1 ASSUMPTIONS AND AN HYPOTHESIS

The relations between igneous activity and other geologic processes were stated in preliminary fashion in Chapter 1. Now, with the benefit of a more sophisticated vocabulary and the detailed observations set out in the intervening chapters, we can examine the relations between specific igneous rock types and their tectonic settings. Fundamental to such an examination are three assumptions leading to a working hypothesis. The assumptions are:

1. Magmas are generated by partial fusion of mantle or crust; there is no permanent, worldwide reservoir of magma.

2. Melting is a response to dynamic processes. Heat cannot be "focused" into a small, high-temperature volume, and radiogenic heat alone is insufficient to cause fusion, so magma must be produced in one of three ways: (a) raising the temperature by introduction of magma from below, or by friction and shear, or by subduction to transport rock into a higher-temperature regime; (b) decreasing the pressure by faulting, uplift, or diapirism; (c) changing composition to lower the solidus by metasomatism. All three are dynamic, not static, processes. A stationary mass of rock of fixed composition and constant temperature will never begin to melt.

3. Once generated, magma moves up into (or through) the crust instantaneously on the scale of geologic time, and even if modified during ascent by differentiation or assimilation, it retains some chemical characterstics that were imposed at the site of magma generation.

TABLE 15-1 NINE TECTONIC REGIMES
OF MAGMATISM

At constructive plate boundaries (spreading centers)
1. Oceanic ridges
2. Back-arc basins
3. Continental rifts

Parallel to destructive plate boundaries (above subduction zones)
4. Ocean–ocean collisions (island arcs)
5. Ocean–continent collisions (arcs and batholiths)
6. Continent–continent collisions

At conservative plate boundaries (transform faults)
7. Volcanoes on oceanic fracture zones

Intraplate magmatism (unrelated to plate boundaries?)
8. Oceanic islands and island chains
9. Intracratonic magmatism

The working hypothesis based on these assumptions is: The compositions of igneous rocks should reflect tectonic conditions at the time of their emplacement. Whether "tectonic conditions" are those in the mantle or in the crust is a question we will postpone. Tectonic environments in which magmatism occurs are classified in Table 15-1 and shown in schematic fashion in Figure 15-1. The kinds of igneous activity associated with each environment will be taken up later, but the classification is needed in the following discussion of some general principles.

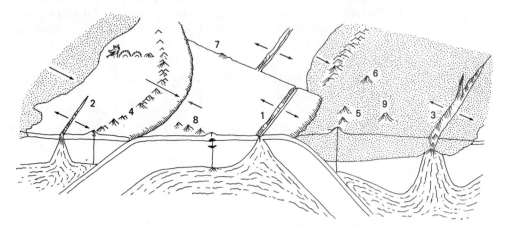

Figure 15-1. Nine types of magmatic environments: 1, oceanic ridge; 2, back-arc basin; 3, continental rift; 4, island arc parallel to ocean–ocean collision boundary; 5, volcanic arc and batholith parallel to ocean–continent collision boundary; 6, continent–continent collision; 7, oceanic fracture zone (transform fault); 8, oceanic islands and island chains; 9, intracratonic. Continental crust is stippled and asthenosphere lined; lithospheric mantle and oceanic crust are unpatterned. Arrows indicate directions of plate motion. The sea serpent is not for scale, but to indicate the degree of credibility that should be accorded this illustration.

15.2 MAJOR AND TRACE ELEMENTS AS TECTONIC INDICATORS

The difficulty in testing the working hypothesis arises from the second part of the third assumption, that all magmas retain some chemical characteristics that were imposed when the parental magmas formed. Obviously, magmatic differentiation, assimilation, and mixing will tend to change the composition of the surviving liquid, especially in terms of those major components that build the rock-forming minerals. However, the persistence of thermal barriers between silica-oversaturated and -undersaturated portions of many systems, especially those pertinent to basaltic rocks, encourages the expectation that a quartz-normative mafic liquid will not evolve into a nepheline-normative one.

As a first test of the hypothesis, we should ask whether the silica saturation of basaltic rocks varies with tectonic setting. This has been examined by Chayes (1972) using 2413 analyzed Cenozoic basalts. Of 507 island–arc basalts, 65% are quartz-normative, 24% contain normative olivine plus hypersthene, and 11% are nepheline-normative. In contrast, of 84 MORB samples, 88% contain normative $ol + hy$, 8% q, and 4% ne. Samples from oceanic islands and from continents were approximately 50% ne-normative, with the remainder nearly equally divided between those with q and those with $ol + hy$. Many more analyses of MORB are now available than could be included in Chayes's 1972 compilation, but the great majority still lack normative quartz or nepheline. The island–arc environment clearly favors silica oversaturation, while intraplate basalts show a weaker tendency toward nepheline in the norm. There is indeed some control exerted on silica saturation by tectonic setting in a broad sense, but this control is not efficient, and silica saturation will hardly serve to permit assignment of a sample of basalt to one tectonic regime. In retrospect, this should not be a surprise, when we recall the many factors (including pressure, degree of partial fusion, ratio of CO_2 to H_2O, and mantle metasomatism) that may determine whether a magma will be nepheline-normative.

Among major oxide components, Chayes (1965) found that TiO_2 is the single best discriminant for distinguishing "oceanic" (MORB and intraplate island) from "circumoceanic" (volcanic arc) basalts. "Oceanic" samples commonly have TiO_2 exceeding 1.75 wt %, while "circumoceanic" basalts have less; there is considerable overlap, however (Figure 15-2).

Other major oxide components do not show such pronounced bimodality, but some combinations of oxides or of normative constituents do. Among these, the Thornton–Tuttle differentiation index shows a distinctly bimodal distribution, for large numbers of samples, in regions that Martin and Piwinskii (1972) called "nonorogenic" (Figure 15-3). Samples from "orogenic" regions (near destructive plate boundaries, in volcanic arcs) show a concentration at an intermediate value, although the total range in differentiation index may be as great as for nonorogenic regions. Martin and Piwinskii also noted that the bimodal population was clearly shown on an

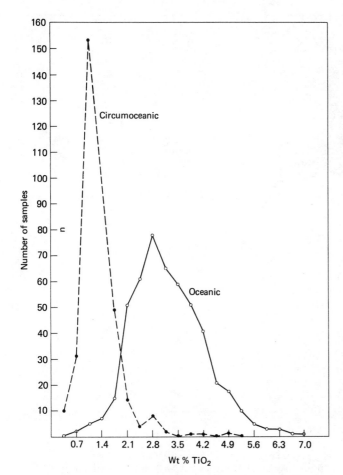

Figure 15-2. Frequency distribution of TiO_2 in 857 samples of Cenozoic lava containing less than 54 wt% SiO_2 and with Thornton–Tuttle differentiation indices of less than 50. (From Chayes, 1965, Fig. 1, with permission of the Mineralogical Society of Great Britain and F. Chayes.)

alkali–iron–magnesium diagram (Figure 15-3), and that the data from non-orogenic regions show pronounced iron enrichment trends. The terms "compressional" and "tensional" were used as synonyms of "orogenic" and "nonorogenic," respectively, by Martin and Piwinskii.

Petro and others (1979) confirmed these observations using a larger set of data, and also showed that $an/(an + ab)$ is bimodal in suites from "extensional plate boundaries" but unimodal in those from "compressional plate boundaries." Extensional suites tend to have higher (Na + K)/Ca ratios at a given silica content, and may include peralkaline rocks. Petro and others (1979, p. 217) conclude that "no single criterion should be used to distinguish tectonic setting."

In the papers just cited, and in others, several sets of contrasts or dualities are named: oceanic versus circumoceanic, extensional versus compressional, anorogenic or nonorogenic versus orogenic, shallow-melting versus

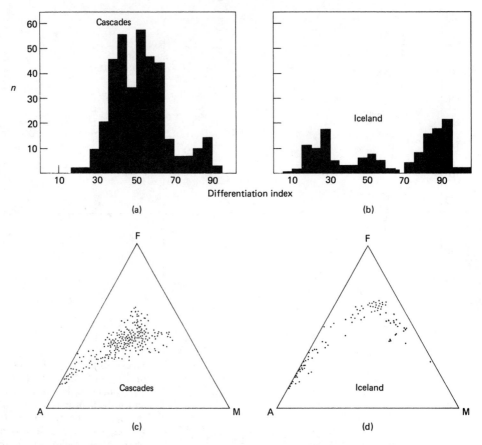

Figure 15-3. (a) and (b) Frequency versus Thornton–Tuttle differentiation index for Cascades (379 analyses) and Iceland (138 analyses), representing orogenic and nonorogenic suites, respectively. (c) and (d) The same samples plotted on alkali (A)–total iron as FeO (F)–magnesium (M) diagrams. (After Martin and Piwinskii, *Journal of Geophysical Research, 77*, 4966–4975, Figs. 1 and 3, 1972, copyrighted by the American Geophysical Union.)

deep-melting. The first members of each pair are not synonymous or geographically congruent, and neither are the second.

Other researchers attempted to discriminate more than two tectonic settings, using the relative proportions of major elements or oxides. For example, Pearce and others (1977) plotted boundaries on a triangular diagram of MgO–FeO–Al_2O_3 delimiting rocks from five tectonic regimes. The five fields correctly classified from 64 to 82% of the samples from a given tectonic setting, but even these unimpressive scores were attained only by restricting the samples to silica-saturated and silica-oversaturated rocks containing from 51 to 56 wt % SiO_2. Silica-undersaturated rocks, and those containing more or less SiO_2, showed much more scatter across the diagram, independent of their tectonic setting. Earlier, Pearce and others (1975) had classified basalts as oceanic or nonoceanic using a TiO_2–P_2O_5–K_2O plot. The mixed success of this approach is less important than the reasoning behind it. These three oxide components are in low concentrations and, most significantly, they do not readily enter the phases (plagioclase, olivine, and pyroxenes) that make up q-normative and $ol + hy$-normative basalts. If all three components remain in the liquid, effects of fractional crystallization upon their proportions can be ignored until the late stages when opaque oxides, apatite, and alkali feldspar begin to form. The ratios among TiO_2, P_2O_5, and K_2O should reflect those in the parental liquid. Use of such components that are rejected by early and abundant minerals, rather than the major components, offers greater promise of seeing through the effects of crystal–liquid fractionation to decipher the composition of the parent magma. Two trace elements with very small distribution coefficients (Section 4.5) should keep a constant ratio during fractional crystallization. Other processes, including liquid immiscibility, assimilation, magma mixing, metasomatism, and weathering, can perturb the ratio by introducing or removing one element more than the other. Many combinations of trace elements are now being assessed as tectonic discriminators. It is unlikely that a single set of all-purpose discriminators, applicable to all rocks and to all settings, will ever be found, for several reasons. One is that most elements have a larger distribution coefficient in some crystalline phase (for example, Sr in plagioclase/liquid). Another is that the distribution coefficient depends on the liquid composition as well as on the identity of the crystalline phase and the temperature. Furthermore, an element may begin to enter a phase as a major component when the liquid becomes saturated with respect to that phase [for example, Zr in zircon, which crystallizes early in most granitic liquids (Watson, 1979), so Zr is not a "residual" element in extremely felsic liquids]. Most trace elements are not distributed uniformly throughout a rock, but tend to concentrate in accessory minerals which may be particularly susceptible to redistribution by crystal settling, flow differentiation, partial fusion, or alteration.

Because the differences between their masses are so small, isotopes of the heavier elements (especially Rb, Sr, Nd, Pb, Th, and U) are not fractionated at high temperatures so, as already described for $^{87}Sr/^{86}Sr$ and $^{143}Nd/^{144}Nd$, their ratios are not changed by crystal–liquid fractionation. Metasomatism and assimilation may, however, change an isotopic ratio by introducing material of a different isotopic ratio or by providing more of the radioactive parent element to decay into the radiogenic daughter isotope during long time intervals between introduction and the sampling. Faure (1977) discusses the application of isotope chemistry to petrology, providing the background needed for comprehension of some important recent papers.

The literature on trace elements and major elements as tectonic discriminators is voluminous and, in general, more contradictory than that concerning isotopic ratios. Some of the successes and failures of all three approaches will be mentioned for specific tectonic settings in the next section.

15.3 CORRELATIONS BETWEEN MAGMA TYPES AND TECTONIC REGIMES

Petrologic differences among the nine environments of Table 15-1 and Figure 15-1 are not clear cut; some are matters of considerable controversy, but the present status of knowledge (and confusion) deserves a summary.

At constructive plate boundaries (spreading centers), seismic data and high heat flow indicate that asthenosphere has risen through lithospheric mantle to approach the base of the crust. The boundary between lithosphere and asthenosphere has been defined on the basis of several criteria. These include decreases in viscosity, seismic wave velocities and density, increase in electrical conductivity, change in the dominant mechanism of heat transfer from conduction to convection, and appearance of interstitial magmatic liquid. These changes are unlikely to occur at the same depth or temperature, so the boundary is vaguely located. The question is not merely a semantic one, because it concerns the role of magma: Is the asthenosphere "weak" because it contains magma, or is magma generated by pressure release as mobile asthenosphere rises? In other words, is magma a propellant, a lubricant, or a byproduct of plate motion?

If the asthenosphere, however defined, reaches the base of the continental crust, a *continental rift* may evolve into an oceanic spreading center. Previously, extension and intrusion took place over a broad zone (Figure 15-4a), but when the top of the asthenosphere impinges directly on the subcontinental Moho, crustal extension and magmatism are likely to become sharply localized. Injection of magma in a narrow zone heats the adjacent crust, so that subsequent intrusions can ascend to the surface without losing most of their heat to the walls of the conduit (Lachenbruch, 1976). Once the lithospheric segments become completely detached along a magma-filled crack, separation becomes more rapid, but magmatism may keep pace with it by building thin oceanic crust on the trailing edges of the separating plates (Figure 15-4b). A reasonable assumption that is frequently made in speculating about magmatism in tectonic settings is especially applicable here; magma should move upward more easily through lithosphere that is under extension than through lithosphere that is under lateral compression.

According to the model in Figure 15-4, although oceanic ridges can evolve out of continental rifts, *back-arc spreading centers* must arise by some other mechanism. Saunders and Tarney (1979) summarize the possibilities, all of which take into account the observed positions of back-arc basins over subducting slabs. Regardless of whether the asthenospheric diapirs that feed back-arc spreading centers are simply displaced by sinking lithosphere or are triggered by shearing, frictional heating, or metasomatism from the subducting slab, the source of back-arc magmas could be more variable than the source of ocean-ridge magmas. The compositions of back-arc magmas are indeed more diverse than those of mid-ocean ridges, but both kinds are com-

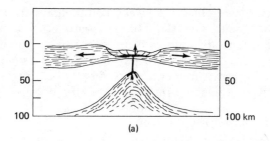

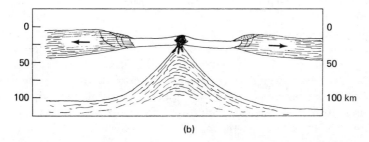

Figure 15-4. Cross sections (no vertical exaggeration). (a) Continental rift; continental crust (shaded) thins by faulting and by ductile necking, being heated and weakened by intrusion of magma (black) in a wide zone. (b) Oceanic ridge; asthenosphere reached the base of the continental crust, allowing more rapid and efficient injection of magma to form oceanic crust (unshaded) at a spreading center. The edges of the former continental rift are now passive margins of continents, partly buried by sediments (stippled). Arrows indicate directions of plate motion. (After Salveson, 1978, with permission of J. O. Salveson, Chevron Resources Co.)

monly *ol* + *hy*-normative basalts with very low K, Rb/Sr, and Fe/Mg. Local variations in back-arc basalts can be attributed to metasomatic introduction of K, Rb, Ba, ^{87}Sr, and H_2O into the mantle from the subducting slab, but "most [back-arc] basin basalts are within the compositional range of basalts produced at normal mid-ocean ridge spreading centres. It is unlikely therefore that geochemical criteria could be erected to unequivocally separate the two" (Saunders and Tarney, 1979, p. 569). This is a significant obstacle in attempts to decide whether a particular ophiolite assemblage was formed at an oceanic ridge or at a back-arc spreading center.

Lavas of continental rifts differ markedly from those of oceanic spreading centers. Basalts range from *q*- to *ne*-normative. Fe/Mg and ^{87}Sr/^{86}Sr ratios are generally higher in mafic lavas of continental rifts, and there are many similarities to intraplate magmas (those of oceanic islands and intracratonic volcanoes). Among these similarities are the presence of phonolites and rhyolites, including peralkaline varieties of both, in some continental rifts, and enrichment in the "residual" trace elements. It is completely unclear whether the chemical resemblance between continental rifts and intraplate magmatism on the one hand and the contrasts with oceanic spreading centers on the other are due to differences in thickness or composition of lithospheric mantle, composition or degree of partial fusion of asthenosphere, or presence of continental crust. Continental rifts show a puzzling variety in attendant magmatism, from tholeiites (Mesozoic rifts of eastern North America) to voluminous alkali basalts and phonolites (Kenya rift) to sparse carbonatites and leucite-rich lavas (Uganda rift).

Magmatism associated with destructive plate boundaries is also poorly understood. Volcanic arcs parallel to collision boundaries between two oceanic plates should involve the smallest number of complications, so they will be considered first. Marshall (1911), following Eduard Seuss's recognition of island arcs, pointed out that an "andesitic line" (delimiting the seaward extent of andesitic volcanism in these arcs) was "the true border of the Pacific area" (p. 32). Figure 15-5 compares Marshall's line with inferred plate boundaries in the southwest Pacific. The near congruence of the two sets of boundaries, one defined by petrology and the other by seismology, indicates that "andesites" (used in a broad sense, including q-normative basalts at one extreme and latites at the other) in volcanic arcs do reflect a specific tectonic setting.

Arc volcanoes form rows on one side of a trench, the side overlying the seismic (Benioff) zone tracing the sinking plate, and define narrow bands, generally about 10 km wide, broken into segments 100 to 500 km long and

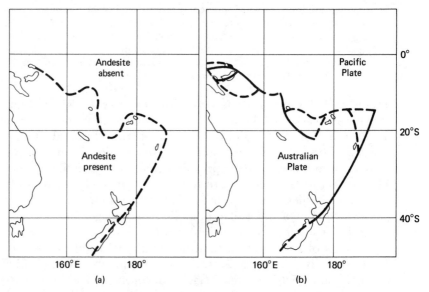

Figure 15-5. Plate boundaries in the southwest Pacific. (a) The "andesitic line" of Marshall (1911, Fig. 3). (b) Plate boundaries, dashed where uncertain, of Johnson and Molnar (*Journal of Geophysical Research, 77,* 5000–5032, Fig. 1, 1972, copyrighted by the American Geophysical Union).

offset from each other, apparently along "breaks or tears in the subducting plate" (Marsh, 1979b, p. 687). Adjacent segments of the sinking slab differ in dip. Marsh interprets the fairly regular spacing of arc volcanoes (at intervals of 30 to 70 km), and their restriction to narrow bands that are closer to the trench over more steeply dipping seismic zones, as indicating detachment of regularly spaced diapirs from a gravitationally unstable ribbon of buoyant material. The ribbon may be derived from the subducting slab itself (Marsh, 1979a), or from the overlying wedge of mantle. Metasomatic introduction of volatile components and "residual" elements into the mantle wedge from dehydrating oceanic crust would produce a parent material capable of yielding arc lavas with much smaller degrees of fusion than either the sinking crust or depleted (sterile) mantle.

Neodymium and strontium isotope ratios indicate that arc magmas are derived from source materials that are mixtures of mantle and altered ocean-ridge basalts (DePaolo and Johnson, 1979). Probably the "mixture" is figurative; a two-stage process appears to satisfy isotopic and trace element constraints if a fluid (magma or aqueous solution) escapes from the subducting slab, carrying residual elements, and metasomatically changes the overlying mantle, which then partially melts to yield arc magmas (Frey, 1979, p. 808). Alternatively, Kay (1980) suggests that diapirs of depleted mantle in the sinking slab detach and rise, acquiring residual elements from subducted sediments and altered basalt; the diapirs continue to rise and melt, feeding arc volcanoes. In either view, crustal materials, especially residual elements, are recycled.

As noted by Kuno (1959), basalts from volcanoes closest to the trench generally are q-normative. If the volcanic belt is broad, or if activity spreads farther from the trench through time, the younger and more distant volcanoes tend to erupt more silica-undersaturated lavas. This might be expected if ne-normative basalts are generated at greater depths than q-normative basalts. Silica-oversaturated basalts are the most abundant lavas in most island arcs associated with ocean–ocean collisions. Compared with basalts of oceanic ridges and back-arc spreading centers, island–arc lavas are generally higher in Al, Rb, Sr, Ba, Pb, CO_2, and H_2O, lower in Mg, Ti, P, Zr, Nb, Ni, and Cr, and have higher $^{87}Sr/^{86}Sr$, K/Rb, Rb/Sr, Fe/Mg, CO_2/H_2O, and Cl/F ratios (Saunders and Tarney, 1979; Perfit and others, 1980; Stern, 1980).

Two examples show the variations among intraoceanic arcs. In the northern Marianas arc (Dixon and Batiza, 1979), two oceanic plates meet, causing a steeply dipping zone of low seismicity. The highly porphyritic lavas are tholeiites and andesites with silica contents ranging from 47 to 61 wt %. In contrast, the central segment of the Sunda arc (Foden and Varne, 1980) contains three volcanoes 150 to 190 km above the Benioff zone that have erupted $ol + hy$-normative basalt, andesite, and dacite from one, but ne-normative lavas from two others. Closer to the trench, leucite-bearing rocks have erupted, and have slightly higher $^{87}Sr/^{86}Sr$ ratios than those from the presumably deeper sources farther from the trench. Magma sources under the Sunda arc appear to have local variations in K, Rb, and Sr.

Volcanic arcs and batholiths associated with ocean–continent collisions are even more complex than arcs parallel to ocean–ocean destructive boundaries. In part, the added complications may be due to the direct involvement of continental crust, but a postulated greater thickness of subcontinental lithosphere, compared with that under oceans, may also play a role. The *arc-trench gap*, the distance between the volcanic arc and the near side of the trench, maintains a fairly constant width along a given plate boundary but varies from 50 to 250 km wide from one boundary to another (Dickinson, 1973). There are exceptions, but the narrower arc–trench gaps generally lie over steeply dipping subduction zones (as in the Solomon Islands, New Britain, New Hebrides, and northern Marianas), usually marking ocean–ocean collisions. The widest gaps tend to lie over oceanic slabs that dip gently under continental crust (as in the Peru–Chile, Eastern Alaska Peninsula, and Japanese Inland Sea systems). The correlation between dip of the subduction zone and width of the arc–trench gap implies that magmas are fed from sources requiring that the subducting plate reach a minimum depth; if the plate goes down steeply, this depth is reached at a lateral distance closer to the trench (Figure 15-6). The minimum depth appears to be about 110 km, according to Marsh and Carmichael (1974). Sinking oceanic crust should have completed its dehydration and conversion to eclogite before reaching that depth.

Keyed to the corresponding letters in Figure 15-6, possible means of generating arc magmas are: partial fusion of (*A*) subducted continental crust and trench sediment; (*B*) subducted oceanic crust; (*C*) hydrated and metasomatized lithospheric mantle, the water probably being supplied by dehydration of the subducting oceanic crust; (*D*) dry lithospheric mantle; (*E*) asthenospheric mantle; and (*F*) lower continental crust. Some specific lines of evidence have been advanced in favor of each possibility. The observations of Barazangi and Isacks (1979), that volcanism is absent above shallow subduction zones lacking an overlying wedge of asthenosphere, favor (*E*).

Among several calculated models of temperature and stress distributions at destructive boundaries, that of Anderson and others (1980) emphasizes a contrast between ocean–ocean and ocean–continent collisions. At ocean–ocean boundaries, temperature is high enough to permit partial fusion only in hydrated asthenosphere over the sinking slab. At ocean–continent boundaries, thick, more highly radioactive continental crust and thick lithospheric mantle permit partial fusion to occur not only in the hydrated asthenosphere wedge but in the subducting oceanic crust itself, in lithospheric mantle, and in lower continental crust. This model "therefore permits immense complexity and variety in continental arc volcanic compositions caused by an intermixing of melts from several regions" (Anderson and others, 1980, p. 448). Although this model is not the only plausible one, the observed variety among lavas associated with ocean–continent collisions does lend it support.

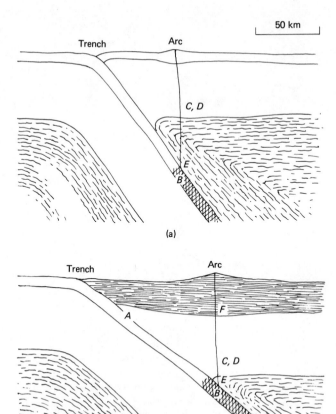

Figure 15-6. Schematic cross sections of two varieties of destructive plate boundaries; (a) ocean–ocean collision, (b) ocean–continent collision. Subducted oceanic crust converts to eclogite (crosshatched). Continental crust is shaded, asthenosphere lined. Letter symbols are discussed in the text.

As an example, active volcanism in the Andes defines three linear segments (Figure 15-7). The northern and central volcanic lines are 140 km above Benioff zones that dip 25 degrees to the east; continental crust under the northern and central segments is 40 to 50 km and 70 km thick, respectively. The southern volcanic line is only 90 km above a Benioff zone that also dips 25 degrees to the east, and lies on continental crust that is 30 to 45 km thick (Thorpe and Francis, 1979; Harmon and others, 1981). The three segments are separated by two zones lacking volcanism and underlain by subducting slabs that dip only 10 degrees to the east (Barazangi and Isacks, 1979), apparently too shallow to permit an asthenospheric wedge between oceanic slab and subcontinental lithosphere. The northern volcanic segment erupts "basaltic andesites" and andesites with 53 to 61 wt % silica and with $^{87}Sr/^{86}Sr$ ratios of 0.7038 to 0.7044. The central segment has erupted voluminous rhyolitic ash-flow tuffs, but now erupts andesites and dacites with 55 to 66 wt % silica, strontium isotopic ratios of 0.7058 to 0.7072, and higher K, Rb, Sr, and Ba than the northern or southern segments. The southern segment

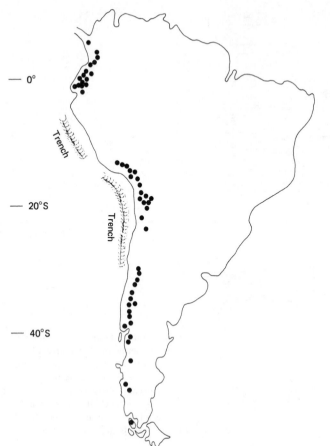

Figure 15-7. Segments of the Andean volcanic chain; black dots indicate volcanoes known or likely to have been active in the last 10,000 years. The two wide gaps in the chain are underlain by more gently dipping subduction zones. (In part after Morris and others, 1979.)

erupts basalts, andesites, and rare dacites with 51 to 68 wt % silica and strontium isotopic ratios of 0.7054 to 0.7112 (Harmon and Thorpe, 1980).

In addition to this variation between segments, there is east–west change in lava compositions and ages within each segment. Younger, more easterly, lavas are richer in K, Rb, and Ba, poorer in Sr. Furthermore, an alkali basalt–trachyte–phonolite suite and a high-potassium andesite and dacite ("shoshonite") suite are scattered up to 400 km east of the main volcanic lines. Lavas in the northern segment appear least affected by crustal contamination, but specific magma sources and the processes that modified the liquid compositions in the three segments remain unclear.

The diversity of rock types in other continental arcs, and in circum-Pacific batholiths (Section 12.7), further testifies to the varied heritage of magmas associated with ocean–continent collisions. A further complication, in addition to the involvement of continental crust, is the apparently common preservation of inhomogeneities of the oceanic plate after its subduction.

Pakiser and Brune (1980) interpret seismologic data as indicating an asymmetric root under the Sierra Nevada batholith, with at least some of the asymmetry being caused by an east-dipping remnant of subducted oceanic crust. Continuity of oceanic plate structure under other continental arcs and batholiths is indicated by segmentation, as in the Andes; segment boundaries in Central America (Carr and others, 1979) seem to coincide with fracture zones offshore. DeLong and others (1975) recognized that in ocean–ocean collisions, also, anomalous occurrences of silica-undersaturated magmatism apparently mark the edges of subducting plates or discontinuities within them. These edges and slits may allow asthenosphere, or magma derived therefrom, to ascend from below a sinking slab. Hughes and others (1980) document the relations between segmentation and lava compositions in the Cascades; volcanoes within segments exhibit restricted ranges of composition (mostly andesite), whereas volcanoes on or near segment boundaries show much more diverse lava compositions with a tendency toward bimodality (basalt plus rhyolite). These differences are yet to be explained.

The excellent book on andesites by Gill (1981) is essential reading. Gill reviews the tectonic setting, petrology, chemistry, and experimental phase relations concerning andesitic arc magmatism, and concludes that most andesite magma forms by fractional crystallization of plagioclase, pyroxenes, olivine, and magnetite from basalt magma at shallow depths. Andesites appear most commonly near destructive plate boundaries, rather than showing up wherever basalts do, because of three peculiarities of collision boundaries. First, the asthenospheric wedge is modified by material ascending from the subducting plate, encouraging the production of more siliceous primary basalt liquid. Second, crystal–liquid fractionation may be enhanced by more tortuous conduit geometry and slower magma ascent in a compressional stress regime. Third, higher oxygen and water fugacities cause the fractionating crystals to be lower in silica.

Continent–continent collisions (Roeder, 1979) are the least understood of the three magmatic regimes on destructive plate boundaries, in large part because we have only one active example, the Himalayas, to study. According to Molnar and Gray (1979), continental crust can be subducted in spite of its greater buoyancy, thickness, and rigidity, but the efficiency of the process is unknown. Perhaps only a small portion of a continental plate may sink at a steep angle, or may stick to the underside of the other plate. At the other extreme, Powell and Conaghan (1975) attribute the twice-normal thickness of continental crust in the Tibetan plateau north of the Himalayas to underthrusting of one continental plate as much as 1000 km along the base of the other. Magmatism within the Himalayas produced S-type granites with very high $^{87}Sr/^{86}Sr$ ratios (around 0.780), almost certainly products of crustal fusion without significant mantle contribution (Allègre and Ben Othman, 1980). On the Tibetan plateau, young potassium-rich andesites, dacites, and rhyolites have erupted (Dewey and Burke, 1973).

Conservative plate boundaries are divided into oceanic transform faults and continental transcurrent faults. Among the latter (for example, the San Andreas and Anatolian faults), no contemporary volcanism is assuredly related to the faults, so this brief discussion centers on the oceanic features. As summarized by Burke and Sengör (1979), the total width of the zone of deformation along a transform fault commonly exceeds 10 km, but maximum displacement is concentrated in a belt less than 1 km wide. The displacement is accomplished by faults that branch and rejoin along strike, isolating splinters that are shifted vertically as well as horizontally. There is little information and less agreement concerning the kinds of magmatism along transform faults. For example, Bryan and others (1980) found no significant difference between basalts extruded in the Kane Fracture Zone and at the crests of Mid-Atlantic Ridge segments where these are offset at least 200 km by the transform fault. On the other hand, Shibata and others (1979) examined the intersection of the Mid-Atlantic Ridge with another transform fault farther north than the Kane Fracture Zone, and found *ne*-normative basalts enriched in residual trace elements. They conclude (p. 139): "Magmas injected into transform faults . . . segregate at greater mantle depths than tholeiites and possibly from a compositionally distinct mantle source. Presumably, it is the tectonics of a fracture zone which enables such alkalic magmas to reach the surface."

Intraplate magmatism far from, and apparently unrelated to, plate boundaries is the most diverse of all kinds in the various tectonic regimes. It is divisible into two groups, one feeding oceanic islands and island chains, the other piercing old, thick, and stable continental crust.

The best-known example of oceanic intraplate volcanism is provided by the Hawaiian–Emperor chain in the Pacific (Figure 15-8). This great parade of islands and submerged seamounts is progressively older farther from the present site of volcanism on the island of Hawaii, indicating that oceanic lithosphere moved north and northwest, relative to a magma source which "oscillates in size and shape with time" (Shaw and others, 1980, p. 691), at about 6 cm/yr until about 1.4 million years ago, then at about 18 cm/yr since. Eruption rates have varied from effectively zero to 0.5 km^3/yr; the present average is 0.1 km^3/yr. Most of the volcanoes in the chain show a sequence of abundant *q*-normative basalt followed, after caldera formation, by *ne*-normative basalts in greatly diminished proportions and finally, after a long erosional interlude, by strongly silica-undersaturated nephelinites as localized flows and cinder cones marking the end of activity at that volcano (Lanphere and Dalrymple, 1980). Some volcanoes omitted the second or third stage, or both. The reasons for this progression are still subjects of speculation and need more study. It is easy to conclude that the sequence indicates tapping of smaller quantities of magma derived from deeper sources by smaller degrees of partial fusion, but this merely restates the problem without explanation.

Some other oceanic volcanoes are arrayed in chains showing an age pro-

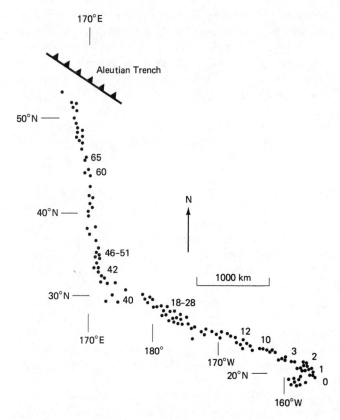

Figure 15-8. Schematic map of volcanic centers in the Hawaii–Emperor Seamount chain. Numbers are rounded potassium–argon ages of volcanism; for example, Midway Island 18 to 28 million years ago, and Mauna Loa and Kilauea, 0 million years ago (active). The ages are from Table 1 of Shaw and others (1980). (Generalized from Shaw and others, 1980, Fig. 1, with permission of the American Journal of Science.)

gression (for example, the Samoan chain; Natland, 1980), but others form clusters with no discernible pattern of development. On many, the parts above sea level contain subordinate q-normative basalt or may entirely lack silica-oversaturated lavas but instead exhibit alkali basalts, nephelinites, or even melilite-bearing rocks and carbonatites, although the most strongly silica-undersaturated rocks are extremely rare in the oceans. Other islands have peralkaline trachytes and rhyolites completing a sequence from hawaiite and mugearite, similar to the suite found in many continental rifts and intraplate volcanoes.

Within continents, intraplate magmatism is scattered through *cratons* (stable areas of older, thicker, and colder crust). Kimberlites, carbonatites, feldspathoidal rocks, and A-type granites are the typical rocks in intraplate continental settings, although each also occurs in other tectonic regimes. Some of the great accumulations of flood basalts, lying unconformably upon much older basement rocks, have also been considered as intraplate cratonic. However, the vent areas for some flood basalts may have been within continental rifts that evolved into oceanic margins (Figure 15-4b) and are now covered by continental shelf sediments. Indeed, the difficulty of assigning all kinds of cratonic volcanic rocks to a rift or an intraplate regime may be insurmountable by the criteria now at our disposal. All of the rocks in question seem to be derived from subcrustal magma sources and, except for those of flood basalts, these sources appear to have been relatively deep in the mantle. The longevity of kimberlitic and carbonatitic magmatism in a region (Chapter 13) indicates that the sources remained fertile for long times. Other intraplate rock types also tend to recur; for example, in southeastern Egypt eight ring complexes of silica-undersaturated rocks yield isotopic ages of 404 to 89 million years (Serencsits and others, 1981).

Intraplate magmatism within continents shows so many similarities to that within oceans (sharing mildly silica-undersaturated basalts to nephelinites, phonolites, trachytes, and peralkaline rhyolites, plus low strontium isotope ratios and enrichment in "residual" trace elements) that a common source has been suspected for intraplate magmas regardless of the kind of crust they invade. To assess this similarity in sources, Allègre and others (1981) compare the isotopic compositions of strontium, neodymium, and lead in *ne*-normative basalts lying on Precambrian rocks in the Sahara with similar basalts of the Canary Islands (Figure 15-9). The isotopic ratios at the continental and oceanic sites (both within the same plate) indicate a common, surprisingly homogeneous, fertile source that contrasts with a more depleted asthenospheric source for MORB and a variably metasomatized lithospheric mantle source for continental flood basalts.

Some intracratonic magmatism may not be independent of plate boundaries, after all. Sykes (1978) reviewed seismicity and igneous activity in the interior of several continents, and observed that earthquakes and magmatism "seem to be located preferentially along old zones of weakness near the ends of major oceanic transform faults that were active in the early opening of adjacent oceans. In several places, alkaline magmatism and earthquakes ex-

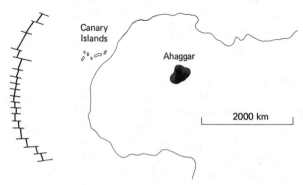

Figure 15-9. Location map of the Canary Islands and the Ahaggar volcanic field, Sahara, two intraplate alkali basalt sites in the same plate but on oceanic and continental crust. The western boundary of the African plate (the Mid-Atlantic Ridge) is shown diagrammatically.

tend several hundred kilometers inland from the ends of oceanic transform faults (but not necessarily with the same strike as the transform fault)" (p. 621). The continental zones of weakness do not show large horizontal displacements and probably are not the actual continuations of oceanic fracture zones. Instead, they seem to be periodically reactivated features, perhaps extending all the way through the lithosphere, that are inherited from older episodes of deformation and are rejuvenated only when they chance to line up with oceanic transform faults.

Rather than being permissively localized along intraplate zones of weakness (and especially at their intersections), magmatism within plates may directly result from narrow convective "plumes" rising from deep in the mantle. Persistent sites of volcanism or "hot spots" on the surface have been attributed (Morgan, 1972) to plumes that are fixed relative to the deep mantle, and over which the lithosphere drifts to produce volcanic chains such as that in Figure 15-8.

Table 15-2 summarizes the relations between igneous rock types and tectonic regimes, and Table 15-3 lists examples. Only the localities are given, because literature references would be too numerous and rapidly outdated. With the geographic names and rock types, a reader will be able to find recent papers listed in *Mineralogical Abstracts* and in the *Bibliography and Index of Geology*. The tables show that no rock type is clearly unique to one tectonic setting, and that no tectonic setting (with the possible exception of continent–continent collisions) has only one rock type. Certain magma compositions are favored by certain tectonic regimes.

TABLE 15-2 QUALITATIVE ASSESSMENT OF THE DISTRIBUTION OF SOME IGNEOUS ROCK TYPES IN NINE TECTONIC SETTINGS

	Tectonic regimes[a]								
	1	2	3	4	5	6	7	8	9
q-normative basalt	x	x	x	X	X			x	x
$ol + hy$-normative basalt	X	X	x	·	·		x	x	x
ne-normative basalt			x	·	·		x	x	x
Nephelinite			x					x	x
Phonolite			x		·			·	x
Peralkaline rhyolite			x		·			·	x
Calcalkaline, S- and I-type granites				X	X	X		·	·
Leucite-bearing rocks			·	·	·		·	·	·
Kimberlite							·	·	·
Carbonatite			·				·	·	·

[a]Numbers are those of Table 15-1. X, abundant in all examples; x, abundant in some examples; ·, rare.

TABLE 15-3 CENOZOIC EXAMPLES OF SOME IGNEOUS ROCK TYPES IN NINE TECTONIC SETTINGS

	q-normative basalts	*ol* + *hy*-normative basalts	*ne*-normative basalts
1.	Mid-Atlantic Ridge, 33 to 53°N	Mid-Atlantic Ridge, 25°S to 33°N	—
2.	East Scotia Sea	Lau Basin	—
3.	Kenya Rift	Kenya Rift	Kenya Rift
4.	Northern Marianas	Sunda Arc	Grenada, Lesser Antilles
5.	Japan	Japan	Japan
6.	—	—	—
7.	—	Kane Fracture Zone, Atlantic	Siqueiros Fracture Zone, Pacific
8.	Hawaii	Hawaii	Hawaii
9.	Columbia Plateau, Washington–Oregon	Snake River Plain, Idaho	Ahaggar, Sahara

	Nephelinites	Phonolites	Peralkaline rhyolites
1.	—	—	—
2.	—	—	—
3.	Kenya Rift	Kenya Rift	Kenya Rift
4.	—	—	—
5.	—	Anahim Volcanic Belt, British Columbia	Anahim Volcanic Belt, British Columbia
6.	—	—	—
7.	—	—	—
8.	Hawaii	Tahiti	Canary Islands
9.	Southeast Australia	Ahaggar, Sahara	McDermitt Caldera, Nevada

	Calcalkaline, S- and I-type granites, rhyolites	Leucite-rich rocks
1.	—	—
2.	—	—
3.	—	Uganda Rift
4.	Northern Marianas	Sunda Arc
5.	Cascades, Washington–Oregon	Celebes
6.	Himalayas	—
7.	—	—
8.	—	Tristan da Cunha
9.	Yellowstone National Park, Wyoming	Gaussberg, Antarctica

	Kimberlites	Carbonatites
1.	—	—
2.	—	—
3.	—	Oldoinyo Lengai, Tanzania
4.	—	—
5.	—	—
6.	—	—
7.	—	—
8.	Malaita, Solomon Islands	Cape Verde Islands
9.	Buell Park, Arizona	—

Note: Numbers are those of Table 15-1.

TABLE 15-4 ESTIMATED ABUNDANCES OF UNERODED, UNMETAMORPHOSED, IGNEOUS ROCKS IN THE EARTH'S CRUST

	Mass, $\times 10^{24}$ g	Percentage of mass in total crust $(28.56 \times 10^{24}$ g)
Continental crust		
Flood basalts	0.02	0.07
Arc basalts	0.18	0.63
Arc andesites	0.15	0.53
Granites, granodiorites, diorites	3.68	12.89
Phonolites, nepheline syenites, syenites	0.03	0.11
Gabbro	0.38	1.33
Ultramafic rocks	0.01	0.04
Oceanic crust		
Tholeiitic basalts	5.39	18.85
Alkali basalts	0.05	0.20

Source: Generalized from Table 3 of Ronov and Yaroshevsky, *Geophysical Monograph 13,* p. 37-57, 1969, copyrighted by the American Geophysical Union.

15.4 RELATIVE ABUNDANCES OF IGNEOUS ROCK TYPES

As indicated in Table 15-2, some igneous rock types are much more abundant than others. Quantitative estimates will be premature until the entire earth has been mapped and drilled, but one widely accepted set of numbers is summarized in Table 15-4. These data show the obvious bimodal distribution of felsic and mafic rocks in continents and oceans, and also emphasize that most of the originally igneous rocks in continental crust have been reworked into sedimentary and metamorphic rocks.

Ronov and others (1980) calculate that, in the last 153 million years, 5.1×10^8 km³ of oceanic lavas have formed and been preserved, compared to 0.25×10^8 km³ of continental lavas in the same time span. Of these totals, 0.04×10^8 km³ were of intermediate and felsic rocks, whereas 5.3×10^8 km³ were mafic. It appears unlikely that the rates, or proportions of rocks extruded, have remained constant. Rates of volcanism on continents and in oceans appear to have had maxima in the Cretaceous and Eocene–Miocene, with a minimum in the Paleocene (Ronov and others, 1980). Farther back in the earth's history, there were also variations in the proportions of igneous rocks. The rarity of komatiites and massif anorthosites in Cambrian and younger rocks has been mentioned in earlier chapters. Silica-undersaturated basalts are increasingly uncommon in progressively older terrains (Engel and others, 1965). The ratio K/Na in igneous rocks has changed through time, increasing most rapidly between 2.5 and 2.0×10^9 years ago (Engel and others, 1974). Similarly, $^{87}Sr/^{86}Sr$ of carbonate sedimentary rocks (reflecting the strontium isotopic ratio of the seawater from which they precipitated) also showed a rapid increase from 2.5 to 2.0×10^9 years ago (Veizer and Compston, 1976). The Archean jumps in K/Na and oceanic $^{87}Sr/^{86}Sr$ both appear to indicate rapid release of residual elements from the mantle, and rapid development of continental crust.

Tarling (1980) proposes that in the earliest stages of the earth's development the lithosphere was thinner because radiogenic heat production, and consequently the geothermal gradient, were greater. According to Tarling, thin oceanic lithosphere was not subducted deeply enough, before dehydrating, to permit enrichment of the overlying mantle wedge in water, alkalis, and, perhaps, silica. Later, after the first 2×10^9 years of cooling, thicker oceanic lithosphere could retain its water until subducted to greater depth. Increased mantle metasomatism then produced the parent from which calcalkaline magmas could arise to build continental crust through volcanic arcs. Such speculations will prompt the gathering of more complete data; eventually, we should be able to answer the questions raised here and in Chapter 1 concerning the rates and efficiency with which crust has evolved from mantle.

15.5 DISTRIBUTION OF HOLOCENE VOLCANOES

Volcanoes known or likely to have been active during the last 10,000 years have been plotted on a world map by Morris and others (1979), with an accompanying text (Simkin and others, 1981). For these young sites of magmatism, we can be reasonably certain of the tectonic regime. Such an instantaneous snapshot (representing only 0.0003% of recorded geologic time) may not be a fully representative sample of the response of magmatism to tectonism, but it is the best available. To ignore this short-term record would be analogous to generalizing about climate without ever considering weather.

However, we are hampered in using the distribution of these Holocene volcanoes, and the relative proportions of magma types among them, by a surprising lack of data. For fully one half of the Holocene volcanoes on land, we have no adequate petrographic descriptions or chemical analyses of even one sample, let alone a representative suite.

15.6 IN PLACE OF A SUMMARY: OUTSTANDING PROBLEMS

Even if we had a complete inventory of the compositions and amounts of lavas and pyroclastic materials emitted by every active volcano, we would still find it difficult to establish exact correspondence between magma composition and tectonic setting, for two reasons. First, we can be certain that the mantle is heterogeneous, but we do not know the scale of this variability, compared to the volume of mantle rock from which a given magma batch is derived. Metasomatism in the mantle may itself reflect stress conditions, but it is likely that compositional change is slow and may long precede the actual generation of magma. Second, there are heterogeneities within a regional stress field (localized extension within the generally compressive regime of an island arc, for example; Johnson, 1979) which can be invoked to explain the anomalous occurrence of "rift-type" volcanic rocks at destructive plate boundaries. This raises an important and unanswered question: What is the minimum size of a stress regime (in crust or mantle) that can influence magmatic compositions?

There is an abundance of models that codify magmatic compositions according to tectonic settings (for example, Jakeš and White, 1972), but evidence is accumulating that these attempts are premature, although necessary and stimulating. Johnson and others (1978) and Arculus and Johnson (1978), among others, point out some of the failures of the available models.

Although the kinds and abundances of igneous rocks within a region may give important evidence concerning the tectonic setting in which they were emplaced, this evidence must not take precedence over that gained from study of the associated sedimentary rocks, which may be far more sensitive environmental indicators (Blatt and others, 1980; Dickinson, 1980).

16

Relations of Magma to Energy and Mineral Resources

16.1 THERMAL ENERGY FROM MAGMA AND LAVA

As a silicate liquid crystallizes, heat equal to the enthalpy of fusion (Section 6.1) is given off. Commonly this amounts to 200 to 300 J per gram of material crystallized. In addition to heat liberated during crystallization, about 1 J (1 W sec) is given off per gram of rock or silicate liquid cooling 1°C. If all of the heat given off by 1 km³ of magma completely crystallizing and cooling 100°C could be converted to electrical energy, 2.4×10^{11} kWhr would be available. This seemingly large number is only about 1% of the total annual energy consumption in the United States. Nevertheless, magma brings abundant thermal energy within our reach in localized areas on the earth's surface; the question is the efficiency with which we can extract it.

The most direct method of tapping magmatic thermal energy is to insert heat exchangers in lava flows, lava lakes, or shallow intrusive magma bodies, so that heat can be conducted from the silicate liquid through the walls of the heat exchangers into a circulating fluid such as water; the resultant hot liquid or gas flowing at high velocity to the surface can turn the vanes of a turbine to generate electric power. Tests with a 1977 Hawaiian basalt flow (Hardee, 1979), at a temperature of 1095°C and moving between 3 and 5 km/hr, showed that conductive heat transfer amounted to 200 kW per square meter of heat-exchanger surface at the beginning of the test and decayed to 10 kW/m² within 5 minutes. Vigorous convection of magma or lava would increase the energy yield several-fold (Hardee, 1981); nevertheless, a large area of heat-exchange surface must be immersed in order to acquire economically significant amounts of thermal energy at a practical rate. The heat source, and the material of which the heat exchanger is fabricated, must last long enough to repay the large capital investment required for such an energy-extraction plant, probably a 10- to 30-year period (Peck, 1975).

Direct extraction of energy from magma or lava may be feasible, but we need much more information on the thermal and mechanical properties of natural silicate melts, on the survival of metals and ceramic materials immersed in these liquids, and on the response of a shallow magma body to human intervention.

16.2 THERMAL ENERGY FROM COOLING ROCK AND NATURALLY CONVECTING WATER

Some of the problems inherent in direct energy extraction from magma or lava can be circumvented using the natural heat exchangers provided by convecting groundwater in and around igneous bodies. Water is a very efficient carrier of heat because of its high heat capacity (five times as much heat is liberated by cooling water as by cooling the same mass of rock through the same temperature interval), as well as its low viscosity and its buoyancy (due to low density and high thermal expansion). The mechanical and thermal effects of intrusion usually cause fractures in the wall rock, increasing its permeability. Convective circulation through and over intrusive bodies (Figure 16-1) has been a topic of much recent research, summarized by Norton and Cathles (1979) and by Parmentier and Schedl (1981).

Above its solidus, magma is unlikely to be sufficiently permeable to allow the flow of groundwater. Furthermore, intense thermal metamorphism of wall rock close to the contact may produce an impermeable seal. Nevertheless, vigorous circulation of water in rock farther from the intrusive contact is to be expected. Later, after the magma body cools, crystallizes, and fractures, convective cells probably invade the igneous material itself. A detailed reconstruction of the buildup and decay of a hydrothermal convection system around the Skaergaard pluton is provided by Norton and Taylor (1979).

Hydrothermal convection systems customarily are divided into *vapor-dominated* and *liquid-dominated* (White and others, 1971), named for the phase that controls the pressure of the circulating fluid. Vapor-dominated systems are less common, but are at present the major sources of geothermal energy, as at Lardarello in Italy and The Geysers in California. Liquid-dominated systems are not only more widespread but contain much more heat, and will undoubtedly be utilized (White and Guffanti, 1979). It is important to remember that most of the heat in a hydrothermal system resides in the rock, not the water, and that the latter acts primarily as the transport medium. It is therefore possible, at least in theory, to extract thermal energy from impermeable dry rock by pumping water down one well, allowing it to travel

Surface

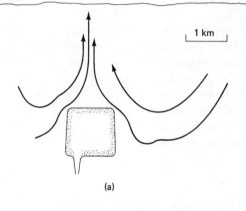

1 km

(a)

Surface

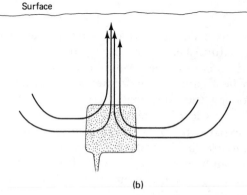

(b)

Figure 16-1. Flow lines of convecting groundwater in and around a cooling igneous body, in idealized cross section. (a) Magmatic crystallization is incomplete, and the intrusive body is impermeable. (b) The intrusive body has cooled below its solidus throughout and is now permeable. Flow strongly converges above the igneous body, where velocity and discharge of water are highest.

through artificially induced fractures in the hot rock, and removing it through one or more wells that intersect the fractures (Kolstad and McGetchin, 1978).

With the exception of Hawaiian lava lakes and hydrothermal systems in Iceland, magma-related geothermal resources are associated with shallow felsic intrusions less than a few million years old (Smith and Shaw, 1975). Apparently, mafic magmas tend to rise through the crust and spend most of their heat at the surface, whereas more viscous felsic magmas stop at shallow levels in the crust and lose more heat by conduction and convection. The generally higher water content of felsic magmas may inhibit their rise to the surface (Chapter 9) and may also provide larger quantities of water for the convective system after the solidus has been reached.

The properties of geothermal systems impose some severe restrictions on their utilization. In addition to sporadic occurrence and possibly rapid depletion through loss of pressure when wells are drilled, geothermal systems have other drawbacks because the "water" is usually a highly concentrated brine (as described in the next section). The high salinity causes fouling of pipes, valves, and turbines, and poses a formidable waste-disposal problem, because the cooled water generally is too rich in dissolved constituents to allow its dumping into lakes and rivers, but not rich enough to permit profitable extraction. One possible solution is to inject the spent hydrothermal fluid, under pressure, back into the hot reservoir rock.

Although localized extraction of energy from geothermal systems has proved practical, there are still many problems, some petrologic, to be addressed. These include changes in permeability due to precipitation or solution along the fractures through which the hot solutions migrate, over a span of years after the natural convective regime has been interrupted by drilling. Finally, there is the possibility that magma may come up through a drill hole; this has already happened once, in Iceland (Larsen and others, 1979).

16.3 ECONOMIC MINERAL DEPOSITS AND MAGMAS

Even for the most abundant metals (aluminum and iron) in the earth's crust, the lower limit of concentration for economically feasible mining is four or five times the average concentration in crustal rocks. For the less abundant elements, concentrations must be increased up to thousands of times the crustal average to produce ore deposits (bodies of rock from which one or more metals can be extracted at a profit). Because the term "ore deposit" is defined on the basis of nongeologic factors, this chapter will use the term *mineral deposit*, defined as a body of rock containing an abnormally high concentration of one or more elements, independent of economic considerations.

Extreme enrichment within a localized volume of the crust requires either physical segregation of one mineral (as by crystal settling), or introduction in the form of a highly concentrated liquid ("ore magma"), or precipitation from a large mass of dilute fluid (hydrothermal solution) that percolates through rock and leaves only a small fraction of its total dissolved load. The cumulate origin of some mineral deposits is demonstrable, as described later. The concept of ore magma is not in favor as an explanation of most mineral deposits, although dikes and lava flows consisting mostly of magnetite have been found (Henriquez and Martin, 1978), and sulfide liquids, immiscible with silicate liquids, carry high concentrations of some metals. In general, however, most mineral deposits associated with igneous bodies have precipitated from convecting hydrothermal fluid, the circulation of which was powered by magmatic heat. Opinions differ widely as to whether magma provides only the heat, or also supplies some or all of the components of the hydrothermal fluid (H_2O, CO_2, F, Cl, S) that was the transporting solvent, or also is the source of the economically important elements. Probably magmas have played all three roles, singly or in combinations, in forming the great variety of mineral deposits associated with igneous rocks.

Petrology and economic geology profit mutually from attempts to understand the genesis of mineral deposits. Areas of economic potential are areas of scientific importance, also, because mineral deposits are by definition anomalies, sites where geologic processes have been carried to extremes, and are therefore places deserving detailed study. Efficient feedback exists between basic research and the minerals extraction industry. For example, the petrologist benefits from having hundreds or thousands of meters of drill core available to supplement the scanty two-dimensional picture provided by surface outcrops, and the mining engineer benefits by being told the direction in which metal concentration is most likely to increase. The test of a scientific hypothesis is its ability to make successful predictions; economic geology provides unrivalled opportunity to test predictions by deeper drilling or excavation.

16.4 HYDROTHERMAL VENTS AT OCEANIC SPREADING CENTERS

Geologists saw mineral deposits in the act of forming when submersible craft explored the crest of the East Pacific Rise at 21°N. Jets of very hot water issued from cracks in the ridge crest, and moved at 1 to 5 m/sec with exit temperatures of up to $380 \pm 30°C$ (kept from boiling by the high pressure under 2.6 km of seawater). The water jets carried small particles of sulfides, sulfates, native sulfur, and silica. Porous "chimneys" up to 5 m in basal diameter and 20 m high accumulated over the hydrothermal vents as the suspended particles settled out and were cemented by hydrous iron oxides and clays.

The sulfide crusts (containing pyrrhotite, pyrite, sphalerite, and chalcopyrite) carry up to 29 wt % Zn, 6 wt % Cu, and 43 wt % Fe locally (Francheteau and others, 1979). The prominent sulfate minerals are anhydrite, gypsum, and barite. Apparently, the deposits rapidly oxidize and disperse unless buried by lava flows, but the total amount of heavy metals concentrated on the seafloor by hydrothermal exhalation must be large.

To enrich these elements by factors of 10^4 or more relative to their concentrations in MORB requires their removal from large volumes of oceanic crust by percolating water, circulating in convection cells over the magma chamber under the spreading ridge crest. Efficient removal of heavy metals from basalts requires reaction with percolating seawater to take place at temperatures higher than 300°C or at water/rock ratios exceeding 10 (Seyfried and Bischoff, 1981). The hot brine ascends rapidly to exit at the vents. Probably the sulfide ions and the native sulfur form by reduction of the abundant sulfate ions in seawater. Insoluble sulfides precipitate as the hydrothermal brine cools and mixes with seawater upon leaving the vents.

16.5 MASSIVE SULFIDE DEPOSITS

In one of the rare unblemished triumphs of petrologic deduction, the process observed at the hydrothermal vents on the East Pacific Rise had already been reconstructed from evidence in much older mineral deposits. The so-called *massive sulfide deposits* are layers or lenses, usually no more than a few hundred meters in lateral extent with sharp boundaries, very rich in pyrite and pyrrhotite and commonly containing chalcopyrite, sphalerite, galena, quartz, anhydrite, gypsum, and barite in varying proportions. Temperatures of crystallization were 200 to 300°C, where these can be estimated from isotopic ratios of sulfur and oxygen, from mineral assemblages, or from study of fluid inclusions. Commonly, however, the sulfide-rich bodies have been completely recrystallized during metamorphism, so the temperatures of initial crystallization can rarely be inferred. Less metamorphosed massive sulfide deposits are underlain by altered rock marking the conduits through which the hydrothermal fluid ascended.

Turner and Gustafson (1978), on the basis of fluid dynamic theory and scale-model experiments, showed that hot brine escaping from a vent on the seafloor would not immediately mix with seawater, but would retain its identity and either rise as a buoyant plume or flow along the bottom as a density current, depending on whether its combined temperature and salinity made it more dense or less dense than the seawater. The buoyant plume would precipitate more highly dispersed sulfides of different composition than those deposited from a density current, which would tend to collect in topographic depressions.

Massive sulfide deposits commonly are zoned, with Zn/Cu increasing upward and outward from the presumed vent. If the vents on the East Pacific Rise are typical of those that formed massive sulfide deposits in the past (and this is by no means certain), variations in velocity and buoyancy of the solution, and in settling velocities of the sulfide particles, may account for the varying proportions of metals in lateral and vertical zones.

Although one kind of massive sulfide deposit, rich in copper but with or without zinc, lead, or gold, does occur with basalt pillows in ophiolite complexes (for example, on Cyprus), there are other kinds, as reviewed by Sawkins (1976) and Hutchinson (1980). In one widespread type, named for the Kuroko deposits of Japan, copper–zinc massive sulfide deposits occur with calcalkaline felsic volcanic rocks (especially on the flanks of rhyolitic to dacitic lava domes) in a volcanic arc setting. Like the Cyprus type, Kuroko deposits show evidence of rapid submarine accumulation and outward zonation from copper-rich to zinc-rich to lead-rich in the parts farthest from the vents. Fluid inclusion studies indicate that the trapped fluids had salinities approximately twice that of normal seawater, and temperatures from 250 to 300°C. The evidence thus suggests that the mechanism was the same in forming both types, although the differences in tectonic setting and in compositions of the associated igneous rocks are profound. Differences in the overall proportions of copper, zinc, lead, and other metals from one kind of massive sulfide deposit to another probably reflect differences in the underlying rocks from which the metals were leached. Further to complicate the picture, there are other classes of massive sulfide deposits occurring in sedimentary rocks with little or no volcanic material.

16.6 FUMAROLE AND HOT SPRING DEPOSITS

When magmatic volatiles and convecting groundwater escape on land rather than on the seafloor, minerals precipitate at gas vents (fumaroles) and in hot springs. During the later 1980 activity at Mount St. Helens, powdery hematite and native sulfur formed localized crusts within the crater. At Vesuvius, unusual chlorides, fluorides, sulfides, and sulfates of heavy metals have long been recognized as vent linings. Stoiber and Rose (1974) described the rich variety of minerals deposited at Central American fumaroles.

Mercury, arsenic, and antimony are among the volatile metals that are mined from fumarole and hot spring deposits. In a geothermal field in New Zealand, dilute hydrothermal brine is depositing high concentrations of gold and silver at the surface (Ewers and Keays, 1977), and, at shallow depths, is altering the volcanic rocks and introducing arsenic and antimony as well as gold. Deeper in the same hydrothermal system, copper, lead, zinc, bismuth, and selenium are precipitating. Cooling of the solution, although important, seems to be subordinate as a control of metal deposition to boiling, which releases CO_2 and H_2S and changes the pH, diminishing the solubility of metals in complexes.

16.7 VEIN DEPOSITS

Veins are tabular bodies formed by partial or complete filling of fractures by minerals precipitated from circulating fluids. They may have parallel walls, but more commonly they branch, pinch, and swell. Textural and mineralogical zoning commonly occurs across the width of a vein, along its horizontal extent, and with depth. This three-dimensional variation reflects changes in the temperature and composition of the fluid through time and space. Because vein deposits by definition are narrow relative to their other two dimensions, and commonly are bounded by low-grade or barren rock, they are only profitable to mine if they contain high values per ton (either high grade or high unit price). Consequently, they are important sources for gold, silver, tin, uranium, fluorite, and barite, but are rarely worked today for copper, lead, or zinc. The most common *gangue* (nonore) minerals in veins are quartz and calcite.

Some vein deposits have formed from low-temperature fluids in areas where no igneous rocks are known. Because of their variability and their often tenuous links to magma, veins will not be considered further. Evans (1980) and the higher-level book edited by Barnes (1979) give much information on these and other kinds of mineral deposits.

16.8 PORPHYRY MINERAL DEPOSITS

In contrast to the other types of mineral deposits discussed in this chapter, economically important minerals may occur as small grains dispersed throughout the rock or concentrated along closely spaced fractures, in and around porphyritic felsic plutons that were emplaced within a few kilometers of the earth's surface. The ratio of rock to economically significant minerals is very large in these so-called porphyry deposits. The fractures that localize mineralization were formed late in the emplacement and crystallization histories of the plutons and are generally conceded to have formed in response to a sudden volume increase in the pluton caused by formation of a gas phase (boiling). Opinions differ as to whether the boiling was resurgent (Chapter 9) and caused by crystallization of anhydrous materials in a closed system, or whether it resulted from the pressure drop when fluid began to escape from the previously impermeable pluton into the much more permeable wall rock (Figure 16-2).

Figure 16-2. Brecciation and boiling in a shallow pluton. On the left is a schematic cross section through the pluton (X pattern) and its wall rock, showing flow lines for convecting groundwater. On the right is a depth (pressure)–temperature diagram showing boiling curves (A; for lithostatic pressure, controlled by the weight of the overlying column of rock; B; for hydrostatic pressure, controlled by the weight of the overlying column of water that is free to escape at the surface). Point m represents the temperature of boiling of a highly saline brine at a depth of 1.6 km in the pluton. Cracking of the impermeable shell of the pluton allows the fluid to begin equilibrating with convecting groundwater, whereupon the boiling curve moves from A to B, allowing rapid formation of vapor at m. The volume increase causes shattering (brecciation) of the pluton and adjacent wall rock. (Simplified from Cunningham, 1978, Figs. 1 and 4.)

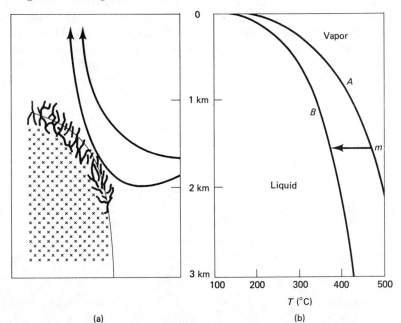

(a)   (b)

The most important porphyry deposits are those of copper-bearing sulfides, yielding more than half of the world's annual copper production (Gustafson, 1979). Commonly molybdenum, gold, silver, lead, and zinc are also extracted, singly or in combinations, as byproducts of copper mining. Molybdenum, tungsten, and tin are the primary products of other porphyry deposits.

Porphyry copper deposits typically are large and of low grade, usually less than 1% Cu, but contain more than 50 million tons of ore-grade rock. More than 100,000 metric tons of rock per day are extracted in a typical operation by a work force of fewer than 300 people; "an average open-pit copper or molybdenum mine, including pit, plant, waste disposal sites and all peripheral lands, occupies less than 900 hectares and will produce at least one billion dollars worth of metal over a period of thirty years" (Sutherland Brown and Cathro, 1976, p. 15). Because such mining operations must move large tonnages of rock, they must be highly mechanized. Subtle variations in metal content and in mineralogy, texture, spacing of fractures, and alteration, all controlling the behavior of the rock during removal, crushing, and metal extraction, must be continually monitored. Consequently, many large porphyry mineral deposits have been documented in detail in three dimensions as mining progressed. Greater understanding of the petrology of these felsic intrusive rocks and their wall rocks is partly a byproduct, partly a prerequisite, of profitable mining.

Nearly all porphyry copper deposits are in or around calcalkaline plutons that do not obviously differ, in major or trace element chemistry, from those abundant calcalkaline rocks that lack copper mineralization (Gustafson, 1979). More alkalic rocks can be hosts for porphyry copper deposits, as in the Canadian Cordillera (Barr and others, 1976); these contain less molybdenum, but higher concentrations of gold and silver, than calcalkaline rocks of the same general age in the same region. There is even one profitable copper sulfide deposit in a carbonatite, at Palabora, South Africa. Regardless of the rock type, porphyry copper deposits show abundant evidence that they formed close to the earth's surface. Nearly all such deposits are less than 200 million years old; probably older ones were formed but destroyed by erosion because of their near-surface genesis.

Alteration in porphyry mineral deposits generally defines concentric zones in the pluton and the wall rock. From the center outward (Figure 16-3), these tend to follow a specific sequence (Lowell and Guilbert, 1970; Rose and Burt, 1979). In the *potassic* zone, K-feldspar and/or biotite replace plagioclase and mafic silicates. Carbonates, white mica, and anhydrite are also prominent additions. In the *sericitic* zone, fine-grained white mica (sericite) replaces alkali feldspar and plagioclase, and is generally accompanied by pyrite and quartz. In the *propylitic* zone, either epidote or chlorite is an abundant alteration product, with or without albite and carbonates. An *argillic* zone may lie within the potassic zone or intervene between the sericitic and propylitic zones; the

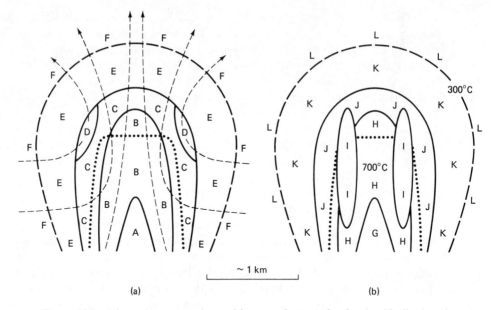

(a)   ~ 1 km   (b)

Figure 16-3. Schematic cross sections, with approximate scale, showing idealized zoning in a porphyry copper deposit. (a) Alteration zones: A, unaltered or deuterically altered intrusive rock; B, potassic zone; C, sericitic zone; D, argillic zone; E, propylitic zone; F, unaltered rock, except for local alteration next to veins. (b) Mineralization zones: G, disseminated magnetite more abundant than sulfides; H, low-grade ore zone, with low concentrations of pyrite, chalcopyrite, and molybdenite, disseminated and in small veinlets; I, high-grade ore zone containing approximately 1% pyrite, 1 to 3% chalcopyrite, 0.01 to 0.1% molybdenite, disseminated and in small veinlets; J, pyrite zone with 10% pyrite, 1 to 3% chalcopyrite, trace of molybdenite, in veinlets dominant over disseminated grains; K, 2% pyrite, plus chalcopyrite, galena, sphalerite, silver sulfides, and gold, in veins; L, larger but more widely spaced veins carrying quartz, native gold, silver, and copper. The dotted line is the intrusive contact, for reference only; it may lie higher or lower, farther in or out, relative to the alteration and mineralization zones. Light dashed lines in (a) indicate flow of convecting fluid. Temperature estimates in (b) are based on fluid inclusion studies. (Modified from Guilbert and Lowell, 1974, Figs. 1, 2, and 3, with permission of the Canadian Institute of Mining and Metallurgy.)

argillic zone takes its name from the abundant clay minerals which replace feldspars. Quartz and alunite, $KAl_3(SO_4)_2(OH)_6$, or other hydrous sulfate minerals accompany the clays. Argillic and sericitic alteration are generally younger than potassic and propylitic, and tend to be confined to the vicinity of fractures. On the other hand, potassic and propylitic alteration tend to pervade rocks more uniformly. These alteration zones are superimposed on, and therefore postdate, the effects of thermal metamorphism at intrusive contacts. Propylitic and argillic alteration also are common in the overlying volcanic rocks.

Like alteration, sulfide mineralization also tends to define concentric zones. Although there are many exceptions and omissions, the typical sequence from the center outward is Mo, Cu, Fe, Zn, Ag, Pb, and Mn. Mineralization may reach its highest intensity in the pluton, the wall rock, or at the contact between them. The symmetrical and concentric zonation of both the alteration and the mineralization suggests that both are related to the convective flow pattern (idealized in Figure 16-3a) that was established after the pluton had cooled enough to become permeable to circulating groundwater. Note also, in Figure 16-3b, the tendency for fracture-filling mineralization to become dominant over disseminated mineralization farther from the center of the hydrothermal system; this implies that the sulfide-depositing fluid in the cooler parts of the convection system became more highly channeled into widely spaced fractures, rather than along grain boundaries and the closely spaced fractures in breccia near the intrusive core.

The composition and temperature of the hydrothermal fluid that produced the alteration and mineralization can be inferred from several lines of evidence. The most direct is the testimony of fluid inclusions trapped as crystals grew. Most fluid inclusions are less than 0.01 mm in diameter and occupy far less than 0.1% of the volume of the host crystal (Roedder, 1979b). When these occur in transparent minerals (quartz or sphalerite, as examples), thin sections can be studied microscopically using transmitted light and a variety of ingenious techniques. Many fluid inclusions contain two phases (a liquid and a vapor bubble) at room temperature. The customary assumption is that the hydrothermal fluid completely filled the cavity at the temperature of crystallization, but shrank upon cooling (the host crystal would shrink much less in cooling through the same temperature interval). Therefore, the temperature at the time the fluid was trapped can be estimated by heating the prepared mineral in a chamber mounted on a microscope stage and observing the temperature at which the vapor bubble disappears (the liquid exactly fills the cavity), as shown in Figure 16-4. Conversely, freezing the sample, then allowing it to warm to room temperature while observing the melting of the ice, permits estimation of the salinity of the fluid. The temperature at which the last ice crystal disappears upon warming is a better estimate of the freezing temperature than is the temperature of ice crystallization upon cooling, because of the reluctance with which ice nucleates in a fluid inclusion. Other methods of analysis are described by Roedder (1979b).

Most fluid inclusions in porphyry and vein sulfide deposits contain aqueous solutions, with salinities ranging up to at least 40%, in which the major solutes are Na^+, K^+, Ca^{2+}, Mg^{2+}, Cl^-, and SO_4^{2-}. Some fluid inclusions contain crystalline phases, the most common of which is halite. Significantly, Cu, Pb, Mo, Zn, Ag, and S^{2-} rarely exceed a few hundred parts per million (Skinner, 1979), and the total of heavy metal ions is far greater than the sulfide anion concentration. The abundance of barite, anhydrite, and alunite as alteration products agrees with the compositions of fluid inclusions in indicating that most of the sulfur in the fluid was present as sulfate, rather than sulfide, anion. Nearly all metal sulfides are extremely insoluble in aqueous solutions,

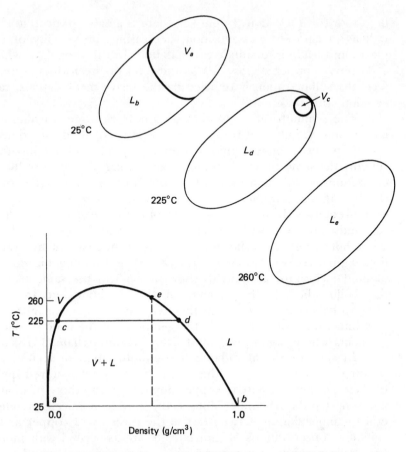

Figure 16-4. Behavior of a two-phase (liquid + vapor) fluid inclusion upon heating or cooling. At 25°C, liquid with density *b* coexists with vapor of density *a*. At 225°C, the vapor bubble, now with greater density *c*, coexists with liquid of decreased density *d*. At 260°C, the vapor bubble disappears, and the inclusion is completely filled with fluid of density *e*. The temperature of entrapment of the inclusion was 260°C if the trapped fluid was entirely one phase (with density *e*), and if the cavity remained closed and did not change volume after the fluid was trapped. Roedder (1979b) discusses the effects of pressure and salinity on the temperature–density diagram.

even at high temperatures. The general conclusion is that the metals are transported as soluble chloride, fluoride, carbonate, or polysulfide complexes in sulfate-bearing fluid; when the sulfate is reduced to sulfide, or the complexing anions are lost, metal sulfides precipitate.

White and others (1971) proposed that porphyry copper deposits probably form in the lower parts of vapor-dominated hydrothermal systems. At the interface between boiling brine (with a salinity exceeding 40%) and the overlying vapor (with less than 1 ppm Cl), there is an abrupt drop in temperature as well as in chloride concentration, decreasing the stability of chloride complexes and leading to sulfide precipitation. In this model, in which the top of a porphyry copper deposit marks the level of the former brine–vapor interface, the sulfate anion is reduced by the heavy metal cations, rather than by reaction with rock.

The symmetrical disposition of concentric zones of alteration, mineralization, and salinity and trapping temperature of fluid inclusions, is around a specific intrusive body within a composite pluton. Other intrusive pulses, of roughly the same composition, age, and setting, did not produce sulfide mineralization or alteration (Nielsen, 1976). This evidence argues strongly against the hypothesis that the copper was leached from wall rocks by convecting groundwater, with magma supplying only the heat. Instead, the copper, sulfur, and chlorine must have been carried up in the magma, then retained in dense hot brine. Maps showing variations in intensity of hydrothermal alteration, copper concentration, and the salinity and homogenization temperatures of fluid inclusions, all tend to show maxima in the same places, suggesting that boiling brine is responsible for the alteration and the mineralization. "The distribution of ore as an inverted cuplike form near the margin of its host intrusion probably reflects the geometry of the zone of steepest pressure gradients where boiling was most likely" (Cunningham, 1978, p. 747).

In spite of the abundance and sophistication of techniques applied to porphyry mineral deposits, there remain many unanswered questions. Many of these pertain not only to copper deposits, but to those of molybdenum, tin, and tungsten. Molybdenite, MoS_2, is the only economically significant ore mineral for molybdenum, in contrast to the diversity of copper and copper–iron sulfides. Concentrations of molybdenite are associated with more felsic rocks (generally granites), whereas porphyry copper deposits tend to occur in and around quartz diorites, granodiorites, and quartz monzonites (Sutherland Brown, 1976). Porphyry molybdenum deposits generally contain higher fluorine concentrations than the copper analogues, and this correlation of F with Mo led Gustafson (1979, p. 461) to conclude that fluorine is "as critical for distribution of molybdenite as chlorine is for copper," by allowing transport as a soluble molybdenum fluoride complex. As with copper, molybdenum mineralization is related to specific intrusive pulses (Figure 16-5).

Sillitoe and others (1975) defined porphyry tin deposits analogous to porphyry copper and molybdenum, although the ore mineral is an oxide, cassiterite, SnO_2, rather than a sulfide. Like chalcopyrite and molybdenite in porphyry copper and molybdenum deposits, cassiterite is subordinate to abundant pyrite. Cassiterite is also known in some porphyry Cu and Mo deposits, where it may be accompanied by tungstates [$CaWO_4$ and $(Fe,Mn)WO_4$] that provide tungsten as a byproduct during extraction of other metals, but most important economic deposits of tungsten are in skarns and veins.

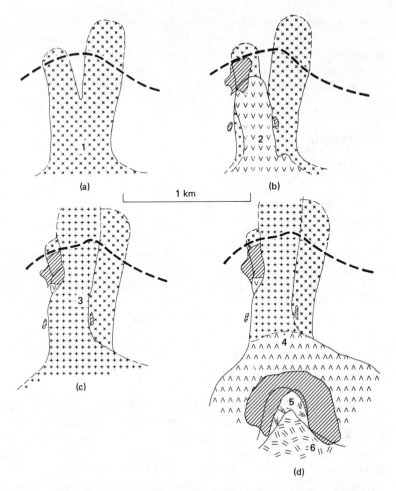

Figure 16-5. Cross sections showing four stages in the development of the Urad and Henderson molybdenum deposits at Red Mountain, Colorado. The heavy dashed line is the topographic surface now. Line pattern represents rock with more than 0.2 wt% MoS_2, formed by only the second and last pulses of granitic magma. (Generalized, after Wallace and others, 1978, Figs. 6, 7, 9, 10, and 21, with permission of Economic Geology Publishing Company.)

16.9 SKARN DEPOSITS

Contact metasomatism, particularly of carbonate wall rocks, has produced deposits enriched in iron, zinc, copper, fluorite, and other commodities. Tungsten and tin have largely been produced from skarns, generally next to I-type and S-type granites, respectively (Chappell and White, 1974). Fluorine is an important component of the metasomatizing fluid in such deposits (Rose and Burt, 1979). On the other hand, chlorine plays an essential role in the genesis of magnetite-rich skarns (Eugster and Chou, 1979); iron is apparently transported as $FeCl_2$ in convecting hydrothermal solution, which cools and reacts with carbonate wall rock to precipitate magnetite.

16.10 PEGMATITES

The mechanisms by which these extremely coarse grained and heterogeneous rock bodies were formed were described in Chapter 9. Many pegmatites, especially those of granitic composition, support small-scale mines for cesium, lithium, beryllium, tantalum, tin, and uranium. The small size and strong zoning prohibit large, highly mechanized operations, with few exceptions. One important exception is the Tanco (Bernic Lake) pegmatite, Manitoba, worked for tantalum but also greatly enriched in lithium, cesium, and beryllium. This complexly zoned body, of still unknown extent, formed by repeated episodes of resurgent boiling, with concentration of potassium, lithium, and silica in the less dense supercritical fluid in the upper parts (Crouse and Černy, 1972).

16.11 LAYERED DEPOSITS IN INTRUSIVE ROCKS

For some metals the most extreme enrichment, compared with their average crustal abundances, occurs when minerals containing these metals as essential components become concentrated as layers within crystallizing magma. Such segregations are thin (less than a centimeter to a few meters) but laterally extensive (from a few meters to more than 100 km). Some are nearly monominerallic (effectively pure magnetite, ilmenite, or chromite). Others contain more than one phase and formed by immiscibility between silicate and sulfide liquids.

Magnetite forms nearly pure layers in the upper parts of some gabbro and anorthosite plutons; chromite forms similar layers lower in gabbro bodies and less continuous pods in ultramafic cumulates of some ophiolites. Chromite layers in mafic and ultramafic rocks are almost the only source of chromium; magnetite layers, as in the upper part of the Bushveld intrusion, South Africa, have been mined for vanadium as well as iron, because the magnetite in some layers contains more than 1 wt % V_2O_5.

The mineralogically more diversified type of segregation results from liquid immiscibility. As mentioned in Chapter 9, the solubility of sulfur in silicate melts decreases with falling temperature. A dense sulfur-rich liquid then can separate from magma, carrying with it large concentrations of metals, and remain mobile because its solidus temperature is appreciably below that of the silicate liquid. In a Kilauea lava lake (Skinner and Peck, 1969), quenched basalt glass contained sulfide spheres up to 0.5 mm in diameter, consisting of approximately 75% pyrrhotite ($Fe_{1-x}S$), 15% magnetite, and 10% chalcopyrite ($CuFeS_2$). The bulk composition of the spheres is roughly 61% Fe, 31% S, 4% Cu, and 4% O, all in sharp contrast to the basalt liquid, which held 100 ppm Cu. Copper enrichment in the sulfide relative to the silicate liquid was therefore 400 times. If the spheres had coalesced and settled, a sulfide layer would have formed deep in the lava lake.

In large mafic and ultramafic plutons, where slow cooling gives opportunity for sulfide droplets to accumulate, sulfide layers are sources of nickel, copper, cobalt, and the platinum group metals. However, by no means all such plutons contain economically important segregations; most, in fact, are barren. Naldrett and Cabri (1976, p. 1142) concluded that "the main criterion for the formation of a rich concentration of magmatic sulfides is that the host magma should be saturated in sulfur"; Naldrett and Macdonald (1980) suggest that some important Ni–Cu sulfide bodies resulted after assimilation of sulfur from wall rock.

In one kind of layered sulfide deposit, nickel and copper are the principal metals extracted; others, including the platinum group, are byproducts. In the other type, the platinum group metals (platinum, palladium, rhodium, ruthenium, iridium, and osmium) are the major products, occurring in solid solution in iron and nickel sulfides but also exsolved or independently precipitated as alloys with iron or as sulfides, arsenides, and antimonides. The distribution coefficient expressing concentration in sulfide liquid relative to mafic silicate liquid is approximately 1000 for platinum (Naldrett and Duke, 1980).

The Bushveld intrusion accounts for nearly half of the world's annual production of platinum group metals. Most comes from one layer, the Merensky Reef, about 1 m thick and extending as a continuous layer for at least 150 km, about one third of the way from the base to the top of the layered ultramafic–mafic sequence. The Merensky Reef (Figure 16-6) contains cumulus orthopyroxene, chromite, and rare olivine with intercumulus plagioclase, clinopyroxene, biotite, amphibole, K-feldspar, quartz, and opaques. The opaque phases are pyrrhotite, chalcopyrite, pentlandite [$(Fe,Ni)_9S_8$], Pt–Fe alloy, and arsenides plus other compounds containing platinum group metals (Vermaak and Hendriks, 1976).

For many years the Merensky Reef and a few other Pt-bearing layers in the Bushveld intrusion were considered unique; then a very similar deposit twice as thick and at least as rich in platinum group metals was recognized in another Precambrian pluton of stratified ultramafic and mafic rocks, the Stillwater complex of Montana. "The fact that there are two such zones is as remarkable as either zone in itself. They would seem to reflect some exceptional event (or combination of events) that happened, not just once, but twice, in two intrusions in widely separated parts of the earth's crust" (Todd and others, 1979, p. 461). What these events were remains a topic for study.

Figure 16-6. The Merensky Reef, Rustenburg Platinum Mine, Bushveld Complex, South Africa. (a) Hand sample through the basal contact of the Merensky Reef. Interval 1 is the underlying anorthosite cumulate; interval 2 is a chromite-rich cumulate in irregular contact with anorthosite; interval 3 is coarse orthopyroxene cumulate (intercumulus sulfides are gray and white). (b) Photomicrograph, in plane-polarized transmitted light, of a thin section from interval 3. Cumulus orthopyroxene surrounds a triangular space occupied partly by clear plagioclase and partly by opaque sulfides.

16.12 METALLOGENIC PROVINCES

Geologists have long recognized that certain areas of continental crust contain more mineral deposits of specific metals than would be expected if the distribution of those deposits were random. There is no reason to believe that mineralization should be random, if there is any relationship between magmatism and tectonic setting. Nevertheless, similar mineral deposits tend to occur in clusters that cut across boundaries between tectonic regimes, and mineral-

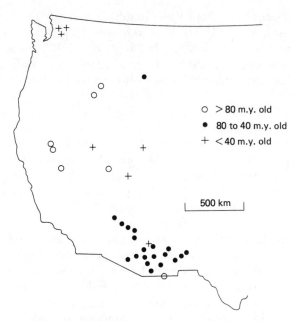

Figure 16-7. Porphyry copper deposits in the western United States, divided into three age groups. (After Guild, 1978, Figs. 5-7, with permission of the Geological Society of London and P. W. Guild.)

ization within these clusters tended to recur at widely separated times. These observations led to the concept of a *metallogenic province*, a region defined by similar mineral deposits that formed within a specific time span (short or long; definitions vary). Examples of such provinces are: the ultramafic and mafic plutons in southern Africa that contain chromite segregations and were emplaced over a span of 1500 million years; the ultramafic and mafic plutons in northeastern North America that contain nickel-rich sulfide layers, also spanning some 1500 million years; the tin-bearing granites of Nigeria, of 500 to 160 million year age (Watson, 1980); the porphyry copper deposits in the western United States (Figure 16-7).

It is unlikely that the recurrence of a specific kind of mineralization is entirely due to recycling of crustal rocks that were abnormally enriched in an element during one initial episode, particularly when the mineralization occurs in mafic and ultramafic rocks derived from the mantle. There is therefore a widely held but tentative conclusion that metallogenic provinces reflect mantle heterogeneities with respect to trace metals and volatile components (Noble, 1974; Watson, 1980).

16.13 IGNEOUS ROCKS AS REPOSITORIES FOR WASTE

At the other end of the economic flowsheet from mining is the problem of toxic and radioactive waste disposal. Toxic wastes require permanent isolation, and radioactive wastes must be sequestered until their radioactivity has decayed to an acceptable level. The problem is compounded because some radioactive elements are also toxic. To hold these materials in a site inaccessible to the atmosphere, hydrosphere, and biosphere requires deep burial in rock that has very low permeability and solubility, has high strength even at high temperatures (such as might be generated by chemical reactions or radioactive decay), and is located in an area unlikely to be disturbed by faulting or igneous activity.

Igneous and high-grade metamorphic rocks consist of minerals that crystallized at high temperature and are, for the most part, relatively insoluble. The textures of many such rocks impart high strength and low permeability, at least on a scale of centimeters, but jointing can result in high permeability through widely spaced fractures. Careful petrological, geophysical, and engineering studies are required of any potential waste repository, even if burial to depths of several kilometers is planned.

Petrologists and mineralogists can also contribute to the solution of waste disposal problems through their knowledge of the chemistry and stability of inorganic compounds at depth. Before burial in any host rock, highly toxic and radioactive elements could be encapsulated, by solid solution, in synthetic minerals that remain stable and insoluble at high temperatures and pressures. There is much that must be learned about phase relations of silicates, phosphates, and oxides that can incorporate as major components such elements as thallium, cadmium, cesium, selenium, and plutonium, although these elements are merely traces in most natural minerals.

16.14 SUMMARY

Large amounts of thermal energy are available in magma and cooling igneous rock. Although direct extraction of heat from magma or lava may be feasible, natural or artificial heat exchange using circulating water is probably more efficient. Young, shallow felsic plutons are the primary targets for geothermal prospecting.

Convecting water in and around cooling igneous bodies not only carries heat, but can transport large quantities of dissolved metals. When the solution encounters changes in physical or chemical environment during its migration, the metals can be precipitated to form mineral deposits, in which the metals are more highly concentrated than in most crustal rocks. Other mechanisms that can efficiently segregate metals include vapor–liquid fractionation (as in pegmatites), sulfide–silicate immiscibility, and the still controversial processes by which cumulus layers form.

The isolation of toxic and radioactive wastes is a severe problem. In the endeavor to solve it, we must understand the interactions between crystalline solids and percolating solutions, applying the same principles that govern the formation of magma-related energy and mineral resources.

References

Allègre, C. J., and D. Ben Othman, 1980, "Nd–Sr isotopic relationship in granitoid rocks and continental crust development: a chemical approach to orogenesis," *Nature, 286,* 335–342.

Allègre, C. J., and J.-F. Minster, 1978, "Quantitative models of trace element behavior in magmatic processes," *Earth and Planetary Science Letters, 38,* 1–25.

Allègre, C. J., B. Dupré, B. Lambret, and P. Richard, 1981, "The subcontinental versus suboceanic debate: 1. Lead–neodymium–strontium isotopes in primary alkali basalts from a shield area: the Ahaggar volcanic suite," *Earth and Planetary Science Letters, 52,* 85–92.

Anderson, A. T., 1974, "Chlorine, sulfur and water in magmas and oceans," *Geol. Soc. America Bulletin, 85,* 1485–1492.

Anderson, A. T., 1975, "Some basaltic and andesitic gases," *Reviews of Geophysics and Space Physics, 13,* 37–55.

Anderson, A. T., 1976, "Magma mixing: petrological process and volcanological tool," *Jour. Volcanol. and Geothermal Research, 1,* 3–33.

Anderson, D. L., 1979, "The deep structure of continents," *Jour. Geophys. Research, 84,* 7555–7560.

Anderson, D. L., 1981, "Rise of deep diapirs," *Geology, 9,* 7–9.

Anderson, O. L., 1979, "The role of fracture dynamics in kimberlite pipe formation," *Proc. Second Internat. Kimberlite Conference, 1,* 344–353.

Anderson, R. N., S. E. DeLong, and W. M. Schwarz, 1980, "Dehydration, asthenospheric convection and seismicity in subduction zones," *Jour. Geology, 88,* 445–451.

Appleyard, E. C., and A. R. Woolley, 1979, "Fenitization: an example of the problems of characterizing mass transfer and volume changes," *Chemical Geology, 26,* 1–15.

Arculus, R. J., and R. W. Johnson, 1978, "Criticism of generalised models for the magmatic evolution of arc–trench systems," *Earth and Planetary Science Letters, 39,* 118–126.

Arndt, N. T., 1977, "The separation of magmas from partially molten peridotite," *Carnegie Inst. Washington Yearbook, 76,* 424–428.

Arndt, N. T., A. J. Naldrett, and D. R. Pyke, 1977, "Komatiitic and iron-rich tholeiitic lavas of Munro Township, northeast Ontario," *Jour. Petrology, 18,* 319–369.

Arth, J. G., 1976, "Behavior of trace elements during magmatic processes—a summary of theoretical models and their applications," *U.S. Geol. Survey Jour. Research, 4,* 41–47.

Atherton, M. P., 1981, "Horizontal and vertical zoning in the Peruvian Coastal Batholith," *Geol. Soc. London Jour., 138,* 343–349.

Atherton, M. P., and J. Tarney, eds., 1979, *Origin of Granite Batholiths: Geochemical Evidence:* Orpington, Kent, Shiva Publishing, Ltd., 148 p.

Avé Lallemant, H. G., J.-C. C. Mercier, N. L. Carter, and J. V. Ross, 1980, "Rheology of the upper mantle: inferences from peridotite xenoliths," *Tectonophysics, 70,* 85–113.

Bailey, D. K., 1974, "Nephelinites and ijolites," p. 53–66 in Sørensen, H., ed., *The Alkaline Rocks:* New York, John Wiley & Sons, Inc., 622 p.

Bailey, D. K., 1976, "Applications of experiments to alkaline rocks," p. 419–469 in Bailey, D. K., and R. Macdonald, eds., *The Evolution of the Crystalline Rocks:* New York, Academic Press, Inc., 484 p.

Bailey, D. K., and J. F. Schairer, 1966, "The system Na_2O–Al_2O_3–Fe_2O_3–SiO_2 at one atmosphere, and the petrogenesis of alkaline rocks," *Jour. Petrology, 7,* 114–170.

Bailey, R. A., 1976, "On the mechanisms of postsubsidence central doming and volcanism in resurgent cauldrons," *Geol. Soc. America Abstracts with Programs, 8,* 567.

Baitis, H. W., and M. M. Lindstrom, 1980, "Geology, petrography, and petrology of Pinzon Island, Galapagos Archipelago," *Contrib. Mineralogy and Petrology, 72,* 367–386.

Baker, B. H., R. Crossley, and G. G. Goles, 1978, "Tectonic and magmatic evolution of the southern part of the Kenya rift valley," p. 29–50 in Neumann, E.-R., and I. B. Ramberg, eds., *Petrology and Geochemistry of Continental Rifts:* Dordrecht, D. Reidel Publishing Co., 296 p.

Baker, B. H., P. A. Mohr, and L. A. J. Williams, 1971, *Geology of the Eastern Rift System of Africa:* Geol. Soc. America Special Paper *136,* 67 p.

Baker, M. C. W., and P. W. Francis, 1978, "Upper Cenozoic volcanism in the central Andes—ages and volumes," *Earth and Planetary Science Letters, 41,* 175–187.

Baksi, A. K., and N. D. Watkins, 1973, "Volcanic production rates: comparisons of oceanic ridges, islands, and the Columbia Plateau basalts," *Science, 180,* 493–496.

Balk, R., 1937, *Structural Behavior of Igneous Rocks:* Geol. Soc. America Memoir *5,* 177 p.

Barazangi, M., and B. L. Isacks, 1979, "Subduction of the Nazca plate beneath Peru: evidence from spatial distribution of earthquakes," *Royal Astron. Soc. Geophys. Jour., 57,* 537–555.

Barker, D. S., 1970, "Compositions of granophyre, myrmekite, and graphic granite," *Geol. Soc. America Bulletin, 81,* 3339–3350.

Barker, D. S., 1976, "Phase relations in the system $NaAlSiO_4$–SiO_2–$NaCl$–H_2O at 400°–800°C and 1 kilobar, and petrologic implications," *Jour. Geology, 84,* 97–106.

Barker, D. S., 1978, "Magmatic trends on alkali–iron–magnesium diagrams," *Amer. Mineralogist, 63,* 531–534.

Barker, D. S., and K. P. Young, 1979, "A marine Cretaceous nepheline basanite volcano at Austin, Texas," *Texas Jour. Science, 31,* 5–24.

Barker, F., 1979, "Trondhjemite: definition, environment and hypotheses of origin," p. 1–12 in Barker, F., ed., *Trondhjemites, Dacites, and Related Rocks*: Amsterdam, Elsevier Scientific Publishing Co., 659 p.

Barker, F., and J. G. Arth, 1976, "Generation of trondhjemitic–tonalitic liquids and Archean bimodal trondhjemite–basalt suites," *Geology, 4,* 596–600.

Barnes, H. L., ed., 1979, *Geochemistry of Hydrothermal Ore Deposits*: New York, John Wiley & Sons, Inc., 798 p.

Barr, D. A., P. E. Fox, K. E. Northcote, and V. A. Treto, 1976, "The alkaline suite porphyry deposits—a summary," p. 359–367, in Sutherland Brown, A., ed., *Porphyry Deposits of the Canadian Cordillera*: Canad. Inst. Mining and Metallurgy Special Volume *15,* 510 p.

Barrière, M., 1976, "Flowage differentiation: limitation of the 'Bagnold effect' to the narrow intrusion," *Contrib. Mineralogy and Petrology, 55,* 139–145.

Barth, T. F. W., and I. B. Ramberg, 1966, "The Fen circular complex," p. 225–257 in Tuttle, O. F., and J. Gittins, eds., *Carbonatites*: New York, John Wiley & Sons, Inc., 591 p.

Bartlett, R. W., 1969, "Magma convection, temperature distribution and differentiation," *Amer. Jour. Science, 267,* 1067–1082.

Basaltic Volcanism Study Project, 1981, *Basaltic Volcanism on the Terrestrial Planets*: New York, Pergamon Press, Inc., 1286 p.

Bateman, P. C., 1979, *Cross Section of the Sierra Nevada from Madera to the White Mountains, Central California*: Geol. Soc. America Map and Chart Series, *MC-28E.*

Bateman, P. C., and B. W. Chappell, 1979, "Crystallization, fractionation and solidification of the Tuolumne Intrusive Series, Yosemite National Park, California," *Geol. Soc. America Bulletin, 90,* 465–482.

Bateman, P. C., and F. C. W. Dodge, 1970, "Variations of major chemical constituents across the central Sierra Nevada batholith," *Geol. Soc. America Bulletin, 81,* 409–420.

Bateman, P. C., L. D. Clark, N. K. Huber, J. G. Moore, and C. D. Rinehart, 1963, *The Sierra Nevada Batholith. A Synthesis of Recent Work across the Central Part*: U.S. Geol. Survey Professional Paper *414-D,* 46 p.

Bates, R. L., and J. A. Jackson, eds., 1980, *Glossary of Geology,* 2nd ed.: Falls Church, Va., American Geological Institute, 749 p.

Bence, A. E., T. L. Grove, and J. J. Papike, 1980, "Basalts as probes of planetary interiors: constraints on the chemistry and mineralogy of their source regions," *Precambrian Research, 10,* 249–279.

Blatt, H., G. V. Middleton, and R. C. Murray, 1980, *Origin of Sedimentary Rocks,* 2nd ed.: Englewood Cliffs, N.J., Prentice-Hall, Inc., 782 p.

Boettcher, A. L., and J. R. O'Neil, 1980, "Stable isotope, chemical and petrographic studies of high-pressure amphiboles and micas: evidence for metasomatism in the mantle source regions of alkali basalts and kimberlites," *Amer. Jour. Science, 280-A,* 594–621.

Boettcher, A. L., and P. J. Wyllie, 1968, "The quartz–coesite transition measured in the presence of a silicate liquid and calibration of piston–cylinder apparatus," *Contrib. Mineralogy and Petrology, 17,* 224–232.

Borley, G. D., 1974, "Oceanic islands," p. 311–330 in Sørensen, H., ed., *The Alkaline Rocks*: New York, John Wiley & Sons, Inc., 622 p.

Boudier, F., and R. G. Coleman, 1981, "Cross section through the peridotite in the Samail ophiolite, southeastern Oman Mountains," *Jour. Geophys. Research, 86,* 2573–2592.

Bowen, N. L., 1913, "The melting phenomena of the plagioclase feldspars," *Amer. Jour. Science,* 4th series, *35,* 577–599.

Bowen, N. L., 1928, *The Evolution of the Igneous Rocks*: Princeton, N.J., Princeton University Press, 334 p. (reprinted 1956 by Dover Publications, Inc., New York).

Bowen, N. L., 1937, "Recent high-temperature research on silicates and its significance in igneous geology," *Amer. Jour. Science, 33,* 1–21.

Bowen, N. L., 1948, "The granite problem and the method of multiple prejudices," *Geol. Soc. America Memoir 28,* 79–90.

Bowen, N. L., and J. F. Schairer, 1935, "The system $MgO–FeO–SiO_2$," *Amer. Jour. Science, 29,* 151–217.

Boyd, F. R., 1973, "A pyroxene geotherm," *Geochim. et Cosmochim. Acta, 37,* 2533–2546.

Boyd, F. R., and A. A. Finnerty, 1980, "Conditions of origin of natural diamonds of peridotite affinity," *Jour. Geophys. Research, 85,* 6911–6918.

Boyd, F. R., and R. H. McCallister, 1976, "Densities of fertile and sterile garnet peridotites," *Geophys. Research Letters, 3,* 509–512.

Brooks, C. K., J. J. Fawcett, J. Gittins, and J. C. Rucklidge, 1981, "The Batbjerg complex, east Greenland: a unique ultrapotassic Caledonian intrusion," *Canad. Jour. Earth Sciences, 18,* 274–285.

Brown, G. M., 1967, "Mineralogy of basaltic rocks," p. 103–162 in Hess, H. H., and A. Poldervaart, eds., *Basalts: The Poldervaart Treatise on Rocks of Basaltic Composition,* Vol. 1: New York, Wiley-Interscience, 482 p.

Bryan, W. B., and J. G. Moore, 1977, "Compositional variations of young basalts in the Mid-Atlantic Ridge rift valley near lat 36°49′ N," *Geol. Soc. America Bulletin, 88,* 556–570.

Bryan, W. B., L. W. Finger, and F. Chayes, 1969, "Estimating proportions in petrographic mixing equations by least-squares approximation," *Science, 163,* 926–927.

Bryan, W. B., G. Thompson, and J. N. Ludden, 1980, "'Normal' basalts from the Kane Fracture Zone and adjacent ridge axes," *Geol. Soc. America Abstracts with Programs, 12,* 394.

Bryson, R. A., and B. M. Goodman, 1980, "Volcanic activity and climatic changes," *Science, 207,* 1041–1044.

Buddington, A. F., 1959, "Granite emplacement with special reference to North America," *Geol. Soc. America Bulletin, 70,* 671–747.

Buddington, A. F., and D. H. Lindsley, 1964, "Iron–titanium oxide minerals and synthetic equivalents," *Jour. Petrology, 5,* 310–357.

Bulau, J. R., H. S. Waff, and J. A. Tyburczy, 1979, "Mechanical and thermodynamic constraints on fluid distribution in partial melts," *Jour. Geophys. Research, 84,* 6102–6108.

Bullard, F. M., 1976, *Volcanoes of the Earth*: Austin, Tex., University of Texas Press, 579 p.

Burke, K., and A. M. C. Sengör, 1979, "Review of plate tectonics," *Reviews of Geophysics and Space Physics, 17,* 1081–1090.

Burnham, C. W., 1975, "Water and magmas: a mixing model," *Geochim. et Cosmochim. Acta, 39,* 1077–1084.

Burnham, C. W., 1979a, "The importance of volatile constituents," p. 439–482 in Yoder, H. S., Jr., ed., *The Evolution of the Igneous Rocks: Fiftieth Anniversary Perspectives*: Princeton, N.J., Princeton University Press, 588 p.

Burnham, C. W., 1979b, "Magmas and hydrothermal fluids," p. 71–136 in Barnes, H. L., ed., *Geochemistry of Hydrothermal Ore Deposits,* 2nd ed.: New York, John Wiley & Sons, Inc., 798 p.

Burnham, C. W., and N. F. Davis, 1969, "Energy relations in water-bearing magmas," *Geol. Soc. America Abstracts with Programs, Part 7,* 26–27.

Burnham, C. W., and N. F. Davis, 1971, "The role of H_2O in silicate melts: I. P–V–T relations in the system $NaAlSi_3O_8$–H_2O to 10 kilobars and 1000°C," *Amer. Jour. Science, 270,* 54–79.

Burnham, C. W., and N. F. Davis, 1974, "The role of H_2O in silicate melts: II. Thermodynamic and phase relations in the system $NaAlSi_3O_8$–H_2O to 10 kilobars, 700° to 1100°C," *Amer. Jour. Science, 274,* 902–940.

Burnham, C. W., J. R. Holloway, and N. F. Davis, 1969, *Thermodynamic Properties of Water to 1000°C and 10,000 bars*: Geol. Soc. America Special Paper *132,* 96 p.

Burns, R. G., 1970, *Mineralogical Applications of Crystal Field Theory*: Cambridge, England, Cambridge University Press, 224 p.

Cameron, E. N., R. H. Jahns, A. H. McNair, and L. R. Page, 1949, *Internal Structure of Granitic Pegmatites*: Econ. Geology Monograph *2,* 115 p.

Campbell, I. H., 1978, "Some problems with the cumulus theory," *Lithos, 11,* 311–323.

Cann, J. R., 1970, "Upward movement of granite magma," *Geol. Magazine, 107,* 335–340.

Cann, J. R., 1974, "A model for oceanic crustal structure developed," *Royal Astron. Soc. Geophys. Jour., 39,* 169–187.

Carl, J. D., and B. B. Van Diver, 1975, "Precambrian Grenville alaskite bodies as ashflow tuffs, northwest Adirondacks, New York," *Geol. Soc. America Bulletin, 86,* 1691–1701.

Carmichael, I. S. E., 1963, "The crystallization of feldspar in volcanic acid liquids," *Geol. Soc. London Quarterly Jour., 119,* 95–131.

Carmichael, I. S. E., 1964, "The petrology of Thingmuli, a Tertiary volcano in eastern Iceland," *Jour. Petrology, 5,* 435–460.

Carmichael, I. S. E., 1967, "The mineralogy and petrology of the volcanic rocks from the Leucite Hills, Wyoming," *Contrib. Mineralogy and Petrology, 15,* 24–66.

Carmichael, I. S. E., 1979, "Glass and the glassy rocks," p. 233–244 in Yoder, H. S., Jr., ed., *The Evolution of the Igneous Rocks: Fiftieth Anniversary Perspectives*: Princeton, N.J., Princeton University Press, 588 p.

Carmichael, I. S. E., F. J. Turner, and J. Verhoogen, 1974, *Igneous Petrology*: New York, McGraw-Hill Book Company, 739 p.

Carmichael, I. S. E., J. Nicholls, F. J. Spera, B. J. Wood, and S. A. Nelson, 1977, "High temperature properties of silicate liquids: applications to the equilibration and ascent of basic magma," *Royal Soc. London Philos. Trans., A, 286,* 373–431.

Carr, M. J., W. I. Rose, and D. G. Mayfield, 1979, "Potassium content of lavas and depth to the seismic zone in Central America," *Jour. Volcanol. and Geothermal Research, 5,* 387–401.

Carswell, D. A., and F. G. F. Gibb, 1980, "Geothermometry of garnet lherzolite nodules with special reference to those from the kimberlites of Northern Lesotho," *Contrib. Mineralogy and Petrology, 74,* 403–416.

Chappell, B. W., and A. J. R. White, 1974, "Two contrasting granite types," *Pacific Geology, 8,* 173–174.

Chayes, F., 1956, *Petrographic Modal Analysis: An Elementary Statistical Appraisal:* New York, John Wiley & Sons, Inc., 113 p.

Chayes, F., 1965, "Titania and alumina content of oceanic and circumoceanic basalt," *Mineral. Magazine, 34,* 126–131.

Chayes, F., 1966, "Alkaline and subalkaline basalts," *Amer. Jour. Science, 264,* 128–145.

Chayes, F., 1969, "The chemical composition of Cenozoic andesite," *Oregon Department of Geology and Mineral Industries Bulletin, 65,* 1–11.

Chayes, F., 1971, *Ratio Correlation: A Manual for Students of Petrology and Geochemistry*: Chicago, University Chicago Press, 99 p.

Chayes, F., 1972, "Silica saturation in Cenozoic basalt," *Royal Soc. London Philos. Trans., A, 271,* 285–296.

Chen, C.-H., and D. C. Presnall, 1975, "The system Mg_2SiO_4–SiO_2 at pressures up to 25 kilobars," *Amer. Mineralogist, 60,* 398–406.

Christensen, N., and J. D. Smewing, 1981, "Geology and seismic structure of the northern section of the Oman ophiolite," *Jour. Geophys. Research, 86,* 2545–2555.

Clague, D. A., 1978, "The oceanic basalt–trachyte association: an explanation of the Daly Gap," *Jour. Geology, 86,* 739–743.

Clarke, D. B., 1981, "The mineralogy of peraluminous granites: a review," *Canad. Mineralogist, 19,* 3–17.

Coleman, R. G., 1977, *Ophiolites: Ancient Oceanic Lithosphere?*: New York, Springer-Verlag, 229 p.

Coleman, R. G., 1981, "Tectonic setting for ophiolite obduction in Oman," *Jour. Geophys. Research, 86,* 2497–2508.

Coleman, R. G., and M. M. Donato, 1979, "Oceanic plagiogranite revisited," p. 149–168 in Barker, F., ed., *Trondhjemites, Dacites, and Related Rocks*: Amsterdam, Elsevier Scientific Publishing Co., 659 p.

Coleman, R. G., D. E. Lee, L. B. Beatty, and W. W. Brannock, 1965, "Eclogites and eclogites: their differences and similarities," *Geol. Soc. America Bulletin, 76,* 483–508.

Cooper, A. F., J. Gittins, and O. F. Tuttle, 1975, "The system Na_2CO_3–K_2CO_3–$CaCO_3$ at 1 kbar and its significance in carbonatite petrogenesis." *Amer. Jour. Science, 275,* 534–560.

Cox, K. G., J. D. Bell, and R. J. Pankhurst, 1979, *The Interpretation of Igneous Rocks:* London, George Allen & Unwin Ltd., 450 p.

Crouse, R. A., and P. Černy, 1972, "The Tanco Pegmatite at Bernic Lake, Manitoba: I. Geology and paragenesis," *Canad. Mineralogist, 11,* 591–608.

Cunningham, C. G., 1978, "Pressure gradients and boiling as mechanisms for localizing ore in porphyry systems," *U.S. Geol. Survey Jour. Research, 6,* 745–754.

Daly, R. A., G. E. Manger, and S. P. Clark, Jr., 1966, "Density of rocks," p. 19–26 in Clark, S. P., Jr., ed., *Handbook of Physical Constants*: Geol. Soc. America Memoir *97,* 587 p.

Davies, G. F., 1980, "Review of oceanic and global heat flow estimates," *Reviews of Geophysics and Space Physics, 18,* 718–722.

Davies, G. F., 1981, "Earth's neodymium budget and structure and evolution of the mantle," *Nature, 290,* 208–213.

Dawson, J. B., 1980, *Kimberlites and Their Xenoliths*: New York, Springer-Verlag, 252 p.

Dawson, J. B., P. Bowden, and G. C. Clark, 1968, "Activity of the carbonatite volcano Oldoinyo Lengai, 1966," *Geol. Rundschau, 57,* 865–879.

Deer, W. A., R. A. Howie, and J. Zussman, 1966, *An Introduction to the Rock-forming Minerals*: New York, John Wiley & Sons, Inc., 528 p.

Deer, W. A., R. A. Howie, and J. Zussman, 1978, *Rock-forming Minerals*: Vol. 2A, *Single-Chain Silicates*: New York, Halsted Press, 668 p.

Delaney, J. S., J. V. Smith, and P. H. Nixon, 1979, "Model for upper mantle below Malaita, Solomon Islands, deduced from chemistry of lherzolite and megacryst minerals," *Contrib. Mineralogy and Petrology, 70,* 209–218.

DeLong, S. E., and M. A. Hoffman, 1975, "Alkali/silica distinction between Hawaiian tholeiites and alkali basalts," *Geol. Soc. America Bulletin, 86,* 1101–1108.

DeLong, S. E., F. N. Hodges, and R. J. Arculus, 1975, "Ultramafic and mafic inclusions, Kanaga Island, Alaska, and the occurrence of alkaline rocks in island arcs," *Jour. Geology, 83,* 721–736.

DePaolo, D. J., 1980a, "Crustal growth and mantle evolution: inferences from models of element transport and Nd and Sr isotopes," *Geochim. et Cosmochim. Acta, 44,* 1185–1196.

DePaolo, D. J., 1980b, "Sources of continental crust: neodymium isotope evidence from the Sierra Nevada and Peninsular Ranges," *Science, 209,* 684–687.

DePaolo, D. J., and R. W. Johnson, 1979, "Magma genesis in the New Britain island–arc: constraints from Nd and Sr isotopes and trace-element patterns," *Contrib. Mineralogy and Petrology, 70,* 367–379.

Dewey, J. F., 1975, "Finite plate implications: some implications for the evolution of rock masses at plate margins," *Amer. Jour. Science, 275-A,* 260–284.

Dewey, J. F., and K. C. A. Burke, 1973, "Tibetan, Variscan, and Precambrian basement reactivation: products of continental collision," *Jour. Geology, 81,* 683–692.

Dickinson, W. R., 1973, "Widths of modern arc–trench gaps proportional to past duration of igneous activity in associated magmatic arcs," *Jour. Geophys. Research, 78,* 3376–3389.

Dickinson, W. R., 1975, "Potash-depth (K-*h*) relations in continental margins and intra oceanic magmatic arcs," *Geology, 3,* 53–56.

Dickinson, W. R., 1980, "Plate tectonics and key petrologic associations," p. 341–360 in Strangway, D. W., ed., *The Continental Crust and Its Mineral Deposits*: Geol. Assoc. Canada Special Paper *20*.

Dixon, T. H., and R. Batiza, 1979, "Petrology and chemistry of Recent lavas in the northern Marianas: implications for the origin of island arc basalts," *Contrib. Mineralogy and Petrology, 70,* 167–181.

Donaldson, C. H., 1976, "An experimental investigation of olivine morphology," *Contrib. Mineralogy and Petrology, 57,* 187–213.

Dowty, E., 1980, "Crystal growth and nucleation theory and the numerical simulation of igneous crystallization," p. 419–458 in Hargraves, R. B., ed., *Physics of Magmatic Processes*: Princeton, N.J., Princeton University Press, 585 p.

Echeverria, L. M., 1980, "Tertiary or Mesozoic komatiites from Gorgona Island, Colombia: field relations and geochemistry," *Contrib. Mineralogy and Petrology, 73,* 253–266.

Eggler, D. H., 1978, "The effect of CO_2 upon partial melting of peridotite in the system $Na_2O–CaO–Al_2O_3–MgO–SiO_2–CO_2$ to 35 kb, with an analysis of melting in a peridotite–H_2O–CO_2 system," *Amer. Jour. Science, 278,* 305–343.

Eggler, D. H., B. O. Mysen, T. C. Hoering, and J. R. Holloway, 1979, "The solubility of carbon monoxide in silicate melts at high pressure and its effect on silicate phase relations," *Earth and Planetary Science Letters, 43,* 321–330.

Ehlers, E. G., 1972, *The Interpretation of Geological Phase Diagrams*: San Francisco, W. H. Freeman and Company, Publishers, 280 p.

Eichelberger, J. C., 1974, "Magma contamination within the volcanic pile: origin of andesite and dacite," *Geology, 2,* 29–33.

Eichelberger, J. C., and R. Gooley, 1977, "Evolution of silicic magma chambers and their relationship to basaltic volcanism," *Amer. Geophys. Union Geophys. Monograph 20,* 57–77.

Ellis, D. J., and D. H. Green, 1979, "An experimental study of the effect of Ca upon garnet–clinopyroxene Fe–Mg exchange equilibria," *Contrib. Mineralogy and Petrology, 71,* 13–22.

Elthon, D., and W. I. Ridley, 1979, "The oxide and silicate mineral chemistry of a kimberlite from the Premier Mine: implications for the evolution of kimberlitic magmas," *Proc. Second Internat. Kimberlite Conference, 1,* 206–216.

Emslie, R. F., 1971, "Liquidus relations and subsolidus reactions in some plagioclase-bearing systems," *Carnegie Inst. Washington Yearbook, 69,* 148–155.

Emslie, R. F., 1975, "Pyroxene megacrysts from anorthositic rocks: new clues to the sources and evolution of the parent magmas," *Canad. Mineralogist, 13,* 138–145.

Emslie, R. F., 1978, "Anorthosite massifs, rapakivi granites, and late Proterozoic rifting of North America," *Precambrian Research, 7,* 61–98.

Engel, A. E. J., C. G. Engel, and R. G. Havens, 1965, "Chemical characteristics of oceanic basalts and the upper mantle," *Geol. Soc. America Bulletin, 76,* 719–734.

Engel, A. E. J., S. P. Itson, C. G. Engel, D. M. Stickney, and E. J. Cray, Jr., 1974, "Crustal evolution and global tectonics: a petrogenetic view," *Geol. Soc. America Bulletin, 85,* 843–858.

Erdosh, G., 1979, "The Ontario carbonatite province and its phosphate potential," *Econ. Geology, 74,* 331–338.

Erikson, E. H., Jr., 1977, "Petrology and petrogenesis of the Mount Stuart Batholith—plutonic equivalent of the high-alumina basalt association?" *Contrib. Mineralogy and Petrology, 60,* 183–207.

Ernst, W. G., 1969, *Earth Materials*: Englewood Cliffs, N.J., Prentice-Hall, Inc., 150 p.

Ernst, W. G., 1981, "Petrogenesis of eclogites and peridotites from the Western and Ligurian Alps," *Amer. Mineralogist, 66,* 443–472.

Eugster, H. P., and I.-M. Chou, 1979, "A model for the deposition of Cornwall-type magnetite deposits," *Econ. Geology, 74,* 763–774.

Evans, A. M., 1980, *An Introduction to Ore Geology*: New York, Elsevier/North-Holland, Inc., 231 p.

Ewart, A., 1976, "Mineralogy and chemistry of modern orogenic lavas—some statistics and implications," *Earth and Planetary Science Letters, 31,* 417–432.

Ewart, A., 1979, "A review of the mineralogy and chemistry of Tertiary-Recent dacitic, latitic, rhyolitic, and related salic volcanic rocks," p. 13–121 in Barker, F., ed., *Trondhjemites, Dacites, and Related Rocks*: Amsterdam, Elsevier Scientific Publishing Co., 659 p.

Ewers, G. R., and R. R. Keays, 1977, "Volatile and precious metal zoning in the Broadlands geothermal field, New Zealand," *Econ. Geology, 72,* 1337–1354.

Fairbairn, H. W., and others, 1951, *A Cooperative Investigation of Precision and Accuracy in Chemical, Spectrochemical and Modal Analysis of Silicate Rocks*: U.S. Geol. Survey Bulletin, *1980.*

Faure, G., 1977, *Principles of Isotope Geology*: New York, John Wiley & Sons, Inc., 464 p.

Feigenson, M. D., and F. J. Spera, 1980, "Melt production by viscous dissipation: role of heat advection by magma transport," *Geophys. Research Letters, 7,* 145–148.

Feininger, T., 1980, "Eclogite and related high-pressure regional metamorphic rocks from the Andes of Ecuador," *Jour. Petrology, 21,* 107–140.

Fenn, P. M., 1977, "The nucleation and growth of alkali feldspar from hydrous melts," *Canad. Mineralogist, 15,* 135–161.

Ferguson, J., and K. L. Currie, 1971, "Evidence of liquid immiscibility in alkaline ultrabasic dikes at Callander Bay, Ontario," *Jour. Petrology, 12,* 561–585.

Ferguson, J., and J. W. Sheraton, 1979, "Petrogenesis of kimberlitic rocks and associated xenoliths of southeastern Australia," *Proc. Second Internat. Kimberlite Conference, 1,* 140–160.

Ferry, J. M., 1978, "Fluid interaction between granite and sediment during metamorphism, south-central Maine," *Amer. Jour. Science, 278,* 1025–1056.

Fink, J. H., and R. C. Fletcher, 1978, "Ropy pahoehoe: surface folding of a viscous fluid," *Jour. Volcanol. and Geothermal Research, 4,* 151–170.

Fletcher, R. C., and A. W. Hofmann, 1974, "Simple models of diffusion and combined diffusion–infiltration metasomatism," p. 243–259 in Hofmann, A. W., B. J. Giletti, H. S. Yoder, Jr., and R. A. Yund, eds., *Geochemical Transport and Kinetics*: Carnegie Inst. Washington Publication *634*, 353 p.

Flower, M. F. J., P. T. Robinson, H.-U. Schmincke, and W. Ohnmacht, 1977, "Magma fractionation systems beneath the Mid-Atlantic Ridge at 36–37°N," *Contrib. Mineralogy and Petrology, 64*, 167–195.

Foden, J. D., and R. Varne, 1980, "The petrology and tectonic setting of Quaternary–Recent volcanic centres of Lombok and Sumbawa, Sunda Arc," *Chemical Geology, 30*, 201–226.

Francheteau, J., and others, 1979, "Massive deepsea sulphide ore deposits discovered on the East Pacific Rise," *Nature, 277*, 523–528.

Freestone, I. C., and D. L. Hamilton, 1980, "The role of liquid immiscibility in the genesis of carbonatites—an experimental study," *Contrib. Mineralogy and Petrology, 73*, 105–117.

Frey, F. A., 1979, "Trace element geochemistry: applications to the igneous petrogenesis of terrestrial rocks," *Reviews of Geophysics and Space Physics, 17*, 803–823.

Frey, F. A., and M. Prinz, 1978, "Ultramafic inclusions from San Carlos, Arizona: petrologic and geochemical data bearing on their petrogenesis," *Earth and Planetary Science Letters, 38*, 129–176.

Frye, K., 1974, *Modern Mineralogy*: Englewood Cliffs, N.J., Prentice-Hall, Inc., 325 p.

Fudali, R. F., and W. G. Melson, 1972, "Ejecta velocities, magma chamber pressure, and kinetic energy associated with the 1968 eruption of Arenal volcano," *Bulletin Volcanologique, 35*, 383–401.

Fyfe, W. S., 1964, *Geochemistry of Solids: An Introduction*: New York, McGraw-Hill Book Company, 199 p.

Gerlach, T. M., 1980a, "Evaluation of volcanic gas analyses from Kilauea volcano," *Jour. Volcanol. and Geothermal Research, 7*, 295–317.

Gerlach, T. M., 1980b, "Chemical characteristics of the volcanic gases from Nyiragongo lava lake and the generation of CH_4-rich fluid inclusions in alkaline rocks," *Jour. Volcanol. and Geothermal Research, 8*, 177–189.

Gill, J. B., 1981, *Orogenic Andesites and Plate Tectonics*: Berlin, Springer-Verlag, 390 p.

Gittins, J., 1979, "The feldspathoidal alkaline rocks," p. 351–390 in Yoder, H. S., Jr., ed., *The Evolution of the Igneous Rocks: Fiftieth Anniversary Perspectives*: Princeton, N.J., Princeton University Press, 588 p.

Gittins, J., and D. McKie, 1980, "Alkalic carbonatite magmas: Oldoinyo Lengai and its wider applicability," *Lithos, 13*, 213–215.

Gittins, J., J. J. Fawcett, C. K. Brooks, and J. C. Rucklidge, 1980, "Intergrowths of nepheline–potassium feldspar and kalsilite–potassium feldspar: a re-evaluation of the 'pseudoleucite problem,'" *Contrib. Mineralogy and Petrology, 73*, 119–126.

Goldich, S. S., 1941, "Evolution of the central Texas granites," *Jour. Geology, 49*, 697–720.

Goles, G. G., 1976, "Some constraints on the origin of phonolites from the Gregory Rift, Kenya, and inferences concerning basaltic magmas in the Rift System," *Lithos, 9*, 1–8.

Greeley, R., and P. D. Spudis, 1981, "Volcanoes on Mars," *Reviews of Geophysics and Space Physics, 19,* 13–41.

Green, D. H., and A. E. Ringwood, 1967, "An experimental investigation of the gabbro to eclogite transformation and its petrological applications," *Geochim. et Cosmochim. Acta, 31,* 767–833.

Green, D. H., I. A. Nicholls, M. Viljoen, and R. Viljoen, 1975, "Experimental demonstration of the existence of peridotitic liquids in earliest Archaean magmatism," *Geology, 3,* 11–14.

Griffin, B. J., and R. Varne, 1980, "The Macquarie Island ophiolite complex: mid-Tertiary oceanic lithosphere from a major ocean basin," *Chemical Geology, 30,* 285–308.

Guilbert, J. M., and J. D. Lowell, 1974, "Variations in zoning patterns in porphyry ore deposits," *Canad. Mining and Metallurgical Bulletin, 67,* no. 742, 99–109.

Guild, P. W., 1978, "Metallogenesis in the western United States," *Geol. Soc. London Jour., 135,* 355–376.

Gupta, A. K., and K. Yagi, 1980, *Petrology and Genesis of Leucite-bearing Rocks*: Berlin, Springer-Verlag, 252 p.

Gustafson, L. B., 1979, "Porphyry copper deposits and calc–alkaline volcanism," p. 427–468 in McElhinny, M. W., ed., *The Earth: Its Origin, Structure and Evolution*: New York, Academic Press, Inc., 597 p.

Haase, C. S., J. Chadam, D. Feinn, and P. Ortoleva, 1980, "Oscillatory zoning in plagioclase feldspar," *Science, 209,* 272–274.

Haggerty, S. E., 1975, "The chemistry and genesis of opaque minerals in kimberlites," *Physics and Chemistry of the Earth, 9,* 295–307.

Haggerty, S. E., 1976, "Opaque mineral oxides in terrestrial igneous rocks," p. Hg101–Hg300 in Rumble, D., III, ed., *Oxide Minerals*: Mineral. Soc. America Reviews in Mineralogy, *3*.

Haggerty, S. E., 1978, "The redox state of planetary basalts," *Geophys. Research Letters, 5,* 443–446.

Hamilton, D. L., and W. S. MacKenzie, 1965, "Phase equilibrium studies in the system $NaAlSiO_4$(nepheline)–$KAlSiO_4$(kalsilite)–SiO_2–H_2O," *Mineral. Magazine, 34,* 214–231.

Hanson, G. N., 1978, "The application of trace elements to the petrogenesis of igneous rocks of granitic composition," *Earth and Planetary Science Letters, 38,* 26–43.

Hardee, H. C., 1979, "Heat transfer measurements in the 1977 Kilauea lava flow, Hawaii," *Jour. Geophys. Research, 84,* 7485–7493.

Hardee, H. C., 1981, "Convective heat extraction from molten magma," *Jour. Volcanol. and Geothermal Research, 10,* 175–193.

Harker, A., 1909, *The Natural History of Igneous Rocks*: New York, Macmillan Publishing Co., Inc., 384 p.

Harmon, R. S., and R. S. Thorpe, 1980, "Petrogenesis of Andean andesites: stable isotope systematics," *Geol. Soc. America Abstracts with Programs, 12,* 442.

Harmon, R. S., R. S. Thorpe, and P. W. Francis, 1981, "Petrogenesis of Andean andesites from combined O–Sr isotope relationships," *Nature, 290,* 396–399.

Harte, B., 1977, "Rock nomenclature with particular relation to deformation and recrystallization textures in olivine-bearing xenoliths," *Jour. Geology, 85,* 279–288.

Harte, B., J. J. Gurney, and J. W. Harris, 1980, "The formation of peridotitic suite inclusions in diamonds," *Contrib. Mineralogy and Petrology, 72,* 181–190.

Haskin, L. A., 1979, "On rare-earth element behavior in igneous rocks," *Physics and Chemistry of the Earth, 11,* 175–189.

Hatch, F. H., A. K. Wells, and M. K. Wells, 1973, *Petrology of the Igneous Rocks,* 13th ed.: New York, Hafner Press, 551 p.

Hawkes, D. D., 1967, "Order of abundant crystal nucleation in a natural magma," *Geol. Magazine, 104,* 473–486.

Head, J. W., III, 1976, "Lunar volcanism in space and time," *Reviews of Geophysics and Space Physics, 14,* 265–300.

Heald, E. F., J. J. Naughton, and I. L. Barnes, Jr., 1963, "The chemistry of volcanic gases: Pt. 2. Uses of equilibrium calculations in the interpretation of volcanic gas samples," *Jour. Geophys. Research, 68,* 545–557.

Hedge, C. E., 1978, "Strontium isotopes in basalts from the Pacific Ocean basin," *Earth and Planetary Science Letters, 38,* 88–94.

Heiken, G., 1972, "Morphology and petrography of volcanic ashes," *Geol. Soc. America Bulletin, 83,* 1961–1988.

Helz, G. R., and P. J. Wyllie, 1979, "Liquidus relationships in the system $CaCO_3$–$Ca(OH)_2$–CaS and the solubility of sulfur in carbonatite magmas," *Geochim. et Cosmochim. Acta, 43,* 259–265.

Henriquez, F., and R. F. Martin, 1978, "Crystal-growth textures in magnetite flows and feeder dikes, El Laco, Chile," *Canad. Mineralogist, 16,* 581–589.

Herbert, F., M. J. Drake, and C. P. Sonett, 1978, "Geophysical and geochemical evolution of the lunar magma ocean," *Proc. Ninth Lunar and Planetary Science Conference, 1,* 249–262.

Herz, N., 1969, "Anorthosite belts, continental drift, and the anorthosite event," *Science, 164,* 944–947.

Hess, H. H., 1962, "History of ocean basins," p. 599–620 in Engel, A. E. J., H. L. James, and B. F. Leonard, eds., *Petrologic Studies: A Volume in Honor of A. F. Buddington*: Geol. Soc. America, 660 p.

Hess, P. C., 1980, "Polymerization model for silicate melts," p. 3–48 in Hargraves, R. B., ed., *Physics of Magmatic Processes*: Princeton, N.J., Princeton University Press, 585 p.

Hildreth, W., 1979, "The Bishop Tuff: evidence for the origin of compositional zonation in silicic magma chambers," *Geol. Soc. America Special Paper 180,* 43–75.

Hildreth, W., 1981, "Gradients in silicic magma chambers: implications for lithospheric magmatism," *Jour. Geophys. Research, 86,* 10153–10192.

Hochella, M. F., Jr., G. E. Brown, and M. Taylor, 1978, "Structural study of basaltic and granitic composition glasses," *Geol. Soc. America Abstracts with Programs, 10,* 422.

Hoffman, A. W., and M. Magaritz, 1977, "Diffusion of Ca, Sr, Ba and Co in a basalt melt: implications for the geochemistry of the mantle," *Jour. Geophys. Research, 82,* 5432–5440.

Holmes, A., 1920, *The Nomenclature of Petrology*: republished 1972 by Hafner Publishing Co., New York, 284 p.

Holmes, A., 1952, "The potash ankaratrite–melaleucitite lavas of Nabugando and Mbuga Craters, southwest Uganda," *Geol. Soc. Edinburgh Trans., 15,* 187–213.

Hoover, J. D., 1978, "Melting relations of a new chilled margin sample from the Skaergaard intrusion," *Carnegie Inst. Washington Yearbook, 77,* 739–743.

Hopson, C. A., R. G. Coleman, R. T. Gregory, J. S. Pallister, and E. H. Bailey, 1981, "Geologic section through the Samail ophiolite and associated rocks along a Muscat–Ibra transect, southeastern Oman Mountains," *Jour. Geophys. Research, 86,* 2527–2544.

Horai, K., and S. Uyeda, 1969, "Terrestrial heat flow in volcanic areas," *Amer. Geophys. Union Geophys. Monograph 13,* 95–109.

Hostetler, C. J., and M. J. Drake, 1980, "Predicting major element mineral/melt equilibria: a statistical approach," *Jour. Geophys. Research, 85,* 3789–3796.

Hughes, C. J., and E. M. Hussey, 1979, "Standardized procedure for presenting corrected Fe_2O_3/FeO ratios in analyses of fine grained mafic rocks," *Neues Jahrbuch für Mineralogie Monatshefte 1979,* 570–572.

Hughes, J. M., R. E. Stoiber, and M. J. Carr, 1980, "Segmentation of the Cascade volcanic chain," *Geology, 8,* 15–17.

Hulme, G., 1974, "The interpretation of lava flow morphology," *Royal Astron. Soc. Geophys. Jour., 39,* 361–383.

Huppert, H. E., and R. S. J. Sparks, 1980, "The fluid dynamics of a basaltic magma chamber replenished by influx of hot, dense ultrabasic magma," *Contrib. Mineralogy and Petrology, 75,* 279–289.

Hutchinson, R. W., 1980, "Massive base metal sulphide deposits as guides to tectonic evolution," p. 659–684 in Strangway, D. W., ed., *The Continental Crust and Its Mineral Deposits*: Geol. Assoc. Canada Special Paper *20.*

Hutchison, C. S., 1974, *Laboratory Handbook of Petrographic Techniques*: New York, John Wiley & Sons, Inc., 527 p.

Irvine, T. N., 1974, *Petrology of the Duke Island Ultramafic Complex, Southeastern Alaska*: Geol. Soc. America Memoir *138.*

Irvine, T. N., 1980, "Magmatic density currents and cumulus processes," *Amer. Jour. Science, 280-A,* 1–58.

Irvine, T. N., and W. R. A. Baragar, 1971, "A guide to the chemical classification of the common volcanic rocks," *Canad. Jour. Earth Sciences, 8,* 523–548.

Irving, A. J., 1978, "A review of experimental studies of crystal/liquid trace element partitioning," *Geochim. et Cosmochim. Acta, 42,* 743–770.

Irving, A. J., 1980, "Petrology and geochemistry of composite ultramafic xenoliths in alkalic basalts and implications for magmatic processes within the mantle," *Amer. Jour. Science, 280-A,* 389–426.

IUGS (International Union of Geological Sciences) Subcommission on the Systematics of Igneous Rocks, 1973, "Plutonic rocks, classification and nomenclature," *Geotimes, 18,* no. 10, 26–30.

Jackson, E. D., 1967, "Ultramafic cumulates in the Stillwater, Great Dyke, and Bushveld intrusions," p. 20–38 in Wyllie, P. J., ed., *Ultramafic and Related Rocks*: New York, John Wiley & Sons, Inc., 464 p.

Jacobsen, S. B., and G. J. Wasserburg, 1979, "The mean age of mantle and crustal reservoirs," *Jour. Geophys. Research, 84,* 7411–7427.

Jaeger, J. C., 1967, "Cooling and solidification of igneous rocks," p. 503–536 in Hess, H. H., and A. Poldervaart, eds., *Basalts: The Poldervaart Treatise on Rocks of Basaltic Composition,* Vol. 2: New York, Wiley-Interscience, 380 p.

Jahns, R. H., 1953, "The genesis of pegmatites: I. Occurrence and origin of giant crystals," *Amer. Mineralogist, 38,* 563–598.

Jahns, R. H., and C. W. Burnham, 1969, "Experimental studies of pegmatite genesis: a model for the derivation and crystallization of granitic pegmatites," *Econ. Geology, 64,* 843–864.

Jahns, R. H., and O. F. Tuttle, 1963, "Layered pegmatite-aplite intrusives," *Mineral. Soc. America Special Paper 1,* 78–92.

Jakeš, P., and A. J. R. White, 1972, "Major and trace-element abundances in volcanic rocks of orogenic areas," *Geol. Soc. America Bulletin, 83,* 29–40.

Jakobsson, S. P., 1972, "Chemistry and distribution pattern of Recent basaltic rocks in Iceland," *Lithos, 5,* 365–386.

Jaques, A. L., and D. H. Green, 1980, "Anhydrous melting of peridotite at 0–15 kb pressure and the genesis of tholeiitic basalts," *Contrib. Mineralogy and Petrology, 73,* 287–310.

Jaupart, C., J. G. Sclater, and G. Simmons, 1981, "Heat flow studies: constraints on the distribution of uranium, thorium and potassium in the continental crust," *Earth and Planetary Science Letters, 52,* 328–344.

Johannsen, A., 1939, *A Descriptive Petrography of the Igneous Rocks:* Vol. I. *Introduction, Textures, Classifications and Glossary:* Chicago, University of Chicago Press, 318 p.

Johnson, R. W., 1979, "Geotectonics and volcanism in Papua New Guinea: a review of the late Cenozoic," *Bureau of Mineral Resources Journal of Australian Geology and Geophysics, 4,* 181–207.

Johnson, R. W., D. E. MacKenzie, and I. E. M. Smith, 1978, "Volcanic rock associations at convergent plate boundaries: reappraisal of the concept using case histories from Papua New Guinea," *Geol. Soc. America Bulletin, 89,* 96–106.

Johnson, T., and P. Molnar, 1972, "Focal mechanisms and plate tectonics of the southwest Pacific," *Jour. Geophys. Research, 77,* 5000–5032.

Johnston, D. A., 1980, "Volcanic contribution of chlorine to the stratosphere: more significant to ozone than previously estimated?" *Science, 209,* 491–493.

Joplin, G. A., 1968, "The shoshonite association: a review," *Geol. Soc. Australia Jour., 15,* 275–294.

Jordan, T. H., 1979, "Mineralogies, densities and seismic velocities of garnet lherzolites and their geophysical implications," *Proc. Second Internat. Kimberlite Conference, 2,* 1–14.

Kay, R. W., 1980, "Volcanic arc magmas: implications of a melting–mixing model for element recycling in the crust-upper mantle system," *Jour. Geology, 88,* 497–522.

Keller, J., 1981, "Carbonatitic volcanism in the Kaiserstuhl alkaline complex: evidence for highly fluid carbonatitic melts at the earth's surface," *Jour. Volcanol. and Geothermal Research, 9,* 423–431.

Kellerhals, R., J. Shaw, and V. K. Arora, 1975, "On grain size from thin sections," *Jour. Geology, 83,* 79–96.

Kelsey, C. H., 1965, "Calculation of the C.I.P.W. norm," *Mineral. Magazine, 34,* 276–282.

Kennedy, C. S., and G. C. Kennedy, 1976, "The equilibrium boundary between graphite and diamond," *Jour. Geophys. Research, 81,* 2467–2470.

Kerrick, D. M., 1977, "The genesis of zoned skarns in the Sierra Nevada, California," *Jour. Petrology, 18,* 144–181.

Kilinc, I. A., and C. W. Burnham, 1972, "Partitioning of chloride between a silicate melt and coexisting aqueous phase from 2 to 8 kilobars," *Econ. Geology, 67,* 231–235.

Kirkpatrick, R. J., 1975, "Crystal growth from the melt: a review," *Amer. Mineralogist, 60,* 798–814.

Kirkpatrick, R. J., G. R. Robinson, and J. F. Hays, 1976, "Kinetics of crystal growth from silicate melts: anorthite and diopside," *Jour. Geophys. Research, 81,* 5715–5720.

Kirkpatrick, R. J., L. Klein, D. R. Uhlmann, and J. F. Hays, 1979, "Rates and processes of crystal growth in the system anorthite–albite," *Jour. Geophys. Research, 84,* 3671–3676.

Knapp, R. B., and J. E. Knight, 1977, "Differential thermal expansion of pore fluids: fracture propagation and microearthquake production in hot pluton environment," *Jour. Geophys. Research, 87,* 2515–2522.

Kolstad, C. D., and T. R. McGetchin, 1978, "Thermal evolution models for the Valles Caldera with reference to hot-dry rock geothermal experiment," *Jour. Volcanol. and Geothermal Research, 3,* 197–218.

Komar, P. D., 1976, "Phenocryst interactions and the velocity profile of magma flowing through dikes or sills," *Geol. Soc. America Bulletin, 87,* 1336–1342.

Komar, P. D., and C. E. Reimers, 1978, "Grain shape effects on settling rates," *Jour. Geology, 86,* 193–209.

Konnerup-Madsen, J., E. Larsen, and J. Rose-Hansen, 1979, "Hydrocarbon-rich fluid inclusions in minerals from the alkaline Ilimaussaq intrusion, South Greenland," *Bulletin de Minéralogie, 102,* 642–653.

Koster van Groos, A. F., and P. J. Wyllie, 1973, "Liquid immiscibility in the join $NaAlSi_3O_8$–$CaAl_2Si_2O_8$–Na_2CO_3–H_2O," *Amer. Jour. Science, 273,* 465–487.

Kuno, H., 1959, "Origin of Cenozoic petrographic provinces of Japan and surrounding areas," *Bulletin Volcanologique, 20,* 37–76.

Kuno, H., 1960, "High-alumina basalt," *Jour. Petrology, 1,* 121–145.

Kushiro, I., 1968, "Compositions of magmas formed by partial zone melting of the earth's upper mantle," *Jour. Geophys. Research, 73,* 619–634.

Kushiro, I., 1969, "Clinopyroxene solid solutions formed by reactions between diopside and plagioclase at high pressures," *Mineral. Soc. America Special Paper 2,* 179–191.

Kushiro, I., 1975, "On the nature of silicate melt and its significance in magma genesis: regularities in the shift of the liquidus boundaries involving olivine, pyroxene, and silica minerals," *Amer. Jour. Science, 275,* 411–436.

Kushiro, I., 1976, "Changes in viscosity and structure of melt of $NaAlSi_2O_6$ composition at high pressures," *Jour. Geophys. Research, 81,* 6347–6350.

Kushiro, I., 1980, "Viscosity, density, and structure of silicate melts at high pressures, and their petrological applications," p. 93–120 in Hargraves, R. B., ed., *Physics of Magmatic Processes:* Princeton, N.J., Princeton University Press, 585 p.

Kushiro, I., H. S. Yoder, Jr., and B. O. Mysen, 1976, "Viscosities of basalt and andesite melts at high pressures," *Jour. Geophys. Research, 81,* 6351–6356.

Lachenbruch, A. H., 1976, "Dynamics of a passive spreading center," *Jour. Geophys. Research, 81,* 1883–1902.

Lane, D. L., and J. Ganguly, 1980, "Al_2O_3 solubility in orthopyroxene in the system MgO–Al_2O_3–SiO_2: a reevaluation, and mantle geotherm," *Jour. Geophys. Research, 85,* 6963–6972.

Langmuir, C. H., and G. N. Hanson, 1980, "An evaluation of major element heterogeneity in the mantle sources of basalts," *Royal Soc. London Philos. Trans., A, 297,* 383–407.

Lanphere, M. A., and G. B. Dalrymple, 1980, "Age and strontium isotopic composition of the Honolulu Volcanic Series, Oahu, Hawaii," *Amer. Jour. Science, 280-A,* 736–751.

Lappin, A. R., and L. S. Hollister, 1980, "Partial melting in the Central Gneiss Complex near Prince Rupert, British Columbia," *Amer. Jour. Science, 280,* 518–545.

Larsen, G., K. Grönvold, and S. Thorarinsson, 1979, "Volcanic eruption through a geothermal borehole at Námafjall, Iceland," *Nature, 278,* 707–710.

Larsen, L. M., 1979, "Distribution of REE and other trace elements between phenocrysts and peralkaline undersaturated magmas, exemplified by rocks from the Gardar igneous province, south Greenland," *Lithos, 12,* 303–315.

Leeman, W. P., J. R. Budahn, D. C. Gerlach, D. R. Smith, and B. N. Powell, 1980, "Origin of Hawaiian tholeiites: trace element constraints," *Amer. Jour. Science, 280-A,* 794–819.

LeMaitre, R. W., 1976, "The chemical variability of some common igneous rocks," *Jour. Petrology, 17,* 589–637.

Levin, E. M., C. R. Robbins, and H. F. McMurdie, 1964, *Phase Diagrams for Ceramists*: Columbus, Ohio, American Ceramic Society, 601 p. (supplement published 1969).

Liebau, F., 1980, "Classification of silicates," *Mineral. Soc. America Reviews in Mineralogy, 5,* 1–24.

Lippard, S. J., 1973a, "Plateau phonolite lava flows, Kenya," *Geol. Magazine, 110,* 543–549.

Lippard, S. J., 1973b, "The petrology of phonolites from the Kenya Rift," *Lithos, 6,* 217–234.

Lofgren, G., 1970, "Experimental devitrification rate of rhyolite glass," *Geol. Soc. America Bulletin, 81,* 553–560.

Lofgren, G., 1971, "Experimentally produced devitrification textures in natural rhyolitic glass," *Geol. Soc. America Bulletin, 82,* 111–124.

Lofgren, G., 1974, "An experimental study of plagioclase morphology," *Amer. Jour. Science, 274,* 243–273.

Lofgren, G., 1979, "The effect of nucleation on basaltic textures," *Geol. Soc. America Abstracts with Programs, 11,* 467–468.

Lofgren, G., 1980, "Experimental studies on the dynamic crystallization of silicate melts," p. 487–551 in Hargraves, R. B., ed., *Physics of Magmatic Processes:* Princeton, N.J., Princeton University Press, 585 p.

Loiselle, M. C., and D. R. Wones, 1979, "Characteristics and origin of anorogenic granites," *Geol. Soc. America Abstracts with Programs, 11,* 468.

Longhi, J., 1978, "Pyroxene stability and the composition of the lunar magma ocean," *Proc. Ninth Lunar and Planetary Science Conference, 1,* 285–306.

Lowell, J. D., and J. M. Guilbert, 1970, "Lateral and vertical alteration–mineralization zoning in porphyry ore deposits," *Econ. Geology, 65,* 378–408.

Luth, W. C., 1976, "Granitic rocks," p. 335–417 in Bailey, D. K., and R. Macdonald, eds., *The Evolution of the Crystalline Rocks:* New York, Academic Press, Inc., 484 p.

Luth, W. C., R. H. Jahns, and O. F. Tuttle, 1964, "The granite system at pressures of 4 to 10 kilobars," *Jour. Geophys. Research, 69,* 759–773.

Maaløe, S., 1978, "The origin of rhythmic layering," *Mineral. Magazine, 42,* 337–345.

Macdonald, G. A., 1967, "Forms and structures of extrusive basaltic rocks", p. 1–61 in Hess, H. H., and A. Poldervaart, eds., *Basalts: The Poldervaart Treatise on Rocks of Basaltic Composition,* Vol. 1: New York, Wiley-Interscience, 482 p.

Macdonald, G. A., 1972, *Volcanoes*: Englewood Cliffs, N.J., Prentice-Hall, Inc., 510 p.

Macdonald, G. A., and T. Katsura, 1964, "Chemical composition of Hawaiian lavas," *Jour. Petrology, 5,* 82–133.

MacKenzie, W. S., and C. Guilford, 1980, *Atlas of Rock-forming Minerals in Thin Section*: New York, John Wiley & Sons, Inc., 98 p.

Manning, J. R., 1974, "Diffusion kinetics and mechanisms in simple crystals," p. 3–13 in Hofmann, A. W., B. J. Giletti, H. S. Yoder, Jr., and R. A. Yund, eds., *Geochemical Transport and Kinetics*: Carnegie Inst. Washington Publication *634,* 353 p.

Marsh, B. D., 1978, "On the cooling of ascending andesitic magma," *Royal Soc. London Philos. Trans., A, 288,* 611–625.

Marsh, B. D., 1979a, "Island-arc volcanism," *Amer. Scientist, 67,* 161–172.

Marsh, B. D., 1979b, "Island arc development: some observations, experiments, and speculations," *Jour. Geology, 87,* 687–713.

Marsh, B. D., and I. S. E. Carmichael, 1974, "Benioff zone magmatism," *Jour. Geophys. Research, 79,* 1196–1206.

Marshall, P., 1911, *Oceania*: Handbuch der regionalen Geologie, 7, no. 2, 36 p.

Martin, R. F., and A. J. Piwinskii, 1972, "Magmatism and tectonic settings," *Jour. Geophys. Research, 77,* 4966–4975.

Mason, B., 1966, *Principles of Geochemistry,* 3rd ed.: New York, John Wiley & Sons, Inc., 329 p.

Mason, B., 1971, "Ytterby, Sweden: a classic mineral locality," *Mineral. Record, 2,* 136–138.

Mathez, E. A., 1976, "Sulfur solubility and magmatic sulfides in submarine basalt glass," *Jour. Geophys. Research, 81,* 4269–4276.

Mathias, M., 1974, "Alkaline rocks of southern Africa," p. 184–202 in Sørensen, H., ed., *The Alkaline Rocks:* New York, John Wiley & Sons, Inc., 622 p.

Mavko, G. M., 1980, "Velocity and attenuation in partially molten rocks," *Jour. Geophys. Research, 85,* 5173–5189.

McBirney, A. R., 1973, "Factors governing the intensity of explosive andesitic eruptions," *Bulletin Volcanologique, 37,* 443–453.

McBirney, A. R., 1975, "Differentiation of the Skaergaard intrusion," *Nature, 253,* 691–694.

McBirney, A. R., 1976, "Some geologic constraints on models for magma generation in orogenic environments," *Canad. Mineralogist, 14,* 245–254.

McBirney, A. R., 1979, "Effects of assimilation," p. 307–338 in Yoder, H. S., Jr., ed., *The Evolution of the Igneous Rocks: Fiftieth Anniversary Perspectives*: Princeton, N.J., Princeton University Press, 588 p.

McBirney, A. R., 1980, "Mixing and unmixing of magmas," *Jour. Volcanol. and Geothermal Research, 7,* 357–371.

McBirney, A. R., and R. M. Noyes, 1979, "Crystallization and layering of the Skaergaard intrusion," *Jour. Petrology, 20,* 487–554.

McCallum, M. E., C. D. Mabarak, and H. G. Coopersmith, 1979, "Diamonds from kimberlites in the Colorado–Wyoming State Line district," *Proc. Second Internat. Kimberlite Conference, 1,* 42–58.

McGetchin, T. R., and B. A. Chouet, 1979, "Energy budget of the volcano Stromboli, Italy," *Geophys. Research Letters, 6,* 317–320.

McGetchin, T. R., and G. W. Ullrich, 1973, "Xenoliths in maars and diatremes with inferences for the Moon, Mars, and Venus," *Jour. Geophys. Research, 78,* 1833–1853.

Melson, W. G., T. L. Vallier, T. L. Wright, G. Byerly, and J. Nelen, 1976, "Chemical diversity of abyssal volcanic glass erupted along Pacific, Atlantic, and Indian Ocean sea-floor spreading centers," *Amer. Geophys. Union Geophys. Monograph 19,* 351–367.

Menzies, M., and V. R. Murthy, 1980, "Mantle metasomatism as a precursor to the genesis of alkaline magmas—isotopic evidence," *Amer. Jour. Science, 280-A,* 622–638.

Mercier, J.-C., 1979, "Peridotite xenoliths and the dynamics of kimberlite intrusion," *Proc. Second Internat. Kimberlite Conference, 2,* 197–212.

Meyer, H. O. A., 1976, "Kimberlites of the continental United States: a review," *Jour. Geology, 84,* 377–403.

Meyer, H. O. A., 1977, "Mineralogy of the upper mantle: a review of the minerals in mantle xenoliths from kimberlites," *Earth-Science Reviews, 13,* 251–281.

Meyer, H. O. A., 1979, "Kimberlites and the mantle," *Reviews of Geophysics and Space Physics, 17,* 776–788.

Meyer, H. O. A., and H.-M. Tsai, 1979, "Inclusions in diamond and the mineral chemistry of the upper mantle," *Physics and Chemistry of the Earth, 11,* 631–644.

Miller, C. F., and L. J. Bradfish, 1980, "An inner Cordilleran belt of muscovite-bearing plutons," *Geology, 8,* 412–416.

Miller, C. F., E. F. Stoddard, L. J. Bradfish, and W. A. Dollase, 1981, "Composition of plutonic muscovite: genetic implications," *Canad. Mineralogist, 19,* 25–34.

Miller, T. P., and R. L. Smith, 1977, "Spectacular mobility of ash flows around Aniakchak and Fisher calderas, Alaska," *Geology, 5,* 173–176.

Minear, J. W., and C. R. Fletcher, 1978, "Crystallization of a lunar magma ocean," *Proc. Ninth Lunar and Planetary Science Conference, 1,* 263–283.

Minster, J. B., T. H. Jordan, P. Molnar, and E. Haines, 1974, "Numerical modelling of instantaneous plate tectonics," *Royal Astron. Soc. Geophys. Jour., 36,* 541–576.

Mitchell, R. H., 1979, "The alleged kimberlite–carbonatite relationship: additional contrary mineralogical evidence," *Amer. Jour. Science, 279,* 570–589.

Mitchell, R. H., D. A. Carswell, and D. B. Clarke, 1980, "Geological implications and validity of calculated equilibration conditions for ultramafic xenoliths from the Pipe 200 kimberlite, northern Lesotho," *Contrib. Mineralogy and Petrology, 72,* 205–217.

Molnar, P., and D. Gray, 1979, "Subduction of continental lithosphere: some constraints and uncertainties," *Geology, 7,* 58–62.

Moore, J. G., 1979, "Vesicularity and CO_2 in mid-oceanic ridge basalt," *Nature, 282,* 250–253.

Moore, J. G., and J. G. Schilling, 1973, "Vesicles, water and sulfur in Reykjanes Ridge basalts," *Contrib. Mineralogy and Petrology, 41,* 105–118.

Moore, J. G., and W. G. Melson, 1969, "Nuées ardentes of the 1968 eruption of Mayon volcano, Philippines," *Bulletin Volcanologique, 33,* 600–620.

Moore, J. G., J. N. Batchelder, and C. G. Cunningham, 1977, "CO_2-filled vesicles in mid-ocean basalt," *Jour. Volcanol. and Geothermal Research, 2,* 309–327.

Moore, J. G., A. Grantz, and M. C. Blake, Jr., 1963, "The quartz diorite line in northwestern North America," *U.S. Geol. Survey Professional Paper 450-E,* p. E89–E93.

Moore, R. B., and others, 1980, "The 1977 eruption of Kilauea volcano, Hawaii," *Jour. Volcanol. and Geothermal Research, 7,* 189–210.

Morgan, W. J., 1972, "Plate motions and deep mantle convection," *Geol. Soc. America Memoir 132,* 7–22.

Morris, L. D., T. Simkin, and H. Meyers, 1979, *Volcanoes of the World,* map published by U.S. National Oceanic and Atmospheric Administration.

Morse, S. A., 1969, "Syenites," *Carnegie Inst. Washington Yearbook, 67,* 112–120.

Morse, S. A., 1970, "Alkali feldspars with water at 5 kb pressure," *Jour. Petrology, 11,* 221–251.

Morse, S. A., 1976, "The lever rule with fractional crystallization and fusion," *Amer. Jour. Science, 276,* 330–346.

Morse, S. A., 1980, *Basalts and Phase Diagrams: An Introduction to the Quantitative Use of Phase Diagrams in Igneous Petrology:* New York, Springer-Verlag, 493 p.

Munhá, J., and R. Kerrich, 1980, "Sea water basalt interaction in spilites from the Iberian Pyrite Belt," *Contrib. Mineralogy and Petrology, 73,* 191–200.

Murase, T., and A. R. McBirney, 1973, "Properties of some common igneous rocks and their melts at high temperatures," *Geol. Soc. America Bulletin, 84,* 3563–3592.

Murck, B. W., R. C. Burruss, and L. S. Hollister, 1978, "Phase equilibria in fluid inclusions in ultramafic xenoliths," *Amer. Mineralogist, 63,* 40–46.

Mutch, T. A., 1979, "Planetary surfaces," *Reviews of Geophysics and Space Physics, 17,* 1694–1722.

Myers, J. S., 1975, "Cauldron subsidence and fluidization: mechanisms of intrusion of the Coastal Batholith of Peru into its own volcanic ejecta," *Geol. Soc. America Bulletin, 86,* 1209–1220.

Mysen, B. O., and D. Virgo, 1980, "Solubility mechanisms of carbon dioxide in silicate melts: a Raman spectroscopic study," *Amer. Mineralogist, 65,* 885–899.

Mysen, B. O., D. Virgo, and C. M. Scarfe, 1980, "Relations between the anionic structure and viscosity of silicate melts—a Raman spectroscopic study," *Amer. Mineralogist, 65,* 690–710.

Naldrett, A. J., and L. J. Cabri, 1976, "Ultramafic and related mafic rocks: their classification and genesis with special reference to the concentration of nickel sulphides and platinum group elements," *Econ. Geology, 71,* 1131–1158.

Naldrett, A. J., and J. M. Duke, 1980, "Platinum metals in magmatic sulfide ores," *Science, 208,* 1417–1424.

Naldrett, A. J., and A. J. Macdonald, 1980, "Tectonic settings of Ni–Cu sulphide ores: their importance in genesis and exploration," p. 633–657 in Strangway, D. W., ed., *The Continental Crust and Its Mineral Deposits*: Geol. Assoc. Canada Special Paper *20.*

Naney, M. T., and S. E. Swanson, 1980, "The effect of Fe and Mg on crystallization in granitic systems," *Amer. Mineralogist, 65,* 639–653.

Nash, W. P., 1972, "Mineralogy and petrology of the Iron Hill carbonatite complex, Colorado," *Geol. Soc. America Bulletin, 83,* 1361–1382.

Naslund, H. R., 1976, "Mineralogical variations in the upper part of the Skaergaard intrusion, east Greenland," *Carnegie Inst. Washington Yearbook, 75,* 640–644.

Natland, J. H., 1980, "The progression of volcanism in the Samoan linear volcanic chain," *Amer. Jour. Science, 280-A,* 709–735.

Nelson, S. A., and I. S. E. Carmichael, 1979, "Partial molar volumes of oxide components in silicate liquids," *Contrib. Mineralogy and Petrology, 71,* 117–124.

Nesbitt, B. E., and W. C. Kelly, 1977, "Magmatic and hydrothermal inclusions in carbonatite of the Magnet Cove complex, Arkansas," *Contrib. Mineralogy and Petrology, 63,* 271–294.

Nesbitt, H. W., 1980, "Genesis of the New Quebec and Adirondack granulites: evidence for their production by partial melting," *Contrib. Mineralogy and Petrology, 72,* 303–310.

Newhall, C. G., 1979, "Temporal variation in the lavas of Mayon volcano, Philippines," *Jour. Volcanol. and Geothermal Research, 6,* 61–83.

Nielsen, R. L., 1976, "Recent developments in the study of porphyry copper geology: a review," p. 487–500 in Sutherland Brown, A., ed., *Porphyry Deposits of the Canadian Cordillera*: Canad. Inst. Mining and Metallurgy Special Volume *15,* 510 p.

Nielsen, R. L., and M. J. Drake, 1979, "Pyroxene-melt equilibria," *Geochim. et Cosmochim. Acta, 43,* 1259–1272.

Noble, J. A., 1974, "Metal provinces and metal finding in the western United States," *Mineralium Deposita, 9,* 1–25.

Nockolds, S. R., 1941, "The Garabal Hill—Glen Fyne igneous complex," *Geol. Soc. London Quart. Jour., 96,* 451–511.

Norton, D., and L. M. Cathles, 1979, "Thermal aspects of ore deposition," p. 611–631 in Barnes, H. L., ed., *Geochemistry of Hydrothermal Ore Deposits,* 2nd ed.: New York, John Wiley & Sons, Inc., 798 p.

Norton, D., and J. Knight, 1977, "Transport phenomena in hydrothermal systems: cooling plutons," *Amer. Jour. Science, 277,* 937–981.

Norton, D., and H. P. Taylor, Jr., 1979, "Quantitative simulation of the hydrothermal systems of crystallizing magmas on the basis of transport theory and oxygen isotope data: an analysis of the Skaergaard Intrusion," *Jour. Petrology, 20,* 421–486.

Norton, J. J., 1970, "Composition of a pegmatite, Keystone, South Dakota," *Amer. Mineralogist, 55,* 981–1002.

O'Hara, M. J., 1968, "The bearing of phase equilibria studies in synthetic and natural systems on the origin and evolution of basic and ultrabasic rocks," *Earth-Science Reviews, 4,* 69–133.

O'Hara, M. J., 1977, "Geochemical evolution during fractional crystallization of a periodically refilled magma chamber," *Nature, 266,* 503–507.

O'Hara, M. J., S. E. Richardson, and G. Wilson, 1971, "Garnet peridotite stability and occurrence in the mantle," *Contrib. Mineralogy and Petrology, 32,* 48–68.

Orville, P. M., 1963, "Alkali ion exchange between vapor and feldspar phases," *Amer. Jour. Science, 261,* 201–237.

Orville, P. M., 1972, "Plagioclase cation exchange equilibria with aqueous chloride solution: results at 700°C and 2000 bars in the presence of quartz," *Amer. Jour. Science, 272,* 234–272.

Pakiser, L. C., and J. N. Brune, 1980, "Seismic models of the root of the Sierra Nevada," *Science, 210,* 1088–1094.

Pallister, J. S., 1981, "Structure of the sheeted dike complex of the Samail ophiolite near Ibra, Oman," *Jour. Geophys. Research, 86,* 2661–2672.

Pallister, J. S., and C. A. Hopson, 1981, "Samail ophiolite plutonic suite: field relations, phase variation, cryptic variation and layering, and a model of a spreading ridge magma chamber," *Jour. Geophys. Research, 86,* 2593–2644.

Papike, J. J., and A. E. Bence, 1978, "Lunar mare versus terrestrial mid-ocean ridge basalts: planetary constraints on basaltic volcanism," *Geophys. Research Letters, 5,* 803–806.

Papike, J. J., and M. Cameron, 1976, "Crystal chemistry of silicate minerals of geophysical interest," *Reviews of Geophysics and Space Physics, 14,* 37–80.

Parmentier, E. M., and A. Schedl, 1981, "Thermal aureoles of igneous intrusions: some possible indications of hydrothermal convective cooling," *Jour. Geology, 89,* 1–22.

Parsons, I., 1979, "The Klokken gabbro-syenite complex, South Greenland: cryptic variation and origin of inversely graded layering," *Jour. Petrology, 20,* 653–694.

Peacock, M. A., 1931, "Classification of igneous rock series," *Jour. Geology, 39,* 54–67.

Pearce, J. A., and M. F. J. Flower, 1977, "The relative importance of petrogenetic variables in magma genesis at accreting plate margins: a preliminary investigation," *Geol. Soc. London Jour., 134,* 103–127.

Pearce, T. H., B. E. Gorman, and T. C. Birkett, 1975, "The TiO_2–K_2O–P_2O_5 diagram: a method of discriminating between oceanic and non-oceanic basalts," *Earth and Planetary Science Letters, 24,* 419–426.

Pearce, T. H., B. E. Gorman, and T. C. Birkett, 1977, "The relationship between major element chemistry and tectonic environment of basic and intermediate volcanic rocks," *Earth and Planetary Science Letters, 36,* 121–132.

Peck, D. L., 1975, "Recoverability of geothermal energy directly from molten igneous systems," *U.S. Geol. Survey Circular 726,* 122–124.

Perfit, M. R., D. A. Gust, A. E. Bence, R. J. Arculus, and S. R. Taylor, 1980, "Chemical characteristics of island-arc basalts: implications for mantle sources," *Chemical Geology, 30,* 227–256.

Petersilje, I. A., and W. A. Pripachkin, 1979, "Hydrogen, carbon, nitrogen and helium in gases of igneous rocks," *Physics and Chemistry of the Earth, 11,* 541–545.

Peterson, D. W., and R. I. Tilling, 1980, "Transition of basaltic lava from pahoehoe to aa, Kilauea Volcano, Hawaii: field observations and key factors," *Jour. Volcanol. and Geothermal Research, 7,* 271–293.

Petro, W. L., T. A. Vogel, and J. T. Wilband, 1979, "Major-element chemistry of plutonic rock suites from compressional and extensional plate boundaries," *Chemical Geology, 26,* 217–235.

Phillips, E. R., 1980, "On polygenetic myrmekite," *Geol. Magazine, 117,* 29–36.

Philpotts, A. R., 1978, "Textural evidence for liquid immiscibility in tholeiites," *Mineral. Magazine, 42,* 417–425.

Philpotts, J. A., 1978, "The law of constant rejection," *Geochim. et Cosmochim. Acta, 42,* 909–920.

Pinkerton, H., and R. S. J. Sparks, 1978, "Field measurements of the rheology of lava," *Nature, 276,* 383–385.

Pitcher, W. S., 1978, "The anatomy of a batholith," *Geol. Soc. London Jour., 135,* 157–182.

Pollack, H. N., and D. S. Chapman, 1977, "On the regional variation of heat flow, geotherms, and lithospheric thickness," *Tectonophysics, 38,* 279–296.

Pollard, D. D., O. H. Muller, and D. R. Dockstader, 1975, "The form and growth of fingered sheet intrusions," *Geol. Soc. America Bulletin, 86,* 351–363.

Powell, C. McA., and P. J. Conaghan, 1975, "Tectonic models of the Tibetan plateau," *Geology, 3,* 727–731.

Powell, J. L., and K. Bell, 1974, "Isotopic composition of strontium in alkalic rocks," p. 412–423 in Sørensen, H., ed., *The Alkaline Rocks*: New York, John Wiley & Sons, Inc., 622 p.

Powell, J. L., P. M. Hurley, and H. W. Fairbairn, 1966, "The strontium isotopic composition and origin of carbonatites," p. 365–378 in Tuttle, O. F., and J. Gittins, eds., *Carbonatites:* New York, John Wiley & Sons, Inc., 591 p.

Presnall, D. C., 1969, "The geometrical analysis of partial fusion," *Amer. Jour. Science, 267,* 1178–1194.

Presnall, D. C., and P. C. Bateman, 1973, "Fusion relationships in the system $NaAlSi_3O_8$–$CaAl_2Si_2O_8$–$KAlSi_3O_8$–SiO_2–H_2O and generation of granitic magmas in the Sierra Nevada batholith," *Geol. Soc. America Bulletin, 84,* 3181–3202.

Propach, G., 1976, "Models of filter differentiation," *Lithos, 9,* 203–209.

Rampino, M. R., S. Self, and R. W. Fairbridge, 1979, "Can rapid climatic change cause volcanic eruptions?" *Science, 206,* 826–829.

Rankin, A. H., 1977, "Fluid-inclusion evidence for the formation conditions of apatite from the Tororo carbonatite complex of eastern Uganda," *Mineral. Magazine, 41,* 155–164.

Read, H. H., 1948, "Granites and granites," *Geol. Soc. America Memoir 28,* 1–19.

Ringwood, A. E., 1975, *Composition and Petrology of the Earth's Mantle*: New York, McGraw-Hill Book Company, 618 p.

Robie, R. A., B. S. Hemingway, and J. R. Fisher, 1978, *Thermodynamic Properties of Minerals and Related Substances at 298.15 K and 1 bar (10^5 Pascals) Pressure and at Higher Temperatures*: U.S. Geol. Survey Bulletin *1452*.

Rock, N. M. S., 1977, "The nature and origin of lamprophyres: some definitions, distinctions, and derivations," *Earth-Science Reviews, 13,* 123–169.

Roedder, E., 1956, "The role of liquid immiscibility in igneous petrogenesis: a discussion," *Jour. Geology, 64,* 84–88.

Roedder, E., 1979a, "Silicate liquid immiscibility in magmas," p. 15–57 in Yoder, H. S., Jr., ed., *The Evolution of the Igneous Rocks: Fiftieth Anniversary Perspectives*: Princeton, N.J., Princeton University Press, 588 p.

Roedder, E., 1979b, "Fluid inclusions as samples of ore fluids," p. 684–737 in Barnes, H. L., ed., *Geochemistry of Hydrothermal Ore Deposits*, 2nd ed.: New York, John Wiley & Sons, Inc., 798 p.

Roedder, E., and D. S. Coombs, 1967, "Immiscibility in granitic melts, indicated by fluid inclusions in ejected blocks from Ascension Island," *Jour. Petrology, 8,* 417–451.

Roedder, E., and P. W. Weiblen, 1971, "Petrology of silicate melt inclusions, Apollo 11 and Apollo 12 and terrestrial equivalents," *Proc. Second Lunar Science Conference,* 507–528.

Roeder, D., 1979, "Continental collisions," *Reviews of Geophysics and Space Physics, 17,* 1098–1109.

Roeder, P. L., 1974, "Paths of crystallization and fusion in systems showing ternary solid solution," *Amer. Jour. Science, 274,* 48–60.

Roeder, P. L. and R. F. Emslie, 1970, "Olivine–liquid equilibrium," *Contrib. Mineralogy and Petrology, 29,* 275–289.

Ronov, A. B., and A. A. Yaroshevsky, 1969, "Chemical composition of the earth's crust," *Amer. Geophys. Union Geophys. Monograph 13,* 37–57.

Ronov, A. B., V. Ye. Khain, and A. N. Balukhovskiy, 1980, "A comparative estimate of volcanism intensity on continents and in oceans," *Internat. Geology Review, 22,* 1383–1389.

Rose, A. W., and D. M. Burt, 1979, "Hydrothermal alteration," p. 173–235 in Barnes, H. L., ed., *Geochemistry of Hydrothermal Ore Deposits*, 2nd ed.: New York, John Wiley & Sons, Inc., 798 p.

Rose, W. I., Jr., and others, 1977a, "The evolution of Santa Maria volcano, Guatemala," *Jour. Geology, 85,* 63–87.

Rose, W. I., Jr., T. Pearson, and S. Bonis, 1977b, "Nuée ardente eruption from the foot of a dacite lava flow, Santiaguito Volcano, Guatemala," *Bulletin Volcanologique, 40,* 23–38.

Rosenhauer, M., and D. H. Eggler, 1975, "Solution of H_2O and CO_2 in diopside melt," *Carnegie Inst. Washington Yearbook, 74,* 474–479.

Ryerson, F. J., and P. C. Hess, 1980, "The role of P_2O_5 in silicate melts," *Geochim. et Cosmochim. Acta, 44,* 611–624.

Sahama, T. G., 1962, "Perthite-like exsolution in the nepheline–kalsilite system," *Norsk Geol. Tidsskrift, 42,* 2nd half-volume, 168–179.

Sakuyama, M., and I. Kushiro, 1979, "Vesiculation of hydrous andesitic melt and transport of alkalies by separated vapor phase," *Contrib. Mineralogy and Petrology, 71,* 61–66.

Salveson, J. O., 1978, "Variations in the geology of rift basins—a tectonic model," p. 82–86 in *1978 International Symposium on the Rio Grande Rift, Program and Abstracts:* Los Alamos Scientific Laboratory, *LA-7487-C.*

Saunders, A. D., and J. Tarney, 1979, "The geochemistry of basalts from a back-arc spreading centre in the East Scotia Sea," *Geochim. et Cosmochim. Acta. 43,* 555–572.

Sawkins, F. J., 1976, "Massive sulfide deposits in relation to geotectonics," *Geol. Assoc. Canada Special Paper 14,* 221–240.

Schairer, J. F., 1950, "The alkali feldspar join in the system NaAlSiO$_4$–KAlSiO$_4$–SiO$_2$," *Jour. Geology, 58,* 512–517.

Schairer, J. F., 1957, "Melting relations of the common rock-forming silicates," *Amer. Ceramic Society Jour., 40,* 215–235.

Schairer, J. F., and Bowen, N. L., 1947, "The system anorthite–leucite–silica," *Soc. Géol. de Finlande Bulletin, 20,* 67–87.

Schilling, J. G., H. Sigurdsson, and R. H. Kingsley, 1978, "Skagi and western neo-volcanic zones in Iceland: 2. Geochemical variations," *Jour. Geophys. Research, 83,* 3983–4002.

Schmid, R., 1981, "Descriptive nomenclature and classification of pyroclastic deposits and fragments: recommendations of the IUGS Subcommission on the Systematics of Igneous Rocks," *Geology, 9,* 41–43.

Schmincke, H.-U., 1974, "Volcanological aspects of peralkaline silicic welded ash-flow tuffs," *Bulletin Volcanologique, 38,* 594–636.

Sclater, J. G., C. Jaupart, and D. Galson, 1980, "The heat flow through oceanic and continental crust and the heat loss of the earth," *Reviews of Geophysics and Space Physics, 18,* 269–311.

Scrope, G. P., 1825, *Considerations on Volcanoes:* London, W. Phillips, 270 p.

Seck, H. A., 1971, "Koexistierende Alkalifeldspäte und Plagioklase im System NaAlSi$_3$O$_8$–KAlSi$_3$O$_8$–CaAl$_2$Si$_2$O$_8$–H$_2$O bei Temperaturen von 650°C bis 900°C," *Neues Jahrbuch für Mineralogie Abhandlungen, 115,* 315–345.

Seifert, F., and W. Schreyer, 1968, "Die Möglichkeit der Entstehung ultrabasischer Magmen bei Gegenwart geringer Alkalimengen," *Geol. Rundschau, 57,* 349–362.

Self, S., L. Wilson, and I. A. Nairn, 1979, "Vulcanian eruption mechanisms," *Nature, 277,* 440–443.

Serencsits, C. M., H. Faul, K. A. Foland, A. A. Hussein, and T. M. Lutz, 1981, "Alka-line ring complexes in Egypt: their ages and relationship in time," *Jour. Geophys. Research, 86,* 3009–3013.

Seyfried, W. E., Jr., and J. L. Bischoff, 1981, "Experimental seawater–basalt interaction at 300°C, 500 bars, chemical exchange, secondary mineral formation and implica-tions for the transport of heavy metals," *Geochim. et Cosmochim. Acta, 45,* 135–147.

Shand, S. J., 1951, *The Study of Rocks:* London, Thomas Murby and Co., 236 p.

Sharp, A. D. L., P. M. Davis, and F. Gray, 1980, "A low velocity zone beneath Mount Etna and magma storage," *Nature, 287,* 587–591.

Shaw, D. M., 1980, "Development of the early continental crust. Part III. Depletion of incompatible elements in the mantle," *Precambrian Research, 10,* 281–299.

Shaw, H. R., 1980, "The fracture mechanisms of magma transport from the mantle to the surface," p. 201–264 in Hargraves, R. B., ed., *Physics of Magmatic Processes*: Princeton, N.J., Princeton University Press, 585 p.

Shaw, H. R., E. D. Jackson, and K. E. Bargar, 1980, "Volcanic periodicity along the Hawaiian–Emperor chain," *Amer. Jour. Science, 280-A,* 667–708.

Sheraton, J. W., and A. Cundari, 1980, "Leucitites from Gaussberg, Antarctica," *Contrib. Mineralogy and Petrology, 71,* 417–427.

Shibata, T., G. Thompson, and F. A. Frey, 1979, "Tholeiitic and alkali basalts from the Mid-Atlantic Ridge at 43°N," *Contrib. Mineralogy and Petrology, 70,* 127–141.

Shima, H., and A. J. Naldrett, 1975, "Solubility of sulfur in an ultramafic melt and the relevance of the system Fe–S–O," *Econ. Geology, 70,* 960–967.

Sigurdsson, H., and R. S. J. Sparks, 1978, "Lateral magma flow within rifted Icelandic crust," *Nature, 274,* 126–130.

Sillitoe, R. H., C. Halls, and J. N. Grant, 1975, "Porphyry tin deposits in Bolivia," *Econ. Geology, 70,* 913–927.

Silver, L. T., H. P. Taylor, Jr., and B. Chappell, 1979, "Some petrological, geochemical and geochronological observations of the Peninsular Ranges Batholith near the international border of the U.S.A. and Mexico," p. 83–110 in Abbott, P. L., and V. R. Todd, eds., *Mesozoic Crystalline Rocks: Peninsular Ranges Batholith and Pegmatites, Point Sal Ophiolite:* San Diego State University, San Diego, Calif., 286 p.

Simkin, T., 1967, "Flow differentiation in the picritic sills of North Skye," p. 64–69 in Wyllie, P. J., ed., *Ultramafic and Related Rocks*: New York, John Wiley & Sons, Inc., 464 p.

Simkin, T., and K. A. Howard, 1970, "Caldera collapse in the Galapagos Islands, 1968," *Science, 169,* 429–437.

Simkin, T., L. Siebert, L. McClelland, D. Bridge, C. Newhall, and J. H. Latter, 1981, *Volcanoes of the World*: Stroudsburg, Pa., Hutchinson Ross Publishing Co., Inc., 233 p.

Simmons, E. C., and G. N. Hanson, 1978, "Geochemistry and origin of massif-type anorthosites," *Contrib. Mineralogy and Petrology, 66,* 119–135.

Simmons, G., 1964, "Gravity survey and geological interpretation, northern New York," *Geol. Soc. America Bulletin, 75,* 81–98.

Skinner, B. J., 1979, "The many origins of hydrothermal mineral deposits," p. 1–21 in Barnes, H. L., ed., *Geochemistry of Hydrothermal Ore Deposits,* 2nd ed.: New York, John Wiley & Sons, Inc., 798 p.

Skinner, B. J., and D. L. Peck, 1969, "An immiscible sulfide melt from Hawaii," *Econ. Geology Monograph 4,* 310–322.

Skinner, E. M. W., and C. R. Clement, 1979, "Mineralogical classification of southern African kimberlites," *Proc. Second Internat. Kimberlite Conference, 1,* 129–139.

Smith, D., 1971, "Stability of the assemblage iron-rich orthopyroxene–olivine–quartz," *Amer. Jour. Science, 271,* 370–382.

Smith, R. L., 1960, "Ash flows," *Geol. Soc. America Bulletin, 71,* 795–841.

Smith, R. L., 1979, "Ash-flow magmatism," *Geol. Soc. America Special Paper 180,* 5–27.

Smith, R. L., and R. A. Bailey, 1968, "Resurgent cauldrons," *Geol. Soc. America Memoir 116,* 613–662.

Smith, R, L., and H. R. Shaw, 1975, "Igneous-related geothermal systems," *U.S. Geol. Survey Circular 726,* 58–83.

Sommer, M. A., 1977, "Volatiles H_2O, CO_2, and CO in silicate melt inclusions in quartz phenocrysts from the rhyolitic Bandelier air-fall and ash-flow tuff, New Mexico," *Jour. Geology, 85,* 423–432.

Sørensen, H., ed., 1974, *The Alkaline Rocks:* New York, John Wiley & Sons, Inc., 622 p.

Sparks, R. S. J., 1978, "The dynamics of bubble formation and growth in magmas: a review and analysis," *Jour. Volcanol. and Geothermal Research, 3,* 1–37.

Sparks, R. S. J., and L. Wilson, 1976, "A model for the formation of ignimbrite by gravitational column collapse," *Geol. Soc. London Jour., 132,* 441–451.

Sparks, R. S. J., H. Pinkerton, and R. Macdonald, 1977, "The transport of xenoliths in magmas," *Earth and Planetary Science Letters, 35,* 234–238.

Sparks, R. S. J., S. Self, and G. P. L. Walker, 1973, "Products of ignimbrite eruptions," *Geology, 1,* 115–118.

Spencer, K. J., and D. H. Lindsley, 1981, "A solution model for coexisting iron-titanium oxides," *Amer. Mineralogist, 66,* 1189–1201.

Spera, F. J., 1980a, "Thermal evolution of plutons: a parameterized approach," *Science, 207,* 299–301.

Spera, F. J., 1980b, "Aspects of magma transport," p. 265–323 in Hargraves, R. B., ed., *Physics of Magmatic Processes:* Princeton, N.J., Princeton University Press, 585 p.

Sprague, D., and H. N. Pollack, 1980, "Heat flow in the Mesozoic and Cenozoic," *Nature, 285,* 393–395.

Spry, A., 1962, "The origin of columnar jointing, particularly in basalt flows," *Geol. Soc. Australia Jour., 8,* 191–216.

Stern, C. R., 1980, "Geochemistry of Chilean ophiolites: evidence for the compositional evolution of the mantle source of back-arc basin basalts," *Jour. Geophys. Research, 85,* 955–966.

Steven, T. A., and P. W. Lipman, 1976, *Calderas of the San Juan Volcanic Field, Southwestern Colorado:* U.S. Geol. Survey Professional Paper *958,* 35 p.

Stewart, D. B., 1978, "Petrogenesis of lithium-rich pegmatites," *Amer. Mineralogist, 63,* 970–980.

Stewart, D. B., 1979, "The formation of siliceous potassic glassy rocks," p. 339–350 in Yoder, H. S., Jr., ed., *The Evolution of the Igneous Rocks: Fiftieth Anniversary Perspectives:* Princeton, N.J., Princeton University Press, 588 p.

Stewart, D. C., and C. P. Thornton, 1975, "Andesite in oceanic regions," *Geology, 3,* 565–568.

Stoiber, R. E., and W. I. Rose, Jr., 1974, "Fumarole incrustations at active Central American volcanoes," *Geochim. et Cosmochim. Acta, 38,* 495–516.

Stoiber, R. E., G. B. Malone, and G. P. Bratton, 1978, "Volcanic emission of SO_2 at Italian and Central American volcanoes," *Geol. Soc. America Abstracts with Programs, 10,* 148.

Stolper, E., 1980, "A phase diagram for mid-oceanic ridge basalts: preliminary results and implications for petrogenesis," *Contrib. Mineralogy and Petrology, 74,* 13–27.

Stolper, E., and D. Walker, 1980, "Melt density and the average composition of basalt," *Contrib. Mineralogy and Petrology, 74,* 7–12.

Stormer, J. C., Jr., 1975, "A practical two-feldspar geothermometer," *Amer. Mineralogist, 60,* 667–674.

Streckeisen, A. L., 1967, "Classification and nomenclature of igneous rocks (final report of an enquiry)," *Neues Jahrbuch für Mineralogie Abhandlungen, 107,* 144–240.

Streckeisen, A. L., 1976, "Classification of the common igneous rocks by means of their chemical composition—a provisional attempt," *Neues Jahrbuch für Mineralogie Monatshefte 1976,* 1–15.

Streckeisen, A. L., 1978, "Classification and nomenclature of volcanic rocks, lamprophyres, carbonatites and melilitic rocks," *Neues Jahrbuch für Mineralogie Abhandlungen, 134,* 1–14.

Streckeisen, A. L., 1980, "Classification and nomenclature of volcanic rocks, lamprophyres, carbonatites and melilitic rocks. IUGS Subcommission on the Systematics of Igneous Rocks, recommendations and suggestions," *Geol. Rundschau, 69,* 194–207.

Streckeisen, A. L., and R. W. LeMaitre, 1979, "A chemical approximation to the modal QAPF classification of the igneous rocks," *Neues Jahrbuch für Mineralogie Abhandlungen, 136,* 169–206.

Sutherland, F. L., 1974, "High-pressure inclusions in tholeiitic basalt and the range of lherzolite-bearing magmas in the Tasmanian volcanic province," *Earth and Planetary Science Letters, 24,* 317–324.

Sutherland Brown, A., 1976, "Morphology and classification," p. 44–51 in Sutherland Brown, A., ed., *Porphyry Deposits of the Canadian Cordillera*: Canad. Institute of Mining and Metallurgy, Special Volume *15,* 510 p.

Sutherland Brown, A., and R. J. Cathro, 1976, "A perspective of porphyry deposits," p. 7–16 in Sutherland Brown, A., ed., *Porphyry Deposits of the Canadian Cordillera*: Canad. Institute of Mining and Metallurgy, Special Volume *15,* 510 p.

Swanson, D. A., and T. L. Wright, 1981, "Guide to geologic field trip between Lewiston, Idaho and Kimberly, Oregon, emphasizing the Columbia River Basalt Group," *U.S. Geol. Survey Circular 838,* 1–14.

Swanson, D. A., T. L. Wright, and R. T. Helz, 1975, "Linear vent systems and estimated rates of magma production and eruption for the Yakima Basalt on the Columbia Plateau," *Amer. Jour. Science, 275,* 877–905.

Swanson, S. E., 1977, "Relation of nucleation and crystal growth rate to the development of granitic textures," *Amer. Mineralogist, 62,* 966–978.

Sykes, L. R., 1978, "Intraplate seismicity, reactivation of preexisting zones of weakness, alkaline magmatism, and other tectonism postdating continental fragmentation," *Reviews of Geophysics and Space Physics, 16,* 621–688.

Takahashi, E., 1978, "Partitioning of Ni^{2+}, Co^{2+}, Fe^{2+}, Mn^{2+} and Mg^{2+} between olivine and silicate melts: compositional dependence of partition coefficients," *Geochim. et Cosmochim. Acta, 42,* 1829–1844.

Tarling, D. H., 1980, "Lithosphere evolution and changing tectonic regimes," *Geol. Soc. London Jour., 137,* 459–467.

Taylor, H. P., Jr., 1978, "Oxygen and hydrogen isotope studies of plutonic granitic rocks," *Earth and Planetary Science Letters, 38,* 177–210.

Taylor, H. P., Jr., and R. W. Forester, 1979, "An oxygen and hydrogen isotope study of the Skaergaard Intrusion and its country rocks: a description of a 55-M.Y. old fossil hydrothermal system," *Jour. Petrology, 20,* 355–419.

Taylor, M., and G. E. Brown, 1979, "Structure of mineral glasses: II. The SiO_2–$NaAlSiO_4$ join," *Geochim. et Cosmochim. Acta, 43,* 1467–1473.

Taylor, M., G. E. Brown, and P. M. Fenn, 1980, "Structure of mineral glasses: III. $NaAlSi_3O_8$ supercooled liquid at 805°C and the effects of thermal history," *Geochim. et Cosmochim. Acta, 44,* 109–117.

Taylor, S. R., 1975, *Lunar Science: A Post-Apollo View*: Elmsford, N.Y., Pergamon Press, 372 p.

Tazieff, H., 1977, "An exceptional eruption: Mt. Niragongo, Jan. 10th, 1977," *Bulletin Volcanologique, 40,* 189–200.

Thorarinsson, S., 1970a, "The Lakagigar eruption of 1783," *Bulletin Volcanologique, 33,* 910–929.

Thorarinsson, S., 1970b, *Hekla, a Notorious Volcano*: Reykjavik, Almenna Bokafelagid, 62 p.

Thornton, C. P., and O. F. Tuttle, 1960, "Chemistry of igneous rocks: I. Differentiation index," *Amer. Jour. Science, 258,* 664–684.

Thorpe, R. S., and P. W. Francis, 1979, "Variations in Andean andesite compositions and their petrogenetic significance," *Tectonophysics, 57,* 53–70.

Todd, S. G., D. J. Schissel, and T. N. Irvine, 1979, "Lithostratigraphic variations associated with the platinum-rich zone of the Stillwater complex," *Carnegie Inst. Washington Yearbook, 78,* 461–468.

Tsai, H.-M., H. O. A. Meyer, J. Moreau, and H. J. Milledge, 1979, "Mineral inclusions in diamond: Premier, Jagersfontein and Finsch kimberlites, South Africa, and Williamson Mine, Tanzania," *Proc. Second Internat. Kimberlite Conference, 1,* 16–26.

Turner, J. S., and L. B. Gustafson, 1978, "The flow of hot saline solutions from vents in the sea floor—some implications for exhalative massive sulfide and other ore deposits," *Econ. Geology, 73,* 1082–1100.

Tuttle, O. F., 1952, "Origin of the contrasting mineralogy of extrusive and plutonic salic rocks," *Jour. Geology, 60,* 107–124.

Tuttle, O. F., and N. L. Bowen, 1958, *Origin of Granite in the Light of Experimental Studies in the System $NaAlSi_3O_8$–$KAlSi_3O_8$–SiO_2–H_2O*: Geol. Soc. America Memoir 74, 153 p.

Upadhyay, H. D., and E. R. W. Neale, 1979, "On the tectonic regimes of ophiolite genesis," *Earth and Planetary Science Letters, 43,* 93–102.

Usselman, T. M., and D. S. Hodge, 1978, "Thermal control of low-pressure fractionation processes," *Jour. Volcanol. and Geothermal Research, 4,* 265–281.

Van Kooten, G. K., 1980, "Mineralogy, petrology, and geochemistry of an ultrapotassic basaltic suite, central Sierra Nevada, California, U.S.A.," *Jour. Petrology, 21,* 651–684.

Vatin-Perignon, N., D. M. Shaw, and J. R. Muysson, 1979, "Abundance of lithium in spilites and its implications for the spilitization process," *Physics and Chemistry of the Earth, 11,* 465–478.

Veizer, J., and W. Compston, 1976, "$^{87}Sr/^{86}Sr$ in Precambrian carbonates as an index of crustal evolution," *Geochim. et Cosmochim. Acta, 40,* 905–914.

Vermaak, C. F., and L. P. Hendriks, 1976, "A review of the mineralogy of the Merensky Reef, with specific reference to new data on the precious metal mineralogy," *Econ. Geology, 71,* 1244–1269.

Viljoen, M. J., and R. P. Viljoen, 1969, "The geology and geochemistry of the lower ultramafic unit of the Onverwacht Group and a proposed new class of igneous rock," *Geol. Soc. South Africa Special Publication 2,* 221–244.

Virgo, D., B. O. Mysen, and I. Kushiro, 1980, "Anionic constitution of 1-atmosphere silicate melts: implications for the structure of igneous melts," *Science, 208,* 1371–1373.

Wadge, G., 1977, "The storage and release of magma on Mount Etna," *Jour. Volcanol. and Geothermal Research, 2,* 361–384.

Wadge, G., 1980, "Output rate of magma from active central volcanoes," *Nature, 288,* 253–255.

Waff, H. S., 1980, "Effects of the gravitational field on liquid distribution in partial melts within the upper mantle," *Jour. Geophys. Research, 85,* 1815–1825.

Waff, H. S., and J. R. Bulau, 1979, "Equilibrium fluid distribution in an ultramafic partial melt under hydrostatic stress conditions," *Jour. Geophys. Research, 84,* 6109–6114.

Wager, L. R., and G. M. Brown, 1967, *Layered Igneous Rocks*: San Francisco, W. H. Freeman and Company, Publishers, 588 p.

Wager, L. R., and W. A. Deer, 1939, "Geological investigations in East Greenland. Part III. The petrology of the Skaergaard intrusion, Kangerdlugssuaq, East Greenland," *Meddelelser om Grønland, 105,* no. 4, 1–352.

Walker, D., T. Shibata, and S. E. DeLong, 1979, "Abyssal tholeiites from the Oceanographer Fracture Zone: II. Phase equilibria and mixing," *Contrib. Mineralogy and Petrology, 70,* 111–125.

Walker, D., E. M. Stolper, and J. F. Hays, 1978, "A numerical treatment of melt/solid segregation: size of the eucrite parent body and stability of the terrestrial low-velocity zone," *Jour. Geophys. Research, 83,* 6005–6013.

Walker, D., J. Longhi, R. J. Kirkpatrick, and J. F. Hays, 1976, "Differentiation of an Apollo 12 picrite magma," *Proc. Seventh Lunar Science Conference,* 1365–1389.

Walker, G. P. L., 1969, "The breaking of magma," *Geol. Magazine, 106,* 166–173.

Walker, G. P. L., 1972, "Crystal concentration in ignimbrites," *Contrib. Mineralogy and Petrology, 36,* 135–146.

Walker, G. P. L., 1973, "Explosive volcanic eruptions—a new classification scheme," *Geol. Rundschau, 62,* 431–446.

Walker, G. P. L., L. Wilson, and E. L. G. Bowell, 1971, "Explosive volcanic eruptions: I. The rate of fall of pyroclasts," *Royal Astron. Soc. Geophys. Jour., 22,* 377–383.

Wallace, S. R., W. B. MacKenzie, R. G. Blair, and N. K. Muncaster, 1978, "Geology of the Urad and Henderson molybdenite deposits, Clear Creek County, Colorado, with a section on a comparison of these deposits with those at Climax, Colorado," *Econ. Geology, 73,* 325–368.

Washington, H. S., 1917, *Chemical Analyses of Igneous Rocks, Published from 1884 to 1913, Inclusive*: U.S. Geol. Survey Professional Paper *99*, 1201 p.

Wass, S. Y., 1980, "Geochemistry and origin of xenolith-bearing and related alkali basaltic rocks from the Southern Highlands, New South Wales, Australia," *Amer. Jour. Science, 280-A,* 639–666.

Watson, E. B., 1979, "Zircon saturation in felsic liquids: experimental results and applications to trace element geochemistry," *Contrib. Mineralogy and Petrology, 70,* 407–419.

Watson, J. V., 1980, "Metallogenesis in relation to mantle heterogeneity," *Royal Soc. London Philos. Trans., A, 297,* 347–352.

Wells, P. R. A., 1977, "Pyroxene thermometry in simple and complex systems," *Contrib. Mineralogy and Petrology, 62,* 129–139.

Wendlandt, R. F., and D. H. Eggler, 1980a, "The origins of potassic magmas: 1. Melting relations in the systems $KAlSiO_4$–Mg_2SiO_4–SiO_2 and $KAlSiO_4$–MgO–SiO_2–CO_2 to 30 kilobars," *Amer. Jour. Science, 280,* 385–420.

Wendlandt, R. F., and D. H. Eggler, 1980b, "The origins of potassic magmas: 2. Stability of phlogopite in natural spinel lherzolite and in the system $KAlSiO_4$–MgO–SiO_2–H_2O–CO_2 at high pressures and high temperatures," *Amer. Jour. Science, 280,* 421–458.

Wendlandt, R. F., and W. J. Harrison, 1979, "Rare earth partitioning between immiscible carbonate and silicate liquids and CO_2 vapor: results and implications for the formation of light rare earth-enriched rocks," *Contrib. Mineralogy and Petrology, 69,* 409–419.

Wendlandt, R. F., and B. O. Mysen, 1978, "Melting phase relations of natural peridotite and CO_2 as a function of degree of partial melting at 15 and 30 kbar," *Carnegie Inst. Washington Yearbook, 77,* 756–761.

White, D. E., and M. Guffanti, 1979, "Geothermal systems and their energy resources," *Reviews of Geophysics and Space Physics, 17,* 887–902.

White, D. E., L. J. P. Muffler, and A. H. Truesdell, 1971, "Vapor-dominated hydrothermal systems compared with hot water systems," *Econ. Geology, 66,* 75–97.

Whitney, J. A., 1975, "Vapor generation in a quartz monzonite magma: a synthetic model with application to porphyry copper deposits," *Econ. Geology, 70,* 346–358.

Whittaker, E. J. W., and R. Muntus, 1970, "Ionic radii for use in geochemistry," *Geochim. et Cosmochim. Acta, 34,* 945–956.

Wicks, F. J., and E. J. W. Whittaker, 1977, "Serpentine textures and serpentinization," *Canad. Mineralogist, 15,* 459–488.

Wiebe, R. A., 1979, "Anorthositic dikes, southern Nairn complex, Labrador," *Amer. Jour. Science, 279,* 394–410.

Wiebe, R. A., 1980, "Anorthositic magmas and the origin of Proterozoic anorthosite massifs," *Nature, 286,* 564–567.

Wilcox, R. E., 1954, *Petrology of Paricutin Volcano, Mexico*: U.S. Geol. Survey Bulletin *965-C*.

Wilcox, R. E., 1979, "The liquid line of descent and variation diagrams," p. 205–232 in Yoder, H. S., Jr., ed., *The Evolution of the Igneous Rocks: Fiftieth Anniversary Perspectives*: Princeton, N.J., Princeton University Press, 588 p.

Wilkinson, J. F. G., 1967, "The petrography of basaltic rocks," p. 163–214 in Hess, H. H., and A. Poldervaart, eds., *Basalts: The Poldervaart Treatise on Rocks of Basaltic Composition,* Vol. 1: New York, Wiley-Interscience, 482 p.

Wilkinson, J. F. G., 1974, "The mineralogy and petrography of alkali basaltic rocks," p. 67–95 in Sørensen, H., ed., *The Alkaline Rocks:* New York, John Wiley & Sons, Inc., 622 p.

Wilkinson, J. F. G., and R. A. Binns, 1977, "Relatively iron-rich lherzolite xenoliths of the Cr-diopside suite: a guide to the primary nature of anorogenic tholeiitic andesite magmas," *Contrib. Mineralogy and Petrology, 65,* 199–212.

Williams, H., 1942, *The Geology of Crater Lake National Park, Oregon*: Carnegie Inst. Washington Publication *540,* 162 p.

Williams, H., and A. R. McBirney, 1979, *Volcanology:* San Francisco, Freeman, Cooper & Company, 397 p.

Williams, H., F. J. Turner, and C. M. Gilbert, 1954, *Petrography: An Introduction to the Study of Rocks in Thin Sections:* San Francisco, W. H. Freeman and Company, Publishers, 406 p.

Williams, L. A. J., 1978, "The volcanological development of the Kenya Rift," p. 101–121 in Neumann, E.-R., and I. B. Ramberg, eds., *Petrology and Geochemistry of Continental Rifts*: Dordrecht, D. Reidel Publishing Co.

Wilshire, H. G., and E. D. Jackson, 1975, "Problems in determining mantle geotherms from pyroxene compositions of ultramafic rocks," *Jour. Geology, 83,* 313–329.

Wilshire, H. G., and J. W. Shervais, 1975, "Al-augite and Cr-diopside ultramafic xenoliths in basaltic rocks from western United States," *Physics and Chemistry of the Earth, 9,* 257–272.

Wilson, L., and J. W. Head, III, 1981, "Ascent and eruption of basaltic magma on the Earth and Moon," *Jour. Geophys. Research, 86,* 2971–3001.

Wilson, L., R. S. J. Sparks, T. C. Huang, and N. D. Watkins, 1978, "The control of volcanic column heights by eruption energetics and dynamics," *Jour. Geophys. Research, 83,* 1829–1836.

Wood, C. A., 1980, "Morphometric evolution of cinder cones," *Jour. Volcanol. and Geothermal Research, 7,* 387–413.

Wood, M. I., and P. C. Hess, 1980, "The structural role of Al_2O_3 and TiO_2 in immiscible silicate liquids in the system SiO_2–MgO–CaO–FeO–TiO_2–Al_2O_3," *Contrib. Mineralogy and Petrology, 72,* 319–328.

Wright, J. V., A. L. Smith, and S. Self, 1980, "A working terminology of pyroclastic deposits," *Jour. Volcanol. and Geothermal Research, 8,* 315–336.

Wright, T. L., 1971, *Chemistry of Kilauea and Mauna Loa Lava in Space and Time*: U.S. Geol. Survey Professional Paper *735,* 40 p.

Wright, T. L., 1974, "Presentation and interpretation of chemical data for igneous rocks," *Contrib. Mineralogy and Petrology, 48,* 233–248.

Wright, T. L., and P. C. Doherty, 1970, "A linear programming and least squares computer method for solving petrologic mixing problems," *Geol. Soc. America Bulletin, 81,* 1995–2008.

Wright, T. L., and R. S. Fiske, 1971, "Origin of the differentiated and hybrid lavas of Kilauea Volcano, Hawaii," *Jour. Petrology, 12,* 1–65.

Wright, T. L., and R. I. Tilling, 1980, "Chemical variation in Kilauea eruptions 1971–1974," *Amer. Jour. Science, 280-A,* 777–793.

Wright, T. L., W. T. Kinoshita, and D. L. Peck, 1968, "March 1965 eruption of Kilauea Volcano and the formation of Makaopuhi lava lake," *Jour. Geophys. Research, 73,* 3181–3205.

Wright, T. L., D. L. Peck, and H. R. Shaw, 1976, "Kilauea lava lakes: natural laboratories for study of cooling, crystallization, and differentiation of basaltic magma," *Amer. Geophys. Union Geophys. Monograph 19,* 375–390.

Wyllie, P. J., 1971, *The Dynamic Earth: Textbook in Geosciences*: New York, John Wiley & Sons, Inc., 416 p.

Wyllie, P. J., 1977, "Effects of H_2O and CO_2 on magma generation in the crust and mantle," *Geol. Soc. London Jour., 134,* 215–234.

Wyllie, P. J., 1979, "Magmas and volatile components," *Amer. Mineralogist, 64,* 469–500.

Wyllie, P. J., 1980, "The origin of kimberlite," *Jour. Geophys. Research, 85,* 6902–6910.

Wyllie, P. J., and O. F. Tuttle, 1960, "The system $CaO–CO_2–H_2O$ and the origin of carbonatites," *Jour. Petrology, 1,* 1–46.

Wyllie, P. J., and O. F. Tuttle, 1964, "Experimental investigation of silicate systems containing two volatile components. Part III. The effects of SO_3, P_2O_5, HCl, and Li_2O, in addition to H_2O, on the melting temperatures of albite and granite," *Amer. Jour. Science, 262,* 930–939.

Yoder, H. S., Jr., 1958, "Effect of water on the melting of silicates," *Carnegie Inst. Washington Yearbook, 57,* 189–191.

Yoder, H. S., Jr., 1969, "Experimental studies bearing on the origin of anorthosite," *New York State Museum and Science Service Memoir 18,* 13–22.

Yoder, H. S., Jr., 1976, *Generation of Basaltic Magma*: Washington, D.C., National Academy of Sciences, 265 p.

Yoder, H. S., Jr., 1978, "Basic magma generation and aggregation," *Bulletin Volcanologique, 41,* 301–316.

Yoder, H. S., Jr., 1979, "Melilite-bearing rocks and related lamprophyres," p. 391–411 in Yoder, H. S., Jr., ed., *The Evolution of the Igneous Rocks: Fiftieth Anniversary Perspectives*: Princeton, N.J., Princeton University Press, 588 p.

Yoder, H. S., Jr., 1980, "Experimental mineralogy: achievements and prospects," *Bulletin de Minéralogie, 103,* 5–26.

Yoder, H. S., Jr., and C. E. Tilley, 1962, "Origin of basalt magmas: an experimental study of natural and synthetic rock systems," *Jour. Petrology, 3,* 342–532.

Yoder, H. S., Jr., D. B. Stewart, and J. R. Smith, 1957, "Ternary feldspars," *Carnegie Inst. Washington Yearbook, 56,* 206–217.

Index